Advances in Industrial Control

Other titles published in this Series:

Digital Controller Implementation and Fragility
Robert S.H. Istepanian and
James F. Whidborne (Eds.)

Optimisation of Industrial Processes at Supervisory Level
Doris Sáez, Aldo Cipriano and
Andrzej W. Ordys

Robust Control of Diesel Ship Propulsion
Nikolaos Xiros

Hydraulic Servo-systems
Mohieddine Jelali and Andreas Kroll

Strategies for Feedback Linearisation
Freddy Garces, Victor M. Becerra,
Chandrasekhar Kambhampati and
Kevin Warwick

Robust Autonomous Guidance
Alberto Isidori, Lorenzo Marconi and
Andrea Serrani

Dynamic Modelling of Gas Turbines
Gennady G. Kulikov and Haydn A.
Thompson (Eds.)

Control of Fuel Cell Power Systems
Jay T. Pukrushpan, Anna G. Stefanopoulou
and Huei Peng

Fuzzy Logic, Identification and Predictive Control
Jairo Espinosa, Joos Vandewalle and
Vincent Wertz

Optimal Real-time Control of Sewer Networks
Magdalene Marinaki and Markos
Papageorgiou

Process Modelling for Control
Benoît Codrons

Computational Intelligence in Time Series Forecasting
Ajoy K. Palit and Dobrivoje Popovic

Modelling and Control of mini-Flying Machines
Pedro Castillo, Rogelio Lozano and
Alejandro Dzul

Rudder and Fin Ship Roll Stabilization
Tristan Perez

Hard Disk Drive Servo Systems (2nd Edition)
Ben M. Chen, Tong H. Lee, Kemao Peng
and Venkatakrishnan Venkataramanan

Measurement, Control, and Communication Using IEEE 1588
John Eidson

Piezoelectric Transducers for Vibration Control and Damping
S.O. Reza Moheimani and Andrew J.
Fleming

Windup in Control
Peter Hippe

Nonlinear H_2/H_∞ Constrained Feedback Control
Murad Abu-Khalaf, Jie Huang and
Frank L. Lewis

Practical Grey-box Process Identification
Torsten Bohlin
Publication due May 2006

Modern Supervisory and Optimal Control
Sandor A. Markon, Hajime Kita, Hiroshi
Kise and Thomas Bartz-Beielstein
Publication due July 2006

Wind Turbine Control Systems
Fernando D. Bianchi, Hernán De Battista
and Ricardo J. Mantz
Publication due August 2006

Soft Sensors for Monitoring and Control of Industrial Processes
Luigi Fortuna, Salvatore Graziani,
Alessandro Rizzo and Maria Gabriella
Xibilia
Publication due August 2006

Practical PID Control
Antonio Visioli
Publication due November 2006

Magnetic Control of Tokamak Plasmas
Marco Ariola and Alfredo Pironti
Publication due May 2007

Stjepan Bogdan, Frank L. Lewis, Zdenko Kovačić
and José Mireles Jr.

Manufacturing Systems Control Design

A Matrix-based Approach

With 152 Figures

Stjepan Bogdan, PhD
Laboratory for Robotics and Intelligent
  Control Systems
Department of Control and
  Computer Engineering
Faculty of Electrical Engineering
  and Computing
University of Zagreb
Zagreb
Croatia

Zdenko Kovačić, PhD
Laboratory for Robotics and Intelligent
  Control Systems
Department of Control and
  Computer Engineering
Faculty of Electrical Engineering
  and Computing
University of Zagreb
Zagreb
Croatia

Frank L. Lewis, PhD
Automation and Robotics Research
  Institute
University of Texas (Arlington)
Fort Worth, Texas
USA

José Mireles Jr., PhD
Instituto de Ingeniería y Tecnología
Universidad Autónoma de Ciudad Juárez
Cd. Juárez
Chihuahua
México

British Library Cataloguing in Publication Data
Manufacturing systems control design : a matrix-based
   approach. - (Advances in industrial control)
   1.Industrial engineering - Automatic control 2.Process
   control 3.Matrices
   I.Bogdan, Stjepan
   629.8
ISBN-13: 9781852339821
ISBN-10: 1852339829

Library of Congress Control Number: 2006924637

Advances in Industrial Control series ISSN 1430-9491
ISBN-10: 1-85233-982-9   e-ISBN 1-84628-334-5   Printed on acid-free paper
ISBN-13: 978-1-85233-982-1

© Springer-Verlag London Limited 2006

MATLAB® and Simulink® are registered trademarks of The MathWorks, Inc., 3 Apple Hill Drive, Natick,
MA 01760-2098, USA. http://www.mathworks.com

Apart from any fair dealing for the purposes of research or private study, or criticism or review, as
permitted under the Copyright, Designs and Patents Act 1988, this publication may only be reproduced,
stored or transmitted, in any form or by any means, with the prior permission in writing of the publishers,
or in the case of reprographic reproduction in accordance with the terms of licences issued by the
Copyright Licensing Agency. Enquiries concerning reproduction outside those terms should be sent to
the publishers.

The use of registered names, trademarks, etc. in this publication does not imply, even in the absence of a
specific statement, that such names are exempt from the relevant laws and regulations and therefore free
for general use.

The publisher makes no representation, express or implied, with regard to the accuracy of the information
contained in this book and cannot accept any legal responsibility or liability for any errors or omissions
that may be made.

Printed in Germany

9 8 7 6 5 4 3 2 1

Springer Science+Business Media
springer.com

Advances in Industrial Control

Series Editors

Professor Michael J. Grimble, Professor of Industrial Systems and Director
Professor Michael A. Johnson, Professor (Emeritus) of Control Systems
and Deputy Director

Industrial Control Centre
Department of Electronic and Electrical Engineering
University of Strathclyde
Graham Hills Building
50 George Street
Glasgow G1 1QE
United Kingdom

Series Advisory Board

Professor E.F. Camacho
Escuela Superior de Ingenieros
Universidad de Sevilla
Camino de los Descobrimientos s/n
41092 Sevilla
Spain

Professor S. Engell
Lehrstuhl für Anlagensteuerungstechnik
Fachbereich Chemietechnik
Universität Dortmund
44221 Dortmund
Germany

Professor G. Goodwin
Department of Electrical and Computer Engineering
The University of Newcastle
Callaghan
NSW 2308
Australia

Professor T.J. Harris
Department of Chemical Engineering
Queen's University
Kingston, Ontario
K7L 3N6
Canada

Professor T.H. Lee
Department of Electrical Engineering
National University of Singapore
4 Engineering Drive 3
Singapore 117576

Professor Emeritus O.P. Malik
Department of Electrical and Computer Engineering
University of Calgary
2500, University Drive, NW
Calgary
Alberta
T2N 1N4
Canada

Professor K.-F. Man
Electronic Engineering Department
City University of Hong Kong
Tat Chee Avenue
Kowloon
Hong Kong

Professor G. Olsson
Department of Industrial Electrical Engineering and Automation
Lund Institute of Technology
Box 118
S-221 00 Lund
Sweden

Professor A. Ray
Pennsylvania State University
Department of Mechanical Engineering
0329 Reber Building
University Park
PA 16802
USA

Professor D.E. Seborg
Chemical Engineering
3335 Engineering II
University of California Santa Barbara
Santa Barbara
CA 93106
USA

Doctor K.K. Tan
Department of Electrical Engineering
National University of Singapore
4 Engineering Drive 3
Singapore 117576

Doctor I. Yamamoto
Technical Headquarters
Nagasaki Research & Development Center
Mitsubishi Heavy Industries Ltd
5-717-1, Fukahori-Machi
Nagasaki 851-0392
Japan

To Jasenka
S.B.

To Chris, my son
F.L.L.

To Dubravka
Z.K.

To Josue, Joel, Alena, and Aaron, my loved kids
J.M.

Series Editors' Foreword

The series *Advances in Industrial Control* aims to report and encourage technology transfer in control engineering. The rapid development of control technology has an impact on all areas of the control discipline. New theory, new controllers, actuators, sensors, new industrial processes, computer methods, new applications, new philosophies..., new challenges. Much of this development work resides in industrial reports, feasibility study papers and the reports of advanced collaborative projects. The series offers an opportunity for researchers to present an extended exposition of such new work in all aspects of industrial control for wider and rapid dissemination.

In some areas of manufacturing, the elements of a flexible manufacturing system form the key components of the process line. These key components are four-fold: a set of programmable robots and machines, an automated materials-handling system that allows parts to be freely routed and re-routed, a buffer storage system where parts and partly-assembled components can wait until required for further processing and assembly and finally, a supervisory control system. The technology employed to coordinate and control all these components as a working system is usually based on programmable logic controllers. The use of this automation hardware and software in manufacturing is designed to yield significant cost reductions and to enhance quality. Economic gains are achieved through the ability of these systems to work continuously (24/7, all year round) and flexibly so that rapid on-line supervisory reprogramming can be performed for the modification and improvement of product parts or the assembly of different products. Unlike human operators, robots and machines do not suffer from biological fatigue so that enhanced product quality can be attained through better repetitive accuracies and consistent mechanical performance.

To achieve the economic and quality goals of flexible manufacturing requires the use of some sophisticated supervisory control algorithms to direct the process scheduling and despatching tasks and to handle any on-line dynamic conflicts that might emerge. In general, these decisions for sequencing operations are constrained by limited resources and shared resources. The other very distinctive characteristic of these supervisory control problems is that the system under supervision is a discrete-event system. In this case, the state-space for the system attains discrete values and the transition from one state to another is caused by an event taking place. This adds to the difficulty in analysing the properties of these systems because the system description is often linguistic unlike the continuous-time equation-based descriptions with which most control engineers are familiar; however, many of these scheduling problems are central issues in manufacturing studies *per se* and as such have already been extensively investigated. For example, many of the methods of operational research were originally driven by the supervisory control problems of manufacturing processes. From this simple background perspective, Professor Stjepan Bogdan and his colleagues present a nicely self-contained treatment of the supervisory control problems of flexible manufacturing systems using recently developed approaches and tools. This new entry to the *Advances in Industrial Control* series has four components. Firstly, the introductory chapters 1 and 2 create the framework for understanding flexible-manufacturing-system concepts and discrete-event system descriptions. Particularly interesting is the discussion of system types – continuous-time, hybrid and discrete – given in Chapter 2 which uses in-depth examples to help the control engineer appreciate the similarities and differences between the three types.

A substantial part of the book, Chapters 3 to 5, pursues matrix models for manufacturing systems; however, it should be noted that because the underlying systems are discrete-event systems, these are matrices defined over an and or (Δ, ∇) algebra. In these chapters, it is fascinating to learn how rule-based systems can be given matrix representation and how there are links to other tools like directed graphs for the analysis of the control of these systems.

Quite often in this text the discussion of manufacturing system properties uses the tools of Petri nets. Petri nets have evolved and developed considerably since their original introduction in the early 1960s. In Chapter 6, the authors present a full introduction to their use in the studies of manufacturing system problems, for example, links between Petri nets and the matrix methods of earlier chapters are established. The authors also usefully describe and demonstrate their own graphical Petri-net-simulation software tool that is available for download for use by the interested reader.

To complete the monograph, Chapter 7 reports on other mainstream simulation tools for virtual factory modelling. The potentially disastrous economic effects of a poorly defined factory layout or an inefficient manufacturing control strategy has ensured the development of a substantial set of factory simulation tools often with advanced graphics for visualisation of operational dynamics and with various analysis tools to assess and compute performance metrics. The thorough survey presented by Professor Bogdan and his colleagues provides a very fitting concluding chapter for this stimulating monograph on flexible manufacturing systems.

M.J. Grimble and M.A. Johnson
Industrial Control Centre
Glasgow, Scotland, U.K.

Foreword

In the late 1980s, the strong needs for modeling, analysis, control, and simulation of complex systems especially computer-integrated manufacturing systems demanded the academic researchers and industrial engineers to seek and investigate better methodologies and tools. Such tools must be able to deal with such system characteristics as asynchronous events, sequences, concurrency, synchronization, mutual exclusion, deadlocks, and choices. While state machines or automata were popular in many applications, they were soon proved to be inadequate since the state explosion problems would be met at the very beginning of system design. Any design flaws or incompleteness may invalidate the entire system design and frequently require rather cumbersome recovery. On the other hand, Petri nets, invented by C. A. Petri in his 1962's doctoral dissertation, are well equipped with the required capabilities to handle the above-mentioned characteristics. They thus gained their popularity among the researchers of discrete event systems and industrial applications in manufacturing automation.

The research group at Rensselaer Polytechnic Institute (RPI) was established and led by Professors Frank DiCesare and Alan Desrochers. It was supported by many leading industrial companies such as IBM, GM, Johnson and Johnson, Sun Microsystems, and Digital Equipment Corporation via an eight-year long Computer Integrated Manufacturing Research Program of the Center for Manufacturing Productivity and Technology Transfer at RPI. They obtained many significant research and application results. Notably, as the first Ph.D. graduate of this group in this area, Dr. Robert Al-Jaar proposed to use generalized stochastic Petri nets for modeling and analysis of production lines. Their work led to their 1994 book Applications of Petri Nets in Manufacturing Systems: Modeling, Control, and Performance Analysis by IEEE Press. As the second Ph.D. graduate of the group, I developed the concepts of parallel and sequential mutual exclusion structures, top-down, bottom-up and hybrid synthesis methods, Petri net-based discrete event controller design and implementation procedures for flexible manufacturing systems (FMS). The results were summarized into the first monograph of its kind, Petri Net Synthesis for Discrete Event Control of Manufacturing Systems, co-authored with Frank DiCesare, Kluwer Academic Publisher in 1993. Dr. Fei-Yue Wang, presently Professor of the University of

Arizona and the Institute of Automation, Chinese Academy of Sciences, pioneered in applying Petri nets to designing intelligent machines and building intelligent control foundation together with his advisor, Dr. George Saridis. He also developed a Petri net method for communication protocol design and performance analysis for manufacturing message specification. From the same group, Dr. Inseon Koh, presently Professor of Hong-Ik University, Korea, perfected a bottom-up method to synthesize Petri nets with desired properties. Dr. Jagdish S. Joshi conducted performance analysis of network and database transactions in a CIM system. Dr. MuDer Jeng, presently Professor of National Taiwan Ocean University, invented a new class of Petri nets suitable for modeling automated manufacturing systems. Dr. Doo Yong Lee, presently Professor of Korea Advanced Institute of Technology, pioneered in using various heuristics to guide optimal or sub-optimal schedule search in timed Petri net models of flexible manufacturing systems. Dr. Alessandro Giua, presently a professor of University of Cagliari, Italy, developed a supervisory control theory in the framework of Petri nets. Dr. Tiehua Cao and Professor Arthur C. Sanderson combined fuzzy logic theory and Petri nets and developed fuzzy Petri nets for intelligent task planning in a robotic system. The research led to the publication of Intelligent Task Planning Using Fuzzy Petri Nets in the Series in Intelligent Control and Intelligent Automation of World Scientific Publisher in 1996. Dr. Hauke Jungnitz developed approximation methods for stochastic timed Petri nets. Dr. James F. Watson formulated a method for performance analysis of discrete event systems with non-exponential random time distributions and state space estimation of a given Petri net model.

The above-mentioned work addressed various issues from model synthesis, performance analysis, simulation, deadlock avoidance, and supervisory control design and made significant contributions to the field of Petri nets and their applications to manufacturing automation. Yet one significant problem remains unsolved: given manufacturing system specifications expressed in Bill of Materials, Assembly Tree, Task Sequencing matrix, and Resource Requirement Matrix, how can one automatically generate a Petri net model and related design for analysis, control, and simulation of a flexible manufacturing system (FMS). This book written by a group of outstanding researchers under the leadership of Dr. Frank Lewis indeed presents an elegant solution to the above long-lasting problem. Their proposed matrix-based approach represents one of the most significant innovations to the area of Petri nets and related discrete-event modeling approaches for manufacturing system control design. The authors are able to identify a unique mapping between the Petri net elements and system specifications and reveal the underlying relations for a number of design and analysis tools used in industrial engineering. More importantly, the research group is able to link what they do to the generation of control code required by Programmable Logic Controllers (PLC). PLC have been the industrial horse in almost every sector of automated manufacturing and packaging industry for three decades.

This present book contributes to the area of manufacturing automation in a number of ways. First, it comprehensively presents a matrix-based modeling and controller design framework. It uses an intelligent material handling workcell to illustrate clearly various steps in matrix-based controller design. Second, the book addresses how to utilize matrices for analyzing structural properties of manufacturing systems. It reveals the underlying relationship among graph descriptions, max-plus algebra, and the proposed matrix models. Third, the book investigates a very important yet difficult class of manufacturing systems, namely, multiple re-entrant flowlines. It answers how deadlocks can be avoided in such systems. PLC-controlled flexible manufacturing systems are used to illustrate various deadlock avoidance strategies. Fourth, the book presents Petri nets and their complementary character with the matrix models. A computer-aided design tool called Petri.NET is developed and presented, allowing researchers and engineers to model and simulate FMS using either Petri nets or matrix models. Finally, the book presents the basics of virtual factory modeling and simulation and a number of CAD tools used in industry. Its contribution includes a web tool called FlexMan that can be used to design and simulate of FMS based on virtual factory models and matrix-based methodologies. Such examples as palletization workcell, FESTO FMS, robotic brick-handling system, Volvo body-manufacturing line, and assembly station are used to demonstrate these tools.

In conclusion, the authors have well presented their innovative manufacturing control design methods based on matrices, Petri nets and other related discrete-event modeling tools. The book clearly advances the state-of-the-art in the area of flexible manufacturing automation and its impact to the area will last long, not only methodologically but also practically.

MengChu Zhou, Ph. D. and Professor
New Jersey Institute of Technology
Newark, NJ
http://web.njit.edu/~zhou

Preface

Being aware that our planet is inhabited with more than six billion human beings having their needs for food, cloths, shelter, medical care, education, transportation, entertainment and many more, efficacious manufacturing of various goods becomes extremely important for the global society.

Even in these days, manufacturing is performed by individuals, and such products, in low volumes, may have a lot of success on the market. Let us just mention exclusive cars, jewellery or some pieces of finest art. High volume production would not be possible without carefully planned production technology, highly automated production processes and engagement of specially designed automation equipment.

The success of one product on the market depends on many factors, how the product looks like, what kind of usefulness it has, how many versions of product exist etc. Investigations of the car market have found long time ago that customers make their decisions based on overall assessment of car manufacturer (e.g., reliability of the vehicle, quality of the service, experience in manufacturing), but they are also judging the individual qualities of the vehicle they want to buy. In this context, sometimes only colour becomes the reason why the customer will pick that car instead of another. What we say about cars holds for any other product.

Many products have very similar constituent parts or the same parts arranged in the slightly different way. The production strategy which has a goal to accommodate to different customers' demands leads naturally to a flexible manufacturing concept. For example, contemporary car industry cannot be imagined without robots, machine tools, belt conveyers, part feeders and other elements. The strength of all single components lies in their integration into a flexible manufacturing system (FMS).

FMS control triggers many parallel worlds - continuous and discrete control loops, as well as many discrete events occurring synchronously or stochastically. In order to be able to control the FMS, sensors, actuators, and controllers, viewing from the lowest to the highest control level, constitute a network. This means that dealing with FMS control actually means dealing with the distributed network-based control. Usually, programmable logic controllers or PC-based solutions are used for implementation of such controllers. Therefore, it is very important that methods and algorithms concerned with analysis and synthesis of FMS control have a form and features which would make them suitable for implementation in a dedicated hardware.

Flexible manufacturing systems are "live" systems composed of a group of allocated resources and tasks being assigned. They can assume different structures and undergo different control strategies depending on the consequences of issued commands (e.g. a new robot has been added to the system) or the states of particular FMS components (e.g. one of the robots is out of order). Under such conditions, sequencing of tasks, parallelism of missions, collision of concurrent decisions, and negligence of planned actions are problems an FMS control designer must confront with.

The authors of this book have been actively involved in the FMS control area more than a decade trying to find such FMS control design methods which would guarantee the stability and the functionality of the control system at one hand, and which would be simple enough, easily implemental and effective in practical engineering applications on the other.

The purpose of this book is to describe the use of matrical approach to the FMS control design. First we introduce the reader with different techniques of FMS control design and then we elaborate the advantages of matrix-based FMS control design, mentioning just one, an ability to convert the matrix-based controller into an effective PLC executable code.

The topics are divided into seven chapters. Chapter 1 is a descriptive introduction into a world of FMS classification, modeling, simulation and design of their controllers. The Chapter 2 gives a brief review of discrete event systems with an emphasis on the time-driven and event-driven systems. Chapter 3 describes the theory and methodology of creating a matrix model and a matrix controller. The reader will also find the example of matrix controller design for an intelligent material handling workcell. Chapter 4 is concerned with an analysis of matrix methods for manufacturing systems. The description of graphs, principles of string composition, and max-plus algebra is given as well as their relations to the matrix model. Manufacturing system structural properties given in the matrix form are the subject of Chapter 5. The focus has been set on the properties of so called multiple re-entrant flowlines (MRF) that are important for the control synthesis, such as circular waits, siphons and traps, and critical subsystems. The discussion is extended also to the free choice multiple re-entrant flowlines (FMRF), whose control properties are even more demanding. Deadlock avoidance strategies are presented and illustrated on the example of a PLC-controlled FMS. Chapter 6 deals with Petri nets that are a widely used tool for MS modeling and control design. The relations between Petri nets and the matrix form are given in order to show their complementary character. A program tool Petri.NET developed at the Laboratory for Robotics and Intelligent Control Systems that utilizes both Petri nets and matrix forms for modeling and simulation of FMS is described. Chapter 7 describes basic principles of virtual factory modeling and simulation as a powerful means of FMS performance visualization. Several commercial program packages for virtual modeling are shortly presented. In addition, a web tool FlexMan developed by the authors and implemented to serve for off-line design and simulation of flexible manufacturing systems using a matrix model and a virtual model of the FMS is described.

Many individuals have contributed to this book. Special credits go to our colleagues Ayla Gurel and Octavian Pastravanu. We are also indebted to the students who contributed by implementing aforementioned program tools and by performing some of the simulation and practical experiments while working on their diploma and masters theses. This list includes, in particular, Bruno Birgmajer, Goran Genter, Krešimir Petrinec, Tomislav Reichenbach, and Nenad Smolić-Ročak.

Zagreb, 10.03.2006.

Stjepan Bogdan
Frank L. Lewis
Zdenko Kovačić
Jose Mireles

Contents

1 Introduction 1
  1.1 Background 2
        1.1.1 Flexible Manufacturing Systems and Their Controllers 2
        1.1.2 Summary of Approaches to Manufacturing System Control 2
  1.2 Flexible Manufacturing Systems 3
        1.2.1 Types of Manufacturing Systems 3
        1.2.2 FMS Design Tools 5
  1.3 Dispatching Rules and Blocking Phenomena 8
  1.4 Models of Discrete Event Manufacturing Systems 9
        1.4.1 Rule-based Expert Systems 9
        1.4.2 Petri Nets 10
        1.4.3 Graphs 14
  1.5 A Matrix-based Discrete Event Controller 15
        1.5.1 Matrix-based Discrete Event Controller Equations 15
  1.6 Simulation of FMS Control Systems 16
        References 17

2 Discrete Event Systems 21
  2.1 Time-driven Systems 22
  2.2 Event-driven Systems 34
        2.2.1 Automaton 36
        2.2.2 Languages and Supervisory Control of DES 45
        References 48

3 Matrix Model and Control of Manufacturing Systems 51
  3.1 System Matrices 53
  3.2 System Equations 58
        3.2.1 Logical State-vector Equation 59
        3.2.2 Job-start Equation 60
        3.2.3 Resource-release and Product-output Equations 61
        3.2.4 Recursive Matrix Model 62
  3.3 Modeling System Dynamics 67

    3.4 Matrix Controller 77
    3.5 A Case Study: Implemetation of the Matrix Controller........................... 86
         3.5.1 Intelligent Material Handling (IMH) Workcell Description......... 86
         3.5.2 IMH Workcell Dispatching Strategy 89
         3.5.3 Implementation of the Matrix Controller on the IMH Workcell .. 91
         3.5.4 The Matrix Controller in LabVIEW Graphical Environment....... 93
    3.6 Excersises 95
         References.. 95

4 **Matrix Methods for Manufacturing Systems Analysis**................................ 97
    4.1 Basic Definitions of Graphs.. 98
         4.1.1 Matrix Representation of the Graph 103
    4.2 String Composition 110
    4.3 Max-plus Algebra 120
         4.3.1 DEDS Model in Max-plus Algebra 124
         4.3.2 Periodic Behavior of DEDS in Max-plus 127
         4.3.3 Buffers in Max-plus Algebra 130
         4.3.4 Deriving Max-plus System Equation from Matrix Model.......... 140
    4.4 Exercises 143
         References.. 144

5 **Manufacturing System Structural Properties in Matrix Form** 147
    5.1 Multiple Re-entrant Flowlines – MRF... 148
         5.1.1 Circular Waits in MRF Systems 150
         5.1.2 Resource Loops in MRF Systems.. 156
         5.1.3 Siphons and Traps in MRF Systems.. 158
         5.1.4 Critical Subsystems in MRF Systems.. 164
         5.1.5 Key Resources and Irregular Systems in MRF.......................... 169
    5.2 Free Choice Multiple Re-entrant Flowlines – FMRF 170
         5.2.1 Structural Properties of FMRF 173
    5.3 Matrix Controller Design in MRF Systems 178
         5.3.1 Deadlock Avoidance in MRF Systems....................................... 178
         5.3.2 Deadlock Avoidance in Irregular Systems 181
         5.3.3 Deadlock Avoidance in FMRF Systems..................................... 184
    5.4 A Case Study: Deadlock Avoidance in PLC-controlled FMS 199
         References.. 208

6 **Petri Nets** 211
    6.1 Basic Definitions 212
    6.2 Manufacturing Systems Modeling... 226
         6.2.1 Petri-net Controller 231
    6.3 Relation Between Petri Nets and Matrix Form 238
    6.4 Petri Nets Simulation and Implementation 242
    6.5 Validation of Implemented Petri Nets 247
         References.. 257

7 Virtual Factory Modeling and Simulation 259
- 7.1 3D Modeling of Manufacturing Systems 261
- 7.2 Modeling FESTO FMS in VRML (X3D) Format 262
  - 7.2.1 Basic VRML Features 263
  - 7.2.2 FESTO FMS VRML Model 265
- 7.3 Modeling in LISA. 267
- 7.4 GRASP2000 (BYG Systems Ltd, UK) 270
- 7.5 Robot Studio (ABB, Sweden) 271
- 7.6 Tecnomatix eM-Plant (UGS, USA) 273
- 7.7 CIMStation Robotics (AC&E, UK) 275
- 7.8 COSIMIR (FESTO, Germany) 275
- 7.9 FlexMan (LARICS, University of Zagreb, Croatia) 276
  - 7.9.1 FlexMan Structure 277
  - 7.9.2 Database 279
  - 7.9.3 Virtual FMS Modeling 279
  - 7.9.4 Functional Modeling of FMS 279
  - 7.9.5 Generating Trajectories in FlexMan 280
  - 7.9.6 Simulation and Visualization of FMS operation 282
  - 7.9.7 Internet-based Multiuser FMS Control with FlexMan 283
  - 7.9.8 A Selection of an FMS Control Method 284
- 7.10 Exercise 290
- References 292

Index 295

1
Introduction

In this book a modern systems theory point of view is offered for the design of supervisory controllers for flexible manufacturing systems (FMS). The supervisory controller is installed on a PLC or on a computer, and sensors situated in the FMS are used to provide information to the controller about the status of the FMS, including job performance information and resource-availability information. Then, the controller performs calculations to determine which jobs should be performed next to achieve the specified performance requirements, such as meeting the product due dates, avoiding blocking phenomena, maximizing machine usage, minimizing time of transit of the product through the FMS, *etc.* Finally, commands are sent back to the FMS to select which jobs should be performed next and which resources should be used. Such a manufacturing controller is called a discrete event (DE) controller since it depends on the events that occur in the FMS.

The DE controller design techniques in this book are based on a *matrix-based* formulation for discrete event systems that streamlines modeling, analysis, simulation, and controller implementation for FMS. The matrices used in the DE controller formulation come from standard industrial engineering data structure techniques including the bill of materials, assembly tree, and resource requirements matrix; they are straightforward to write down for large-scale interconnected manufacturing systems using notions of block matrices.

In this chapter we give a preview of the philosophy behind supervisory control design. We outline some well-known tools in manufacturing industrial engineering, including the bill of materials, assembly tree, and job-sequencing matrix. We outline Petri nets and rule-based systems for DE controller design. We introduce a DE controller that has a very special and convenient form based on matrices and has close connections with all these background tools. Finally, we summarize methods for computer simulation of supervisory FMS controllers, and then techniques for their actual implementation on installed FMS.

Although this book focuses on manufacturing systems, this DE controller formulation is also applicable for other DE systems including autonomous guided vehicles (AGV), communication networks, wireless sensor networks, and computer operating systems.

1.1 Background

Some background is given here on FMS and their control techniques. This lays the foundation for the controller-design philosophy presented in this book.

1.1.1 Flexible Manufacturing Systems and Their Controllers

New developments in FMS, telecommunications systems, wireless networks, multiagent battlefield scenarios, computer operating systems, intelligent highway/vehicle systems, and elsewhere place severe demands on the design of decision-making supervisory control systems. The Internet and wireless communication mechanisms hold out the possibility of large-scale distributed systems spanning physically remote sites. The large-scale interconnected nature of such *discrete event* systems requires controllers/supervisors with increased capabilities for scheduling with optimality and capacity constraints, shared-resource dispatching, conflict resolution and deadlock avoidance, routing, failure handling, and other decisions. Many such DE systems are known to suffer from problems of computational complexity [1], where adding increased computer power will not significantly improve system performance, though performance can be improved through judicious choice of flow and command protocols, as well as improvements in system structure. Therefore, there are heightened demands for advanced supervisory controllers that include efficient organizational schemes, task protocols, and communications network protocols that impose increased structure on the system without detriment to strategic system objectives.

The concept of FMS emerged with the Ingersoll-Rand factory in Roanoke in the 1960s. An FMS consists of (1) programmable machines and robots, (2) an automated material handling system, and (3) a supervisory control system [2]. With the advent of FMS, the importance of the supervisory controller increases. The controller must be capable of quickly reprogramming the FMS to handle different parts and produce different products, and of dynamically handling contention and conflict decisions.

1.1.2 Summary of Approaches to Manufacturing System Control

Standard manufacturing engineering tools for heuristic analysis of decision-making supervisory controllers for flow shops and job shops include the bill of materials (BOM), Steward's task-sequencing matrix, the assembly tree, and the resource requirements matrix [3-5]. A body of work exists for shared-resource conflict resolution in industrial engineering, namely, the work on dispatching and scheduling (*e.g.* [6]). Standard dispatching rules show how to operate manufacturing cells in the presence of limited resources such as pallets, transport robots, machines. Results on kanban, CONWIP [7], and other pull techniques show how to avoid blocking phenomena by limiting the numbers of jobs in certain subsystems. Deadlock avoidance methods appear in [8-10], where the circular wait relations and circular blockings of an FMS are studied. A thorough treatment is given in [11].

Operations Research approaches to manufacturing modeling, analysis and control include mathematical optimization [4, 38], queuing network analysis [12, 13], and discrete event simulation [14, 15]. Mathematical programming models have been extensively studied and are suitable for open-loop planning and scheduling, though rarely for closed-loop controller design. Techniques include linear programming (LP), integer programming (IP), and quadratic programming (QP). Many algorithms exist to solve problems including simplex, and dual simplex. Algorithms that afford complexity reduction include Karmarkar (LP), branch and bound (IP), cut algorithms (IP), Hungarian algorithm (IP), and Fibonacci Search (QP). Dynamic programming [16] has been used to solve various controller design problems. Mathematical programming algorithms have been developed with emphasis on a hierarchical approach to modeling and control [17, 18].

Graphs and Petri nets (PN) [19, 20] afford a popular approach for analysis of FMS and computer operating systems [21]. PN are important as they provide insight into task sequencing and resource assignment, with analytical results on reachability, liveness, conservativeness, and other important system properties. If the PN is a (decision-free) event graph, it can equivalently be written as a linear system over the max/plus or dioid algebra [22], which affords even more analysis tools. Several researchers (*e.g.* [23, 24]) extended PN by using colored PN, stochastic PN, hierarchical PN, *etc.* "Top-down" and "bottom-up" design algorithms were proposed [25] along with improved techniques for the shared-resource allocation problem. Formal techniques for the design of PN supervisors or controllers are developed in [26, 27].

In perturbation analysis (PA) of discrete event systems [28] a dynamic system point of view is employed to study DE system behavior and analyze its performance. Many DE systems suffer from problems of computational complexity [1]. Therefore, the objective of PA is to obtain performance sensitivities with respect to system parameters by analyzing a single sample path of a discrete event system. Other work such as [29, 30-32] brings a system theory flavor into manufacturing dispatching, with the desired performance and bounded buffers being guaranteed via mathematical proofs including Lagrangian relaxation and Lyapunov stability techniques. Some work on fuzzy logic dispatching is available [33]. Supervisory control theory techniques for analyzing DE systems involve language-based approaches [27, 32], which offer effective analysis and design results for DE systems. Other work [34] has by now studied properties of hybrid and DE systems including stability, reachability, and so on.

1.2 Flexible Manufacturing Systems

1.2.1 Types of Manufacturing Systems

To meet competition in a global marketplace and provide flexible manufacturing in today's high-mix low-volume manufacturing environment, manufacturing systems have moved away from old-style fixed-hardware sequential processing lines with

dedicated workstations. The trend for years has been towards flexible manufacturing. An FMS has four major components [35]:
- a set of machines, robots, fixtures, or work stations,
- an automated material handling system that allows flexible part routing,
- distributed buffer storage sites where the parts may be temporarily placed during processing,
- a computer-based supervisory controller for monitoring the status of jobs and directing part routing and machine-job selections.

In order to allow fast setup of the FMS for new parts and product types, an advanced decision-making controller should be used. Proper design of the controller can allow one to program the FMS as easily as one does a personal computer. Such controllers are described in this book. The controllers are called *discrete event controllers* (DEC) since they make decisions based on the current *events* occurring in the FMS.

The controller should be distinguished from the physical portion of the FMS. The physical portion of an FMS is the *manufacturing facility*, comprised of its *resources*: the set of machines or work stations (including also robots, fixtures, tools, *etc.*), the automated material handling system, and the distributed buffers. Each resource type has a distinct function, though resource pools of more than one machine of a type may perform the same function (*e.g.* drill, press fit, paint, *etc.*). The resources serve the *parts*, and parts of the same class or type are grouped together, flowing through the facility on distinct *part paths*. The *job sequence* for each part type is the sequence of jobs required to produce a finished product.

There are several standard structures of manufacturing systems, including the re-entrant flowline, the assembly line, and the job shop. In the general *job shop* the sequencing of jobs is not fixed, or the assignment of resources to the jobs is not fixed. Parts of the same type may visit different machines in different orders to produce the same final product. The effect is that part *routing decisions* must be made during processing. This significantly complicates decision making and control in a manufacturing system and leads to problems with complexity issues.

In the *flowline* greater organization is imposed, and the sequence of jobs for each part type is fixed and the assignment of resource pools to the jobs is fixed. This results in a streamlined protocol that is easier to manage to provide guaranteed performance in the FMS. The result is that parts of each type visit the resources in the same sequence, though different part types may have different sequences. A flowline is said to be *re-entrant* if any part type revisits the same resource pool more than once in its job sequence [32, 30]. This occurs if the same resource is assigned to different jobs in the part's sequence. For instance, the same drilling operation may need to be performed twice at different stages in the part's processing.

An FMS at The University of Texas at Arlington (UTA) Automation & Robotics Research Institute (ARRI) is shown in Figure 1.1. This facility has three robots, an IBM, a PUMA, and an Adept. These robots have been connected through serial ports to allow central coordinated control from a single PC using a LabVIEW© user interface developed at UTA. A *matrix-based discrete event con-

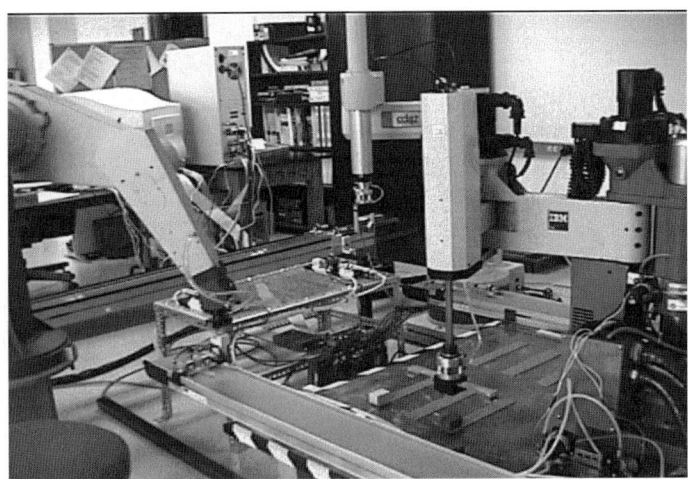

Figure 1.1. An FMS at UTA's Automation & Robotics Research Institute, showing three robots and three conveyor belts

troller such as those discussed in this book has been implemented, allowing for very fast and easy reprogramming of the FMS for new part types and products.

Figure 1.2 shows the re-entrant flowline structure of this FMS. Part type A is processed by the Puma robot twice, part type B is processed by the Adept twice, and both part types visit the IBM twice. Moreover, all three robots are used to process both part types. Thus, the three robots are all *shared resources*, which are visited several times by different parts. The flowline is *re-entrant* since parts of each type revisit the same resource more than once.

Since in the re-entrant flowline certain resources may be shared, either by parts of the same type at different stages of their processing, or across parts of different types, one is faced with a decision at each shared resource involving which part to process next. For instance, robot 2 has three queues where the parts enter - part A for the first time, part A for the second time, and part B for the first time. Parts may arrive at all these points simultaneously. Deciding which part to select next for processing at each shared resource is known as the *dispatching problem* [6]. The dispatching decision is a crucial one that can cause severe problems in a manufacturing system if not properly made.

1.2.2 FMS Design Tools

There are numerous tools available in industrial engineering usage for the design and analysis of manufacturing systems. We shall discuss here the bill of materials (BOM), the assembly tree, the task-sequencing matrix, and the resource-requirements matrix. In this book, these tools are combined into an overall design and analysis technique that results in rigorous algorithms, computer simulation techniques, and supervisory controllers with guaranteed performance. These tools are unified through a matrix-based DEC formulation presented in this book.

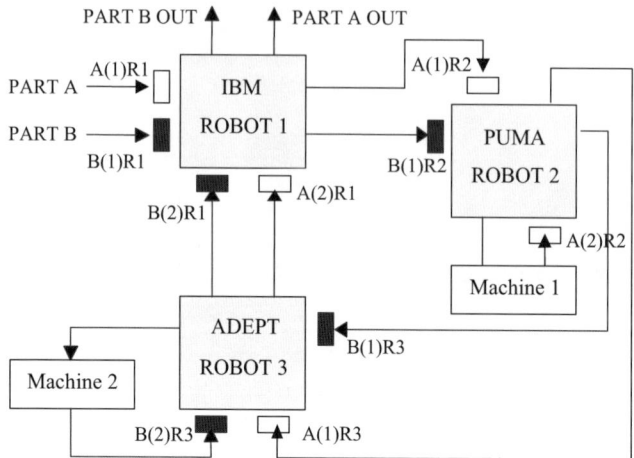

Figure 1.2. Re-entrant flowline structure of the FMS

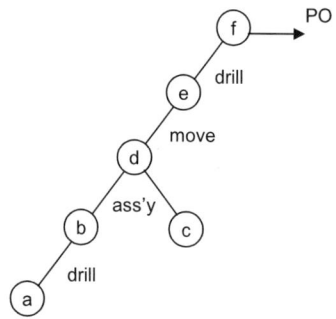

Figure 1.3. An assembly tree

The *bill of materials (BOM)* is a document containing the assembly - subassembly relationships for a specified product line [36]. It may be considered as a lookup table or a matrix in which the (i,j) entry is equal to the number of subassemblies/parts of type j needed to produce one subassembly/part of type i. Thus, row i contains the materials required to form the ith subassembly. The BOM is known for a given product line or part path. BOM information is an integral part of the specifications for all manufactured products.

The information in the BOM may be depicted in graphical form for easy visualization in the *assembly tree* [37], which shows the task decomposition of jobs needed to manufacture a product. A sample assembly tree is given in Figure 1.3. This tree shows that part a enters the workcell, where it is drilled. Then, part a and part c are assembled, moved, drilled again, and finally put out as the finished product, part out *(PO)*.

The sequence of events in an assembly tree can be captured in matrix form by defining the *job or task-sequencing matrix*, which for this example is

$$\mathbf{F}_v = \begin{array}{c} \\ a \\ b \\ c \\ d \\ e \\ f \\ PO \end{array} \begin{array}{c} \begin{array}{cccccc} a & b & c & d & e & f \end{array} \\ \left[\begin{array}{cccccc} 0 & 0 & 0 & 0 & 0 & 0 \\ 1 & 0 & 0 & 0 & 0 & 0 \\ 0 & 0 & 0 & 0 & 0 & 0 \\ 0 & 1 & 1 & 0 & 0 & 0 \\ 0 & 0 & 0 & 1 & 0 & 0 \\ 0 & 0 & 0 & 0 & 1 & 0 \\ 0 & 0 & 0 & 0 & 0 & 1 \end{array} \right] \end{array}$$

Each row i indicates which jobs are required as immediate precursors for job i. For instance, row 4 shows that jobs b and c are needed to perform the assembly job d. This matrix is effectively the BOM, and it was studied by Steward and others [5, 39, 40]. In this matrix, the columns and rows correspond to jobs, and an (i,j) entry of 1 indicates that job j is an immediate prerequisite for job i. The task-sequencing matrix is very useful for representing the partial orderings needed for sequencing manufacturing jobs. In fact, note that the causal sequencing of jobs d, e, f is seen in the diagonal 1s, showing that each job is an immediate prerequisite for the next job. It has been shown that a lower triangular job-sequencing matrix corresponds to a causal ordering of jobs, and that information on the hierarchical subsystem structure of a process can be extracted by raising this matrix to various powers.

The *resource-requirements matrix* (RRM) shows which resources are needed to perform which tasks or jobs, as reflected graphically, for instance, in the *subassembly tree*, which is an assembly tree annotated to indicate the resources assigned to the jobs [37]. The subassembly tree for this example is shown in Figure 1.4, where information has been added to show which resources are assigned to perform which jobs. Note that first the task sequence is prescribed, and then after that the resources are added. The task-sequencing information may come from the BOM or from computer science planning programs. On the other hand, the resource information might be assigned by a factory floor manager.

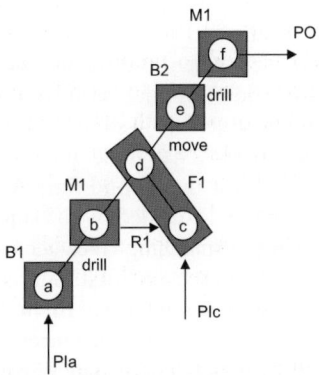

Figure 1.4. A subassembly tree

This subassembly tree shows that part a enters the FSM and is stored in buffer B1 then drilled in machine M1. Part c enters and is sent to fixture F1, where robot R1 assembles part a to it and puts the resulting subassembly in buffer B2. The subassembly is drilled again by machine M1, and finally sent out as the finished product PO. The resource requirements matrix for this example is given by

$$\mathbf{F}_r = \begin{array}{c} a \\ b \\ c \\ d \\ e \\ f \\ PO \end{array} \begin{bmatrix} R1 & F1 & B1 & B2 & M1 \\ 0 & 0 & 1 & 0 & 0 \\ 0 & 0 & 0 & 0 & 1 \\ 0 & 1 & 0 & 0 & 0 \\ 1 & 0 & 0 & 0 & 0 \\ 0 & 0 & 0 & 1 & 0 \\ 0 & 0 & 0 & 0 & 1 \\ 0 & 0 & 0 & 0 & 0 \end{bmatrix}$$

In the resource-requirements matrix, the columns correspond to resources (tools, fixtures, machines, robots) and rows correspond to jobs; an (i,j) entry of 1 indicates that resource j is needed for job i. Row 4, for instance, shows that robot R1 is needed to perform job d. In this example, note that the last column contains two 1s. This indicates that resource M1 is needed for two jobs, and hence it is a shared resource. Kusiak [3] has shown that RRM provides the basis for decision making while assigning or dispatching shared resources.

1.3 Dispatching Rules and Blocking Phenomena

If there is a shared resource, it is important to assign the correct jobs next to accomplish the *performance requirements* that are prescribed for the workcell. Performance requirements might include meeting product-due dates, keeping machine per cent utilization high, guaranteeing that all parts are processed through the FSM within a maximum allowed time, *etc*. The issues involve problems of assignment of shared resources when the same resources are simultaneously requested by more than one job. Similar issues occur in computer systems, communication systems, highway/vehicle systems, and elsewhere.

In Figure 1.2, for instance, robot 2 has three queues where the parts enter - part A for the first time A(1), part A for the second time A(2), and part B for the first time B(1). The dispatching strategy, executed by the supervisor, assigns which jobs to process next. There are many dispatching rules such as first-in-first-out (FIFO), where the parts arriving first are processed first, and earliest due date, where the part with the earliest due date is served first. In first-buffer-first-serve (FBFS) dispatching, the resource serves first the buffer corresponding to the first passage of a given part through the resource. For instance, robot 2 in Figure 1.2 would serve A(1), part type 1 entering for the first time, before serving A(2), the part type

1 entering for the second time. Correspondingly, another dispatching rule is last-buffer-first-serve (LBFS), wherein A(2) is given preference over A(1).

Failure to dispatch shared resources properly can result in blocking phenomena including *deadlock*, where all the resources in the FMS are busy, each waiting for the others to release a part before it can proceed. In this case, all activity in the FMS seizes up and no jobs can proceed. It has been shown that LBFS dispatching avoids deadlock in re-entrant flowlines with only one part type, while with FBFS deadlock may occur. In fact, LBFS is a *pull* technique thatattempts to clear jobs out of a FMS, while FBFS is a *push* technique that tries to load jobs into the workcell.

Deadlock research in computer systems has focused on four main areas [41]. Deadlock prevention is involved with removing any possibility of system deadlocks; the result is often overconservative policies resulting in poor utilization of resources. Deadlock detection focuses on detecting imminent or current deadlocks, and is required for deadlock recovery and avoidance strategies. Deadlock recovery methods are used to clear deadlocks once they occur, often by placing jobs in buffers, by manually removing some parts from machines, or by completely flushing one or more of the deadlocked processes, resulting in lost work. In deadlock avoidance the possibility of system deadlock is not totally removed, but whenever deadlock is imminent, it is sidestepped by a real-time decision-making procedure. Later in this book we shall be interested in online intelligent deadlock avoidance.

1.4 Models of Discrete Event Manufacturing Systems

There are several mathematical models for discrete event manufacturing systems. In manufacturing system control, one should discriminate between the workcell with its resources, and the supervisory controller that sequences the jobs and dispatches those resources. We shall now discuss the methods that are close to the matrix-based controller we will introduce in this book, and that also relate closely to the FMS design tools just discussed. These include rule-based expert systems and Petri nets.

1.4.1 Rule-based Expert Systems

One may describe the task-sequencing conditions and the resource assignments using a *rule-based system*. The task-sequencing rules may be derived from the bill of materials or assembly tree, and the resource assignment rules from the shop-floor supervisor, exactly as detailed above. The term *expert system* refers to the fact that the rules in the rule base are derived from advice and consultation with experts in the domain of interest. The product specialist specifies the task sequencing, while the factory manager specifies the resource assignments.

By examining the assembly tree in Figure 1.3, the partial assembly tree in Figure 1.4, and their associated task-sequencing matrix $\mathbf{F}_v$ and resource assignment matrix $\mathbf{F}_r$, one may directly write down the following rules for implementation of the assembly tree on an FSM. Each rule corresponds to one row in $\mathbf{F}_v$ and $\mathbf{F}_r$.

IF (buffer B1 is available) **THEN** (input part a)
IF (part a is input) **AND** (machine M1 is available) **THEN** (drill part a)
IF (fixture F1 is available) **THEN** (input part c)
IF (job b and job c have just been done) **AND** (robot R1 is available) **THEN** (assemble to form d)
IF (d has just been formed) **AND** (buffer B2 is available) **THEN** (move d to B2)
IF (part e is available in buffer B2) **AND** (machine M1 is available) **THEN** (drill the part)
IF (part has been drilled by M1) **THEN** (send final product PO out)

Note that this rule base implements the controller that generates products based on the given assembly tree. In each rule, the phrases to the left of the "THEN" are termed the rule *antecedent (prerequisites)*, and those following the "THEN" are termed the *consequent*. The antecedent has two parts, one coming from the task-sequencing matrix F_v and one from the resource-assignment matrix F_r.

Rule-based systems are very useful for programming programmable logic controllers (PLC) to implement the controllers for FSM, as we shall see. However, it is difficult to see the structure of a rule-based system, which means it is difficult to ensure that the rules are not conflicting and that they yield a causal job sequencing. It is difficult to use expert systems for computer simulation of FMS since they are difficult to interface with any description of the jobs and resources in the workcell. If some jobs change or some resources change, it is not easy to modify the corresponding rules in a large rule-based system. Finally, it is almost impossible to perform mathematical analysis of FMS performance or blocking phenomena in terms of rule-based systems.

1.4.2 Petri Nets

Event-driven systems are growing in popularity and complexity, and can be used to describe systems in manufacturing, vehicle-traffic systems, communication systems, computers, and wireless-sensor networks. This is motivating the use of well-organized design methodologies to avoid failures and to optimize performance. These systems usually have characteristics such as concurrence, conflicts, priorities, mutual exclusions, shared resources, and many others. These properties are difficult to handle, however, the analysis and design of these systems can be carried out using Petri nets (PN) [19, 20]. There are many varieties of Petri nets from binary PN, which are simple to analyze, to colored nets, which allow the modeling of more complex systems but have fewer analytic results.

Petri nets and their relations with matrix-based modeling and analysis are described in detail in Chapter 6. Here we give just a brief introduction to the topic without formal definitions of terms. A Petri net (PN) is simply a bipartite (*e.g.*, having two sorts of nodes) digraph (*e.g.* directed graph, which has arrows as arcs) described by (P, T, I, O), where P is a set of *places* and T is a set of *transitions* (later in the book we show the very important property that in fact each PN transition corresponds to a rule). These are both nodes in the graph. There are two types of arcs, namely I and O, where I is a set of (input) arcs from places to transi-

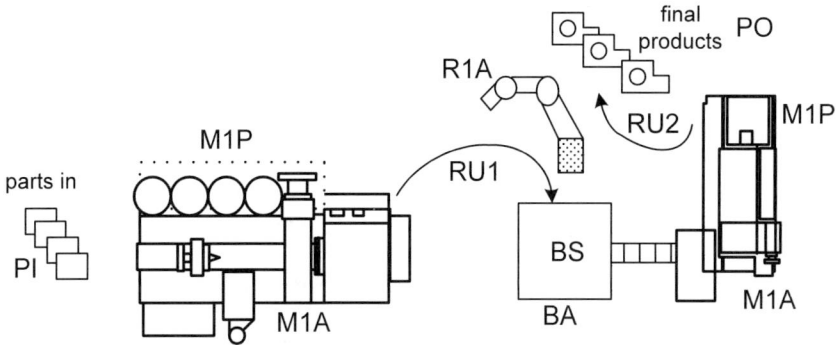

Figure 1.5. Flexible manufacturing system example

tions, and O is a set of (output) arcs from transitions to places. In our application, the PN places represent manufacturing resources and jobs, and the transitions represent decisions or rules for resource assignment/release and starting jobs.

An example of FMS is given in Figure 1.5. This shows one flowline for one part type, the required job sequence, and the required resources for each job. Robot R1 is a shared resource since it is responsible for performing two part moves - RU1 and RU2. Pallets have been added to carry the parts through the workcell; each pallet carries one part. Endings in A denote resource "available", and endings in P or S denote jobs in progress with those resources (buffer storage (S) or job in process (P)). The associated PN is given in Figure 1.6. In this figure, circles represent places, which correspond to jobs or resources, while vertical bars represent transitions, which fire under certain conditions.

The places along the part path denote jobs, while the places off the part path denote resources available. Along the part path, places and transitions alternate.

To denote the numbers of resources available and the numbers of jobs in process, one uses *tokens,* which are represented with black circle inside PN places, as shown in Figure 1.6. This PN shows that initially one has available 4 pallets (*e.g.* 4 parts can be in the workcell simultaneously), 2 machines M1, one robot R1, and so on.

Petri net dynamics is represented by the so-called *token game.* When a transition fires, a particular number of tokens is removed from each input place,

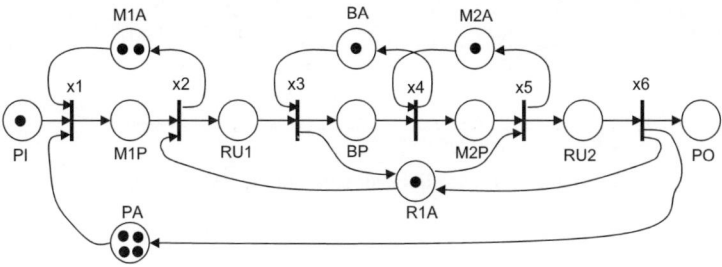

Figure 1.6. Petri net for the FMS example

and added to each output place. By keeping track of tokens one is able to simulate and analyze the behavior of the system described by the PN. The job sequencing, the resources needed to perform the jobs, the resource availability and utilization, jobs currently in progress and many other properties can be easily studied by following the tokens routes. Two already-mentioned phenomena, conflict and deadlock, which are particularly important in FMS supervisor design and most of the book is dedicated to their analysis, also can be allocated by tracking PN tokens.

Using the FMS example, we can illustrate the meaning of conflict and deadlock. In Figure 1.7 we see a setup of the current situation in the FMS. There are two jobs waiting for the pick & place robot resource R1, namely the token in M1P shows that there is a job waiting for robot R1 to clear it from machine M1 in move RU1, while the token in M2P shows that there is a job waiting for R1 to remove it in move RU2. Unfortunately, there is only one robot in the resource pool R1A, and it must select only one of these two jobs to perform, hence, R1A is in a conflict.

Suppose R1 is dispatched to perform move RU1. Then, transition x_2 fires and the situation now moves to that shown in Figure 1.8. This is quite a bad situation. Clearly, there is no way that any transition can now fire in this figure. The problem is that each of the resources is waiting for another resource to become available. However, this will never happen. Therefore, all activity along the part path ceases and can never resume. Some thought can reveal that if the shared-resource robot R1 had elected to perform the *downstream* move RU2 instead of RU1, that would not have resulted in deadlock. This example illustrates the notions in Section 1.3, namely, dispatching using first-buffer-first-serve (FBFS) results in deadlock, but last-buffer-first-serve (LBFS) avoids deadlock. Pull policies generally are safer than push policies.

This short depiction demonstrated that PNs are a powerful graphical tool for discrete event systems modeling. However, in order to be able to provide thorough analysis one needs an appropriate mathematical framework. Since the PN is a graph with two types of nodes, the arcs in the PN are described by two matrices, the PN *input incidence matrix* **I** and the *output incidence matrix* **O**.

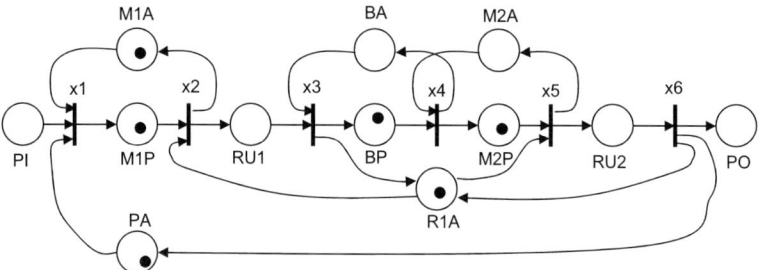

Figure 1.7. Predeadlock situation in the FMS example

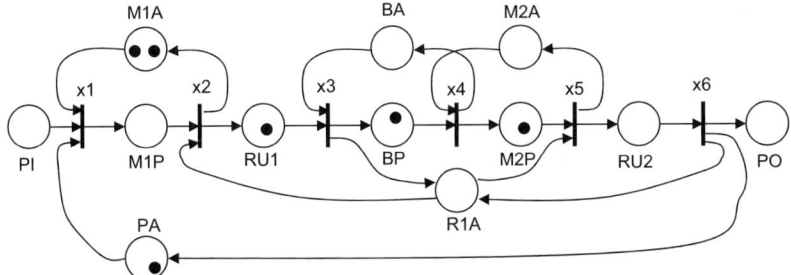

Figure 1.8. Deadlock situation in the FMS example

The PN input incidence matrix **I** has element (i,j) equal to 1 if place j is an input to transition i. The PN output incidence matrix **O** has element (i,j) equal to 1 if place j is an output from transition i. The input incidence matrix for the assembly PN in Figure 1.9 is given by

$$
\mathbf{I} = \begin{array}{c} \\ x1 \\ x2 \\ x3 \\ x4 \\ x5 \\ x6 \\ x7 \end{array} \begin{array}{c} a \; b \; c \; d \; e \; f \end{array} \left[\begin{array}{cccccc|ccccc} 0 & 0 & 0 & 0 & 0 & 0 & 1 & 0 & 0 & 1 & 0 & 0 \\ 1 & 0 & 0 & 0 & 0 & 0 & 0 & 0 & 0 & 0 & 0 & 1 \\ 0 & 0 & 0 & 0 & 0 & 0 & 0 & 0 & 1 & 0 & 0 & 0 \\ 0 & 1 & 1 & 0 & 0 & 0 & 0 & 1 & 0 & 0 & 0 & 0 \\ 0 & 0 & 0 & 1 & 0 & 0 & 0 & 0 & 0 & 0 & 1 & 0 \\ 0 & 0 & 0 & 0 & 1 & 0 & 0 & 0 & 0 & 0 & 0 & 1 \\ 0 & 0 & 0 & 0 & 0 & 1 & 0 & 0 & 0 & 0 & 0 & 0 \end{array} \right]
$$

with column headers: $a\ b\ c\ d\ e\ f\ PA\ R1A\ F1A\ B1A\ B2A\ M1A$.

It is highly interesting to note that the first block of this matrix is simply $\mathbf{F}_v$, the task-sequencing matrix, and the second block simply $\mathbf{F}_r$, the resource-requirements matrix, from Section 1.2.2 (pallets PA have been added). This will be a central theme later in the book. Likewise, the output incidence matrix **O** can be written for this PN very easily.

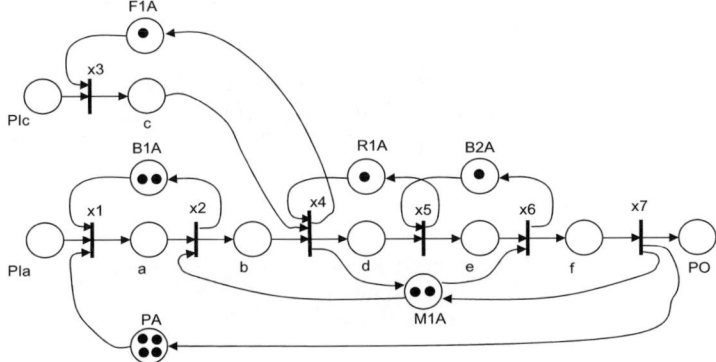

Figure 1.9. PN for the subassembly tree example

In terms of the PN incidence matrices, one can describe the update of the tokens when transitions fire. In fact, if $\mathbf{m}(t_1)$ is a vector whose components correspond with the number of tokens in PN places at a prescribed time t_1, then the updated token placement is described by the vector $\mathbf{m}$ at the next time t_2 as

$$\mathbf{m}(t_2) = \mathbf{m}(t_1) + \mathbf{W}^T \boldsymbol{\tau}$$

where $\mathbf{W} = \mathbf{O} - \mathbf{I}$, and $\boldsymbol{\tau}$ is a vector comprised of integers that correspond to the number of firings of transitions in PN in the time interval $t_2 - t_1$. We shall return to this equation and its relation with matrix-based formalism in Chapter 6.

Unfortunately, this equation is not a complete description of a PN since it does not take into account the order of firing of the transitions, nor whether a given transition can actually fire at any point in time. That is, there is no way known in PN theory to compute the transition firing vector $\boldsymbol{\tau}$.

1.4.3 Graphs

As we see in the previous section, graphs are quite important in manufacturing system analysis and control; not only in PN theory but also, as we shall present through the chapters that follow, in other modeling and design tools. They indeed provide some rigorous techniques for the analysis of discrete event control systems. A *graph* is a set of *nodes* and the *arcs* connecting them. A directed graph, or *digraph*, associates directions with the arcs so that they effectively become arrows.

For example, a special graph can be constructed from a Petri Net by considering only the resource places. Refer to Figure 1.6, and start at any resource. Proceed *backwards* along the arcs until you come to another resource, then backwards from that resource to the next resource, and so on until you have traversed all the arcs. Draw arrows through the resources traversed. The result is the digraph shown in Figure 1.10.

This graph is known as the *wait relation graph* [8]. Note from the PN that resource R1 cannot become available until BA is available and performs job BP. That is, R1 *waits for* the buffer BA to become available before it can become available.

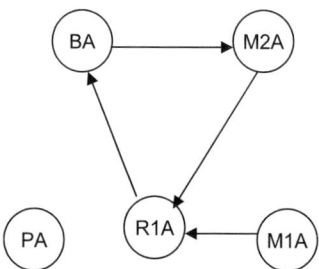

Figure 1.10. Wait relation graph

In the wait relation graph depicted, there is a looming potential problem. Namely, there is a *circular wait relation* (R1A, BA, M2A), wherein each resource waits for another. Unless extreme care is taken in dispatching the jobs within this circular wait, one will arrive at deadlock, exactly as has occurred in Figure 1.8, where all the resources in the circular wait are busy and waiting for each other. Methods for avoiding deadlock, as well as dispatching to achieve performance specifications in FMS, thus hinge on understanding the *structure* of FMS, namely the circular wait relations and other important structural elements.

1.5 A Matrix-based Discrete Event Controller

A rule-based discrete event supervisory controller that is based on matrices is now briefly described. This matrix-based DE controller plays a central theme in this book, and it allows fast programming of FMS for assembly/job sequencing, resource dispatching, and blocking and performance analysis, and facilitates dispatching and routing design. The controller provides a framework for rigorous analysis of the structure and performance capabilities of an FMS. Furthermore, the controller also allows a very convenient method for computer simulation and implementation on actual FMS.

1.5.1 Matrix-based Discrete Event Controller Equations

The DE controller is based on a matrix formulation where each matrix has a well-defined function for job sequencing, resource assignment, and resource release. The matrix-based model of a discrete event system is described by the set of equations. Since each equation is thoroughly described in separate sections later in the book, they are introduced herein with no further explanations:

Logical state-vector equation

$$\bar{x} = \mathbf{F}_v \Delta \bar{\mathbf{v}}_c \nabla \mathbf{F}_r \Delta \bar{\mathbf{r}}_c \nabla \mathbf{F}_u \Delta \bar{\mathbf{u}} \nabla \mathbf{F}_d \Delta \bar{\mathbf{u}}_d$$

Job-start equation

$$\mathbf{v}_s = \mathbf{S}_v \Delta \mathbf{x}$$

Resource-release equation

$$\mathbf{r}_s = \mathbf{S}_r \Delta \mathbf{x}$$

Product-output equation

$$\mathbf{y} = \mathbf{S}_y \Delta \mathbf{x}$$

In these equations, $\mathbf{F}_v$ is exactly the job sequencing matrix, and $\mathbf{F}_r$ the resource-requirements matrix discussed above. Each of the other matrices also has a specific function and meaning, which is explained in the chapter that follows.

The four equations given above are the central part of the matrix-based controller. They are not computed using standard matrix operations of multiplication and addition. In fact, all the matrices and vectors are *logical variables* that have entries of either "0" or "1", exactly as for the job sequencing and resource assignment matrices. Therefore, all matrix operations are defined to be in so-called *and/or algebra* (see section 3.2), denoted $\triangle$ and $\triangledown$, where standard addition is replaced with logical *or* and standard multiplication with logical *and*. The overbar denotes logical *negation*.

Input **u** represents raw parts entering the cell and **y** completed tasks, or products, leaving the cell. The controller, implemented on a PLC or a computer, *dynamically observes, in real time, the workcell status* by looking at the *status outputs* of the DE system or workcell using installed sensors, represented by a job vector $\mathbf{v}_c$, and a resource vector $\mathbf{r}_c$. Hence, the matrix-based supervisor has a *dynamic feedback control* structure. On top of it, higher-level dispatching and/or routing decisions are needed to determine vector $\mathbf{u}_d$ that selects which jobs to initiate. This dispatching input is selected in higher-level control loops using priority assignment techniques (*e.g.* [6]) in accordance with prescribed performance objectives such as minimum resource idle time, task priority orderings, task due dates, minimum time of task accomplishment, and so on as prescribed by the user. Then, the controller sends *commands* to the FMS workcell, namely, vector $\mathbf{v}_s$ for jobs to be started, and vector $\mathbf{r}_s$ for resources to be released.

Since the matrix DE controller is a rule base, it can be *directly used to program a programmable logic controller*. This means that PLC can easily be programmed to control actual industrial processes directly and simply from the matrix DE controller. Two case studies related to the implementation of the matrix controller in FMS are presented in the book with references to other applications of matrix-based approaches.

The matrix-based DE controller unifies tools from different aspects of manufacturing, computer science, and discrete event systems. It uses the BOM, task-sequencing matrix, and resource-requirements matrix. Moreover, it can be shown that the complete task plan ($\mathbf{F}_v$, $\mathbf{S}_v$, $\mathbf{F}_r$, $\mathbf{S}_r$) generates a Petri net. In fact, as one might surmise from the discussion at the end of Section 1.4.2, $\mathbf{F}_v$ and $\mathbf{F}_r$ generate the PN input incidence matrix, while $\mathbf{S}_v$ and $\mathbf{S}_r$ generate the PN output incidence matrix. As we shall show, this means that for any speciation of the matrices in equations one can draw a PN.

1.6 Simulation of FMS Control Systems

A comprehensive approach for analysis of computational complexity in FMS (and elsewhere) is the theory of *NP- (nonpolynomial)-completeness* [1]. Mathematical programming approaches to scheduling were mainly based on combinatorial optimization methods until the development of the theory of NP-completeness in the 1970s. Many traditional scheduling and sequencing problems have been found to be in the NP class, however there is no formal theory describing how to impose *structured flow and command protocols* on an FMS to simplify the complexity.

Since analytical results are often difficult to obtain for DE systems, particularly for transient analysis, the performance of FMS, including scheduling and dispatching rules and other algorithms, has often been studied using simulation [14,

15]. There are available many packages for simulation of manufacturing systems (WITNESS, SIMFACTORY, Gert [42], etc.), Petri nets (Design/CPN, Grafcet [43], TORA, etc.), and general DE systems (SIMAN, Simscript, Simula, Smalltalk-80, GPSS, Extend). In these packages various programming methods are used; object-oriented techniques, knowledge-based approaches, Lotus 1-2-3, Prolog and many others. Various efficient simulation techniques may be based on perturbation analysis or system theory approaches. Many of these tools use brute force approaches that do not take advantage of the protocol structures of manufacturing flowlines, assembly lines, and job-shop systems. The large number of techniques available show the complications arising from simulation of DE systems.

The use of virtual models has become a standard characteristic of modern program tools for virtual modeling and simulation of FMS (*e.g.* Grasp 2000, eM-Plant [44], RobotStudio, Cimstation Robotics, Cosimir, *etc.*). Virtual models provide a very convenient and inexpensive way for the complete factory design, allowing a clear visualization of all potential problems in an FMS caused by a factory layout, job sequencing or resource requirements.

Besides physical modeling that relies on the virtual models of all constituent FMS objects, an important task is functional testing that connects a physical setup with the organization of the simulated FMS. Functional testing comprises several tasks including a definition of a job sequence, setting of FMS parameters, local (at the robot workcell or robot station level), and global (at the whole FMS level) conflict and deadlock analysis, synthesis of control logic, investigation of different job-scheduling strategies, simulation and visualization of dynamic phenomena that occur during FMS operation. Although most of the above-mentioned program tools contain some types of DE simulation tools and DE controller design techniques, and automatically generate downloadable programs for particular FMS elements (*e.g.* robots) and accompanying programmable logic controllers, *they do not allow for ease of computer simulation and do not support direct generation of code needed for FMS controller implementation on actual industrial systems.*

A matrix-based approach to the FMS controller design can be easily integrated in the virtual-reality environment, and the result of simulation with a selected dispatching policy can be effectively visualized and analyzed by observing an animated performance of the FMS. In the book, we describe FlexMan [45], an Internet-based virtual-factory simulator with an integrated matrix-based FMS controller and automatic FMS controller code generator for an industrial PLC (Siemens PLC S 216).

References

[1]　Garey MR, Johnson DS. Computers and Intractability: a Guide to the Theory of NP-completeness, W.H. Freeman, 1979.
[2]　Groover MP. Automation, Production Systems, and Computer-Integrated Manufacturing, 2nd edn, Prentice Hall, 2001.
[3]　Kusiak A. Intelligent scheduling of automated machining systems, in Intelligent Design and Manufacturing, ed. A. Kusiak. New York: John Wiley & Sons, 1992.
[4]　Baker KK. Introduction to Sequencing and Scheduling. New York: John Wiley & Sons, 1974.

[5] Steward DV. The design structure system: a method for managing the design of complex systems, IEEE Trans. Eng. Manag. 1981;71–74.
[6] Panwalker SS, Iskander W. A survey of scheduling rules, Operations Research 1977;26;1:45–61.
[7] Spearman ML, Woodruff DL, Hopp WJ. CONWIP: a pull alternative to kanban, Int. J. Prod. Res. 1990;28;5:879–894.
[8] Wysk RA, Yang NS, Joshi S. Detection of deadlocks in flexible manufacturing systems, IEEE Trans. Rob. Automat. 1991;7:853–859.
[9] Ezpeleta J, Colom JM, Martinez J. A Petri net based deadlock prevention policy for flexible manufacturing systems, IEEE Trans. Rob. Automat. 1995;11;2:173–184.
[10] Lewis FL, Gurel A, Bogdan S, Doganalp A, Pastravanu O. Analysis of deadlock and circular waits using a matrix model for flexible manufacturing systems, Automatica 1998;34;9:1083–1100.
[11] Zhou M. Deadlock Resolution in Computer–Integrated Systems. New York: Marcel Dekker/CRC Press, 2004.
[12] Gross D, Harris CM. Fundamentals of Queuing Theory. New York: John Wiley & Sons, 1985.
[13] Jackson JR. Networks of waiting lines, Operations Research 1957;5:518–521.
[14] Banks J, Carson JS. Discrete Event System Simulation, Prentice Hall, 1984.
[15] Law AM, and Kelton WD. Simulation Modeling and Analysis. New York: McGraw Hill, 1991.
[16] Lewis FL, Syrmos VL. Optimal Control, 2nd edn. NewYork: John Wiley and Sons, 1995.
[17] Bitran GR, Haas EA, Hax AC. Hierarchical production planning: a two stage system, Operations Research 1982;30;2:232–251.
[18] Silver EA, Peterson R. Decision Systems for Inventory Management and Production Planning. New York: John Wiley & Sons, 1985.
[19] Desrochers AA. Modeling and Control of Automated Manufacturing Systems, IEEE Comp. Soc. Press, 1990.
[20] Murata T. Petri nets: properties, analysis and applications, Proc. IEEE 1989;77;4:541–580.
[21] Balbo G, Bruell SC, Ghanta S. Combining queueing network and generalized Petri net models for the analysis of some software blocking phenomena, IEEE Trans. Soft. Eng. 1986;12;4:561–576.
[22] Cohen G, Dubois D, Quadrat JP, Viot M. A linear-system-theoretic view of discrete-event processes and its use for performance evaluation in manufacturing, IEEE Trans. Aut. Contr. 1985;AC-30;3:210–220.
[23] Kasturia E, DiCesare F, Desrochers A. Real time control of multilevel manufacturing systems using colored Petri nets, Proc. IEEE Conf. Rob. Automat. 1988:1114–1119.
[24] Murata T, Komoda N, Matsumoto K, Haruna K. A Petri net-based controller for flexible and maintanable sequence control and its applications in factory automation, IEEE Trans. Ind. Electr. 1986;IE-33;1:1–8.
[25] Zhou MC, DiCesare F. Petri Net Synthesis for Discrete Event Control of Manufacturing Systems. Boston: Kluwer, 1993.
[26] Krogh BH, Holloway LE. Synthesis of feedback control logic for discrete manufacturing systems, Automatica 1991;27;4:641–651.
[27] Ramadge PJ, Wonham WM. The control of discrete event systems, Proc. IEEE 1989;77:81–98.
[28] Ho YC, Cao XR. Perturbation analysis and optimization of queueing networks, J. Optim. Th. and Appl. 1983;40;4:559–582.

[29] Burgess KL, Passino KM. Stability analysis of load balancing systems, Proc. Amer. Contr. Conf. 1993:2415–2419.
[30] Lu SH, Kumar PR. Distributed scheduling based on due dates and buffer priorities, IEEE Trans. Aut. Contr. 1991;36;12:1406–1416.
[31] Luh PB, Hoitomt DJ. Scheduling of manufacturing systems using the Lagrangian relaxation technique, IEEE Trans. Aut. Contr. 1993;38;7:
[32] Kumar PR, Meyn SP. Stability of queueing networks and scheduling policies, Proc. IEEE Conf. Dec. and Contr. 1993:2730–2735.
[33] Angsana A, Passino KM. Distributed fuzzy control of flexible manufacturing systems, IEEE Trans. Contr. Syst. Tech. 1994;2;4:423–435.
[34] Antsaklis P, Kohn W, Nerode A, Sastry S. Hybrid Systems II, Springer-Verlag, 1995.
[35] Buzacott JA, Yao DD. Flexible manufacturing systems: a review of analytical models, Management Sci. 1986;32;7:890–905.
[36] Elsayed EA, Boucher TO. Analysis and Control of Production Systems, 2nd edn, Prentice Hall, 1994.
[37] Wolter J, Chakrabarty S, Tsao J. Methods of knowledge representation for assembly planning, Proc. NSF Design and Manuf. Sys. Conf. 1992:463–468.
[38] Graves SC. A review of production scheduling, Operations Research 1981;29:646–675.
[39] Warfield JN. Binary matrices in system modeling, IEEE Trans. Sys., Man, Cyb. 1973;SMC-3:441–449.
[40] Eppinger SD, Whitney DE, Smith RP. Organizing the tasks in complex design projects, Proc. ASME Int. Conf. Design Theory and Methodology 1990:39–46.
[41] Deitel HM. An Introduction to Operating Systems, Addison-Wesley, 1984.
[42] Cash CR, Wilhelm WE. Simulation modeling approach for analyzing robotic assembly cells, Proc. 18th Conference on Winter Simulation 1986:594–596.
[43] David R, Alla H. Petri Nets and Grafcet: Tools for Modeling Discrete Event Systems. New York: Prentice Hall, 1992.
[44] Heinicke M. U. and Hickman A., Eliminate bottlenecks with integrated analysis tools in eM-Plant, Proc. 2000 Winter Simulation Conference, pp. 229–231, 2000.
[45] Bogdan S., Kovačić Z., Smolić-Ročak N. and Birgmajer B. A Matrix Approach to an FMS Control Design – From Virtual Modeling to a Practical Implementation, IEEE Rob. Aut. Mag., Vol. 11, No. 4, pp. 92–109, December 2004.

2

Discrete Event Systems

From the moment when a human being became aware of its existence, until the present, one question has dominated through a long history of ups and downs: how to predict the future? This question attained many forms; how to predict winds on high seas? How to predict the floods of Nile? How to find the probability that an electron would appear at a particular place in an atom? How to predict the way the Universe ends? Step by step we have found some methods and piece by piece the future revealed its secrets. In the foundations of all of these techniques, whose purpose was to foresee future events, appeared *a model*. Based on the experience gained from observations, people build models and then, by setting these models into various conditions, they are able to predict future events. When these particular conditions, already tested on the model, occur in real life, we can know more or less accurately, what will be the outcome.

Establishing an appropriate model for some general problem might be very demanding. To help ourselves, we separate particular entities from the surroundings. These isolated entities should have the property that they operate together in a way not possible by any one of them individually, *i.e.* they should form *a system*. Where the boundary line between the system and the surroundings is set depends on the problem we want to analyze. The boundary of the system defines *inputs* and *outputs* of the system – its connections to the environment and to the other systems. If an airplane's attitude is a subject of our study, then we treat the whole plane as a system and investigate the influence that airplane *parameters* (mass, wing span), *variables* (speed, elevation) and environment properties (air pressure, wind) have on the attitude. Although an airplane is a very complex system, built of many *subsystems,* we can intentionally ignore the influence of some observable facts (jet engine r.p.m.) in order to make the model feasible. On the other hand, if we focus our intention on the jet engine, which is a subsystem of the plane, then the boundary line is set in such a way that the plane becomes the environment and the jet engine becomes the system under investigation.

Once the boundary line is drawn and the system is determined, its behavior can be described with a model. The above definition of the system, as a set of entities that form a whole and act together, is the broadest one and as such it requires various types of models. It is clear that a political system cannot be described with the same type of model as an airplane. Whilst the latter belongs to the class of *technical systems* and can be modeled by mathematical equations, a model of the former is represented by a set of words, sentences and paragraphs. In the text that follows we are concerned with technical systems, *i.e.* systems that encompass physical devices built by a human.

In the control-engineering literature technical systems are usually divided into two major groups: *time-driven systems* and *event-driven systems*. In this chapter we describe the basic concepts of these two groups. First, we give a brief description of the basic properties of time-driven systems. The well-known facts associated with these systems are given in order to be able to compare their activities with the behavior of the second group, event-driven systems, which are the major topic of the chapter. A reader acquainted with the time-driven systems may proceed directly to Section 2.2.

The notions of *an event, a system state, a clock* and others are presented and described herein. A brief presentation of automata, as the modeling tool that is most frequently used in the analysis of the event-driven system, is given.

2.1 Time-driven Systems

In a mathematical description (a mathematical model) of a dynamic system, quantities that change with time are associated with *variables*, while quantities that describe system properties and generally remain constant are called *parameters* (systems with constant parameters are called *time invariant*; if parameters are changing with time we talk about *a time-variant system*). The role of time in system modeling is interesting and important. Due to its unique property – no matter where the boundary between the system and the environment is drawn and no matter how the variables of the system model are chosen, the time remains *independent* (we do not consider systems that include theory of relativity phenomenon) – each system variable can be represented as a function of time, *i.e.* time is the argument of all functions that describe the system. In this way any change in time causes a change of the system variables. We say that the system is *time driven* and write a system model as

$$\mathbf{y}(t) = G[\mathbf{u}(t)] \quad (2.1)$$

where $\mathbf{u}(t)$ is a system input vector, $\mathbf{y}(t)$ is a system output vector and G is an *operator* that describes how the system *transforms (maps)* the input vector into the output vector. It is common in the literature that relation (2.1) (the same holds for other forms of mathematical descriptions) is referred to as "a system" although it actually represents more or less accurately a "model of the system" (in very few situations the complete model of the system is known).

When that operator G changes with time, a system model is written as

$$\mathbf{y}(t) = G[\mathbf{u}(t), t] \qquad (2.2)$$

For a system described with the model (2.1), input **u**(*t*), applied at the moment *t* = t_0, would have the same effect on the system as if it were applied at any other moment *t* = t_0 + *τ*. This is not the case for a system described with Equation (2.2); as operator **G** changes with time, the output of that system depends not only on the form of the input vector **u**(*t*) but also on the moment in which the input is applied to the system.

Further classification of time-driven systems is closely related to the principle of proportionality, *i.e.* an increase of the system input value by factor *b* will increase the system output value by factor *b*. Systems with this property are called *linear systems*. We say that operator **G** is linear if and only if

$$G[b \cdot u(t)] = b \cdot G[u(t)] \qquad (2.3)$$

Although most technical systems do not behave in accordance with the principal of proportionality (they are *nonlinear*), linear models are usually used for their description since the mathematical tools for the analysis and design of linear models are much easier to understand and implement than these methods used in the nonlinear systems theory [1–3]. Furthermore, in most cases the linear models describe the real systems to the extent that is considered satisfactory from the practical point of view.

So far we were concerned with the so-called input-output representation of the system. This representation can be expanded by the notion of *system state*. The output of a dynamic system depends not only on the current input value but also on the past values of the input, *i.e.* dynamic systems have a "memory". The memory is in the form of conservation of energy and/or information. This means that some kind of internal properties that are not explicitly seen from the model (2.1) are present in the system. Thus, in order to obtain the model that would demonstrate the internal phenomenon of the system, the modeling process should take into account not only the system input and output vectors and their relationships, but the system states should be incorporated into the model also. In the systems theory this kind of model is usually represented in the form of set of differential equations [4]:

$$\begin{aligned} \dot{\mathbf{x}}(t) &= f[\mathbf{x}(t), \mathbf{u}(t), t], \quad \mathbf{x}(t_0) = \mathbf{x}_0 \\ \mathbf{y}(t) &= g[\mathbf{x}(t), \mathbf{u}(t), t] \end{aligned} \qquad (2.4)$$

where **x**(*t*) is *the state vector* containing system states and **f** and **g** are functions. Equations (2.4) are called *state equations* and they uniquely describe the system state at any time instant $t \geq t_0$.

As an example, let us consider the system shown in Figure 2.1. The system represents a DC motor drive with mechanical load. Differential equations that des-

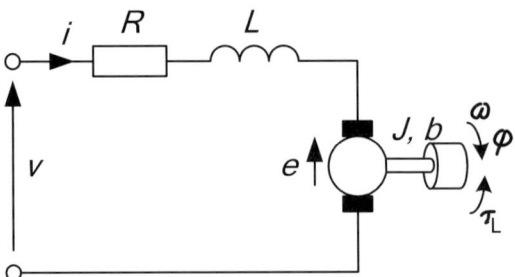

Figure 2.1. A DC motor drive with mechanical load

cribe the electrical and mechanical dynamics of the system have a well-known form:

$$R \cdot i(t) + L\frac{di(t)}{dt} + e(t) = v(t), \quad e(t) = K \cdot \omega(t)$$

$$J\frac{d\omega(t)}{dt} + b \cdot \omega(t) = \tau_M(t) - \tau_L(t), \quad \tau_M(t) = K \cdot i(t)$$

where K is a motor constant, τ_L is a load torque and τ_M is a motor torque. Which physical variable will be defined as an input and which one as an output depends on the purpose of the model. When one is investigating the influence of voltage $v(t)$ on the motor torque $\tau_M(t)$, then the voltage should be considered as the system input and motor torque as the output. On the other hand, if one is concerned with the influence that the voltage has on the rotor position $\varphi(t)$ then this physical variable should be treated as the system output.

For the latter case the state equations obtain the following form:

$$\frac{di(t)}{dt} = \frac{1}{L}\left[-R \cdot i(t) - K \cdot \omega(t) + v(t)\right]$$

$$\frac{d\omega(t)}{dt} = \frac{1}{J}\left[K \cdot i(t) - b \cdot \omega(t) - \tau_L(t)\right] \qquad (2.5)$$

$$\frac{d\varphi(t)}{dt} = \omega(t)$$

$$y(t) = \varphi(t)$$

where the system input $u(t) = v(t)$.

The solutions of state Equations (2.4) are functions that describe evaluation in time of each system state. These functions are called the *state trajectories*. State trajectories for Equations (2.5) together with input $u(t)$ are shown in Figure 2.2. We can see that the system state φ changed after the input (in the form of a short pulse) has been applied. Initial value $\varphi(t_0) = \varphi_0$ changed to $\varphi(t \to \infty) = \varphi_F$, while in the same time the system states i and ω returned to their initial values after the transition period has finished.

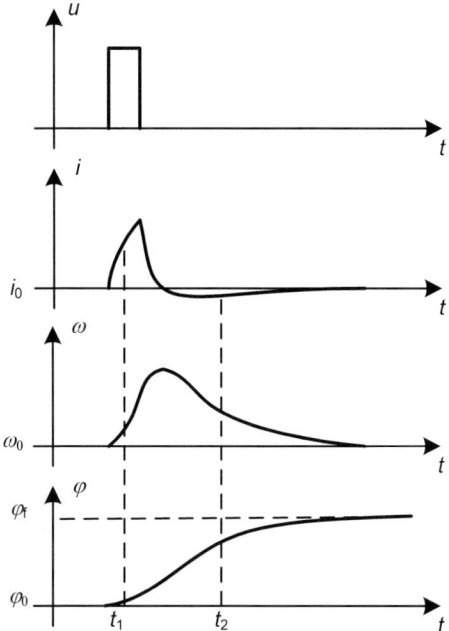

Figure 2.2. Trajectories of the state Equations (2.5)

A very interesting presentation of the state trajectories can be attained if time as a variable is eliminated from functions that correspond with the state trajectories. A new function $q(\mathbf{x})=0$, obtained in this way, represents movement of the state vector $\mathbf{x}$ in the *state space*. The state space of a system is the set of all possible values that the system state vector may attain. For systems with up to 3 states, state-vector trajectories can be presented graphically as in Figure 2.3. Points on the state-vector trajectory correspond to the points on trajectories of system states. As time changes from t_1 to t_2 the system states are changing, thus causing the state vector to travel through the state space from point $\mathbf{x}(t_1)$ to point $\mathbf{x}(t_2)$, which is shown in Figures 2.2 and 2.3.

The state-vector trajectories are very important in system analysis and design. Many interesting system properties, such as stability, can be examined directly from the state-vector trajectories. This is particularly important in the case of nonlinear systems, since, as already mentioned, mathematical tools for this type of systems are complicated and demanding.

As we said earlier, the main purpose of the model is its ability to more or less accurately predict the future states of the system once the current state and the system inputs are known. If we define the system states that have to be reached from the current state and when we are able to manage the system inputs, then we can define these inputs in a way that they guide the state vector directly to the desired state. We say that the system is *controlled*. The question is how to determine the system inputs? Usually there exist at least two control objectives; a) the system should be conducted to the desired state and b) this state should be reached in a particular way, *i.e.* the state vector should follow a predefined trajectory.

Based on the objectives and the system model we can determine an input vector **u**(*t*) that would fulfill both requirements:

$$\mathbf{u}(t) = h\big[\mathbf{u}_\mathbf{r}(t)\big] \quad (2.6)$$

where $\mathbf{u}_\mathbf{r}(t)$ is a *reference input* and *h* is a *control function* that maps the desired system behavior, described by the reference input $\mathbf{u}_\mathbf{r}(t)$, to the system input **u**(*t*). One concept is especially important when the definition of the control function is concerned. This is the concept of *controllability*. In order to determine a control law and apply it to the system, the structure of the system input vector should be known in advance. As a first step in the controller design one must identify system states that can be influenced by the input vector components. Such states are called *controllable states* and outside signals can be supplied into the system only through these states.

A problem with the control law (2.6) lies in the fact that the control function does not take into account possible changes that may happen in the system during implementation of the control law. In our example with a DC motor drive, a change in the load torque, τ_L, would modify the second state equation in Equations (2.5), thus causing different state-vector trajectory and, as a consequence, the desired system state will not be reached. A solution to that problem is one of the fundamental principles in nature; the principle of *feedback*. In biological systems, as well as in social systems, most actions are based on feedback; if we get a fever our body starts to sweat in order to enhance heat exchange with the surroundings, an insecure political situation lowers prices on the stock markets, and so on. In many cases the feedback is an inherent property of the system (our body), while on the other hand it can be artificially added in order to enable the system to cope with various *disturbances* that influence its behavior.

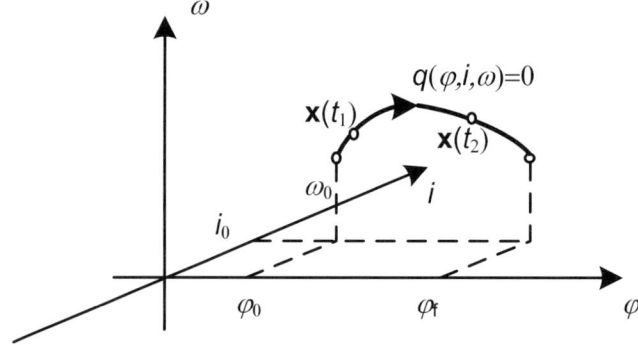

Figure 2.3. The state-vector trajectory $q(i, \omega, \varphi)=0$ of the state Equations (2.5)

In technical systems, the information regarding the current state of the system (feedback) is usually acquired by measurement of the system output (or the system states) and then fed back into the system by the control law,

$$\mathbf{u}(t) = h\left[\mathbf{u}_r(t), \mathbf{x}(t)\right] \tag{2.7}$$

The primary constraint that is related to determination of Equation (2.7) is associated with the notion of *observability*. Specifically, some parts of the system (system states) might be inaccessible for measurement. Such unobservable states cannot be used as feedback for the controller (2.7).

Let us demonstrate the feedback principal on our DC motor drive example. A feedback, which will compensate the influence of load torque, is introduced by measuring the position $\varphi(t)$. In that case the system input, $v(t)$, may be calculated as

$$v(t) = K_\varphi \cdot \left[\varphi_r(t) - \varphi(t)\right] \tag{2.8}$$

where K_φ is the proportionality factor (or *gain*) and $\varphi_r(t)$ is the reference position. The *proportional control law* (2.8) is the simplest form of control function h. Inclusion of Equation (2.8) in the first equation of Equations (2.5) gives

$$\frac{di(t)}{dt} = \frac{1}{L}\left[-R \cdot i(t) - K \cdot \omega(t) + K_\varphi \cdot \left[\varphi_r(t) - \varphi(t)\right]\right]$$

It can be seen that, as long as there is a difference between the reference and current positions, the motor current $i(t)$ will change. As a result, the state vector moves around the state space until the final (reference) state is reached. Although the applied control law leads the system to the desired state despite the changes in load torque, it can not accomplish the second control objective – the way in which the system gets into the desired state (the state-vector trajectory) is changed (certainly, a more complex control law could handle both objectives, but that analysis is beyond the scope of this book).

Now we introduce the other form of time-driven systems. As a start we can consider our everyday experience – parking a car (Figure 2.4).

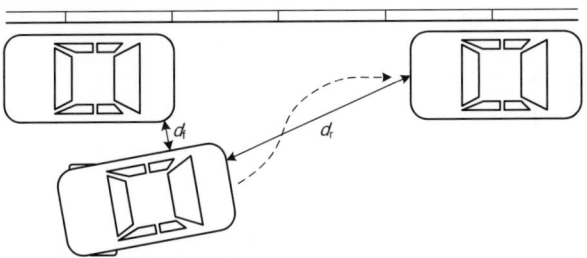

Figure 2.4. A parallel parking example

It is well known that parallel parking is a challenging task (not only for new drivers). Two actions have to be taken simultaneously and very carefully - steering the wheel and balancing the clutch and the break. These actions are generally based on the feedback that is in the form of two variables - a distance from the front car, d_f, and a distance from the rear car, d_r. Since we have only one distance measurement sensor (eyes) that has to deal with two variables, we have to concentrate our attention into two directions. This can be done only if we are toggling our view from one feedback variable (front car distance) to the other (the rear car distance) during the parking. Instead of getting the whole information regarding distances d_f and d_r (Figure 2.5), only a sequence of partial data, taken in *discrete-time intervals* as shown in Figure 2.6, is processed during the parking (it should be noted that as we are getting closer to the front and rear cars we have to acquire data more often). This partial information, collection of *samples* of continuous variables, is sufficient for more or less successful completion of the car-parking maneuver.

The question is how to fit sampled variables into the system model (2.4)? First, to make things easier from the mathematical point of view, instead of stochastically samples would be taken in equal time intervals, $t_1, t_2, t_3, \ldots, t_k, \ldots$, with $T_d = t_k - t_{k-1}$. As a second modification, the sampled value taken in t_k would be "frozen" during the sample interval $t_k + T_d$. Given these modifications, continuous variables shown in Figure 2.5, attain the form presented in Figure 2.7.

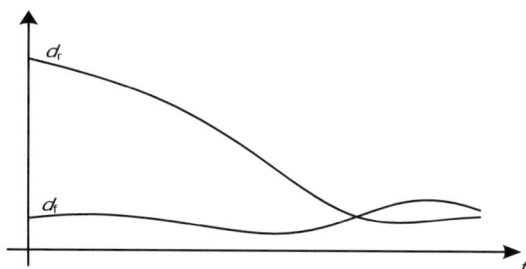

Figure 2.5. The continuous-time variables representing distances from the front and rear cars

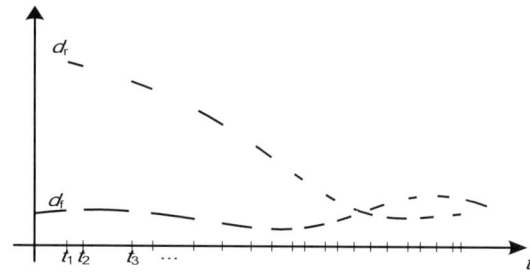

Figure 2.6. Samples taken by the driver

Now, we can rewrite the continuous-time model (2.4) in the discrete-time form:

$$\mathbf{x}(t_{k+1}) = \Phi\left[\mathbf{x}(t_k), \mathbf{u}(t_k), t_k\right], \quad \mathbf{x}(t_0) = \mathbf{x}_0$$
$$\mathbf{y}(t_k) = \Gamma\left[\mathbf{x}(t_k), \mathbf{u}(t_k), t_k\right]$$
(2.9)

As $t_k = k \cdot T_d$, $k = 0, 1, 2, \ldots$, sometimes Equation (2.9) is written as [5]:

$$\mathbf{x}(k+1) = \Phi\left[\mathbf{x}(k), \mathbf{u}(k), k\right], \quad \mathbf{x}(0) = \mathbf{x}_0$$
$$\mathbf{y}(k) = \Gamma\left[\mathbf{x}(k), \mathbf{u}(k), k\right]$$

The evaluation of system states and system output in a *discrete-time model* is obtained recursively by the *difference equations* (2.9) [6]. Since the value of variable k increases as time evolves, the system-state changes are synchronized with time, hence the discrete-time systems are time driven. When Φ and Γ are linear functions, the discrete-time model becomes the system of nonhomogeneous linear equations.

Although in the parking example the nature of the system (one sensor that should monitor two feedback states) was the reason for the sampling of continuous variables, usually the implementation of the control law is why the system has to be represented in the discrete-time form. Nowadays, advances in computer technology provide low-cost solutions for very sophisticated and computationally demanding control algorithms. As the execution of the control laws in the form of a computer program is performed in discrete-time intervals, discrete-time models are needed for appropriate design and investigation of control algorithms. In every sampling interval continuous system variables are sampled and converted by analog-to-digital (A/D) conversion into the numerical values in order to be processed by the computer. Upon execution of the control law, results, in numerical form, are returned into the system by digital-to-analog (D/A) conversion. During the execution cycle (which takes some time), the computer is

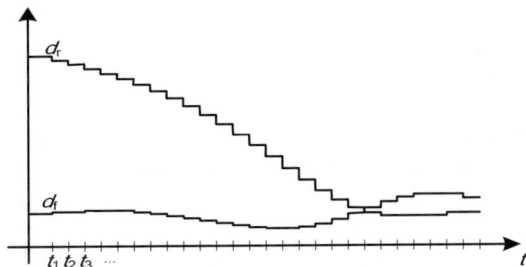

Figure 2.7. Discrete-time form of continuous-time variables

"not aware" of the system development; only at particular moments in time are the system conditions sampled and conveyed into the computer.

Even though this issue is not a subject of the book, it should be mentioned that data exchange between the system and the digital computer requires not only sampling in time but also sampling of state space, as the numerical resolution of the computer is finite. Actually, this requirement is related to the limited resolution of A/D and D/A converters (usually 10 to 16 bits). For example, the continuous system variable that takes values from the set of real numbers, when processed by a 12-bit A/D converter, is mapped into 4096 integers. Usually, the range of the converters is considered sufficient so that discretization of the state space is ignored in the process of discrete-time modeling.

So far we were concerned with the models of systems that change their states in synchronization with time. As an introduction to the concept of an *event* and an *event-driven state* we start with an example of the system that belongs to the class of so-called *hybrid systems* - a broad class of technical systems that integrate both time-driven and event-driven states [7]. Hence, some of the hybrid system states change in synchronization with time, while the change of others is caused by events that occur asynchronously (here we do not elaborate what an event is; it is a primitive concept intuitively understandable). Although very complex and difficult to analyze, during the last decade hybrid systems have become an important topic for researchers and engineers, which is expected, since most of the industrial control and automation solutions fit into this class of systems. However, the hybrid system design is still done mostly by a heuristic approach as current theory is complicated and requires time-consuming methods [8, 9].

The concept of an event-driven state becomes apparent if we bear in mind that in many cases (especially in practical control implementations) the status of an actuator or a sensor is described with only two categories, *i.e.* valve opened-valve closed, motor is running-motor is idle, sensor is active-sensor is idle, *etc*. If the status of a motor is considered as a state, then the event "switch on the motor" changes the system state. Furthermore, in order to make a control algorithm synthesis simple, continuous states of hybrid systems are partitioned in regions that are treated separately during the system design. Then, the goal of the hybrid system controller synthesis is to find an algorithm intended to fulfill the desired requirements for a particular region. The switching between regions and controllers and the binary nature of actuators and sensors make some of the system states event driven.

Example 2.1.1 (event-driven states in hybrid system)

The system under consideration is the longitudinal tunnel ventilation. Vehicles passing through a tunnel produce various types of poisonous gases as well as soot, especially in the case of heavy vehicles with diesel engines. High standards for air quality and the need for good visibility require an advanced ventilation system for management and control of pollution. Two objectives, opposite in nature, have to be fulfilled simultaneously by the ventilation system: a) the system should keep visibility (opacity) at a required level and make certain that pollutants (mainly carbon monoxide - CO) remain within admissible margins and b) energy (costs) used for objective a) should be minimal.

Here we are not concerned with the design of control algorithms that use continuous system states and meet both objectives concurrently (that might be a very complex and demanding issue). Our aim is to design a controller based on event-driven system states. The controller should utilize the carbon monoxide concentration as a feedback signal and the number of vehicles per hour per kilometer as a feedforward signal (variations of this type of controller are used in many practical implementations of the tunnel ventilation). As a first step in achieving our goal both continuous time signals are divided into three regions – low (L), moderate (M) and high (H) (three regions are chosen for simplicity – typical tunnel ventilation controllers use more than 7). Transitions over predefined threshold values between two neighboring regions are considered as *events*. These events will be the driving force of the system model. We assume that low traffic produces a low level of CO, a moderate number of vehicles a moderate level of CO, and a high number of vehicles a high level of CO.

In the tunnel ventilation systems jet fans are usually used as actuators. It is presumed that actuators can be described as active (1) and idle (0). The system is designed in the way that active jet fans reduce carbon monoxide concentration so that it moves down into the neighboring region. The actuator's state is changed by two actions, switching on (ON) and switching off (OFF). These actions, treated as events, can be triggered by an operator or by the controller.

At this point it is important to note that some events, such as jet-fan breakdown or sensor failure for example, are *uncontrollable*. Since this type of event cannot be influenced by the supervisor, modeling and design of the systems regularly starts by neglecting potential effects that uncontrollable events might have on the system performance. Then, as the second step in the system synthesis, the states forced by uncontrollable events are analyzed and additional features (fault tolerance) are incorporated into the supervisor. In our example, we disregard jet-fan failure and concentrate only on events that are necessary for the description of the event-driven state concept.

Having defined the system in that way, we can express the set of events as

$$E = \{V_{inc}, V_{dec}, CO_{inc}, CO_{dec}, ON, OFF\}$$

where V_{inc} and V_{dec} are events related to transitions of thresholds defined for the number of vehicles and CO_{inc} and CO_{dec} are events related to transitions of thresholds defined for the carbon monoxide.

Given that the supervisor has two discrete inputs, a number of vehicles, N_v, and a carbon monoxide concentration, CO, and one output, actuators status, A, the system state x can be represented with triples (N_v, CO, A). For given values, there exist 18 discrete states (for example, state (M, L, 1) stands for "moderate number of vehicles", "low CO concentration" and "actuators running"). It is evident that due to the system nature some states are *unreachable*. Such a state, for example, is (L, H, 1), *i.e.* low number of vehicles cannot produce a high level of CO with jet fans running (actually, such situation can happen in the case of fire in the tunnel, but our model does not consider this catastrophic incident). In a vector form the system state is represented with the system state vector $\mathbf{x} = [N_v\ CO\ A]^T$.

The occurrence of an event from E changes the system state, thus causing a movement of the system state vector in the system state space X, as shown in Figure 2.8. Upon event OFF, the system that resided in $\mathbf{x}^1 = [M\ L\ 1]^T$ goes to the new state $\mathbf{x}^2 = [M\ L\ 0]^T$. Since jet fans have been turned off, the concentration of CO increases, *i.e.* CO_{inc} occurs, and the system state vector becomes $\mathbf{x}^3 = [M\ M\ 0]^T$. Then, a new event V_{inc} forces the system into $\mathbf{x}^4 = [H\ M\ 0]^T$. As the number of vehicles increases the level of carbon monoxide starts to rise, thus causing occurrence of event CO_{inc}, which forces the system into state $\mathbf{x}^5 = [H\ H\ 0]^T$. High concentration of CO can be reduced by switching on the jet fans (event "ON"), that leads the system into state $\mathbf{x}^6 = [H\ H\ 1]^T$. When the jet fans start to dilute CO, CO_{dec} takes place and the system state vector attains its final value $\mathbf{x}^7 = [H\ M\ 1]^T$. Continuous system variables with corresponding events, are shown in Figure 2.9.

For a given event-driven model of the system one is able to determine the event-driven controller. Depending on the control goal and the system characteristics the discrete state space can be partitioned into several regions. For example, one of the regions has already been mentioned, *i.e.* a region of unreachable states. From the control point of view particularly interesting is a *forbidden region*, *i.e.* the region that localizes the states that must be avoided. Once defined, these states are a basis for the system-controller design. For example, if in the tunnel ventilation system all states with a high level of CO are not allowed (states of the form $\mathbf{x} = [\cdot\ H\ \cdot]^T$), then the supervisory control actions might be defined as follows: switch on jet fans each time system arrives at states $\mathbf{x} = [H\ \cdot\ 0]^T$ and switch off jet fans when the system leaves states $\mathbf{x} = [H\ \cdot\ 1]^T$. Continuous system variables of a so-controlled system, together with corresponding events, are shown in Figure 2.10.

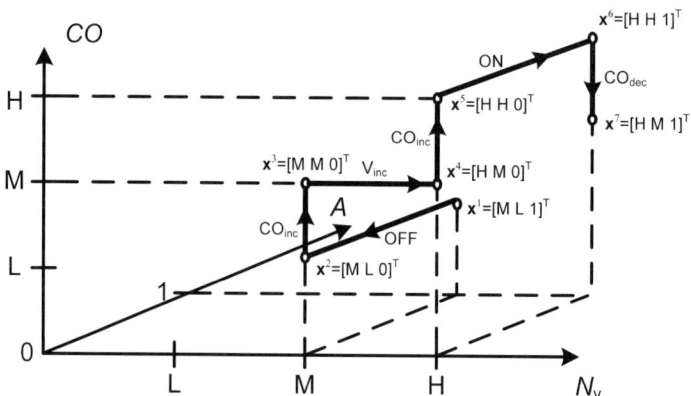

Figure 2.8. Movement of the system state vector in discrete state space

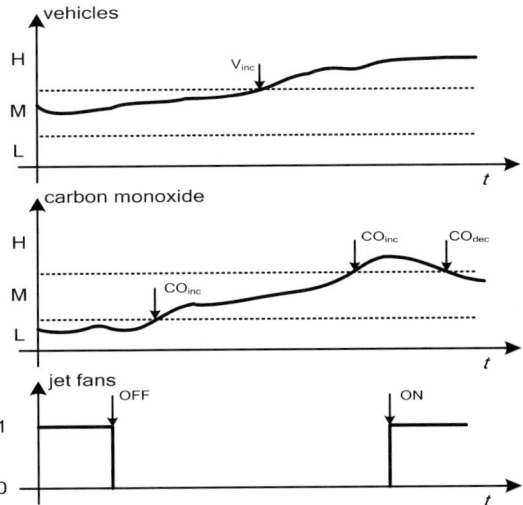

Figure 2.9. Continuous system variables with corresponding events

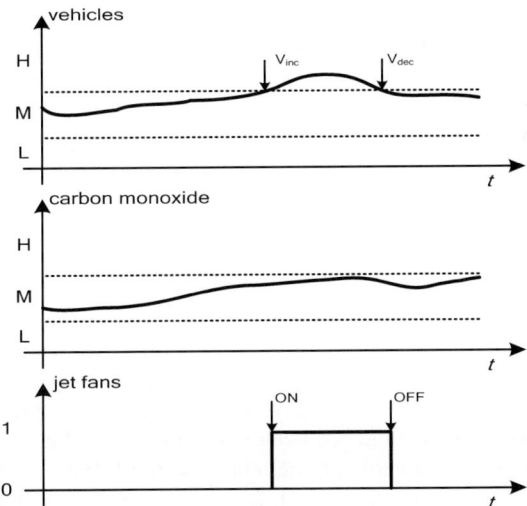

Figure 2.10. Continuous variables of the controlled system (with corresponding events)

It can be seen that event V_{inc} drives the system into state $\mathbf{x} = [\text{H M } 0]^T$. As a reaction, the supervisor triggers the event ON and the system reaches state $\mathbf{x} = [\text{H M } 1]^T$. Without control the system will settle into $\mathbf{x} = [\text{H H } 0]^T$ (see Figure 2.9). As the number of vehicles decreases, event V_{dec} causes a change of the system state to $\mathbf{x} = [\text{M M } 1]^T$. Hence, the supervisor generates the event OFF forcing the system into state $\mathbf{x} = [\text{M M } 0]^T$.

♦

2.2 Event-driven Systems

In the previous example event-driven states were created from time-driven states, and then used for the supervisor design. A simple control objective was accomplished by triggering events that kept the system out of the undesired states. Now we move further and introduce the systems that encompass only event-driven phenomena, i.e. event-driven states are an inherent property of the systems.

Let us now consider a game in which two persons are placed at the table. There is a panel between them so that they are not able to see each other, although, there is an open space below the panel. Person A has in his/her hands a set of 10 cards, each card marked with different letter: a, b, c, d, f, g, h, m, n, p. Person B has a pen, a paper and a watch. The watch has an alarm that is set to be active every 10 seconds. The game is as follows. Person B has closed eyes and on the alarm signal he/she should open eyes and write down the letter displayed on the card placed on the table. If there is no card on the table, a letter x should be written on the paper. Once the letter is marked, person B closes his/her eyes and waits for the next alarm. Person A has headphones and can not hear the alarm. He/she randomly, in various time instants, picks the card, writes down the displayed letter and leaves the card on the table below the panel for a few moments for it to be visible to the other person. Then the card is put aside. A possible outcome of the game is shown in Table 2.1. The table contains the letters written by the players.

Table 2.1. A possible outcome of the card game

Player A	a	h	d	f	n	b	c	g	m	p
Player B	x	x		f		x	c		x	x

From Table 2.1 it can be seen that person B "caught" only 2 cards out of 10 that have been placed on the table by person A; 80% of the information has been lost. After the third alarm person B saw a card on the table for the first time. From his/her perspective person A still had 9 cards in hand while actually only 6 cards remained. Obviously, actions performed by person A can not be accurately measured by the technique used by person B. One can argue that much better results could be achieved if the alarm was set to 1 second, but this is not the point, that is, the observation method has been synchronized by the watch alarms, while the process of placing the cards on the table was random and asynchronous. From the person A's point of view the system state (cards remaining in hand) changed 10 times, while from person B's perspective the system state changed only 2 times. Clearly, this system can not be described with a model in which the system state evolves in synchronization with time, since the evolution of the system state is caused by *asynchronous events* (placement the cards on the table).

The second aspect that has to be taken into account, when one considers modeling the systems like the one described above, is the system state space. In our example the set of cards remaining in hand was regarded as the system state. Since the status of a particular card may take only two values, "in hand" or "put aside", it is apparent that the system state can attain only *discrete values*, thus the state space is *discrete*, in contrast to time-driven systems where the state space is continuous (compare Figures 2.3 and 2.8).

Asynchronous events that cause the change of system state vector in discrete state space characterize *event-driven systems*, also called *discrete event systems* (*DES*) [10, 11].

Let us define a set E as the set that comprises all events e_i that can occur in the system. In the card game example we have

$$E = \{a, b, c, d, f, g, h, m, n, p\}$$

i.e. event b corresponds with placing the card marked with the letter b on the table, event c corresponds with placing the card marked with the letter c on the table, and so on.

In our example, each time an event takes place the system state changes (it should be noted that in some discrete event systems there exist events that do not change the system state). In most systems a simultaneous occurrence of two (or more) events is not allowed (this can be enforced by the system design or it can be its inherent property), thus events arise in some *order* or *sequence*. From Table 2.1 we see that the sequence of events in the game was $s = (a, h, d, f, n, b, c, g, m, p)$ (in Example 2.1.1, Figure 2.9, the sequence was (OFF, CO_{inc}, V_{inc}, CO_{inc}, ON, CO_{dec})). If we associate vector $\mathbf{x}$ with the system state in the way that each component of the vector corresponds to the card in hand, 1 if the card is in hand, 0 if the card is put aside, then, at the beginning of the game, we shall have

$$\mathbf{x}_0 = \begin{bmatrix} 1 & 1 & 1 & 1 & 1 & 1 & 1 & 1 & 1 \end{bmatrix}^T$$

where the first component of the vector stands for the card marked with the letter a, the second component for the card marked with b, and so on.

Given the system state vector $\mathbf{x}$, and the system initial condition $\mathbf{x}_0$, we can express the state of the system after a sequence of events $s_1 = (b, d, f, p)$ as

$$\mathbf{x}^1 = \begin{bmatrix} 1 & 0 & 1 & 1 & 1 & 1 & 1 & 1 & 1 \end{bmatrix}^T$$
$$\mathbf{x}^2 = \begin{bmatrix} 1 & 0 & 1 & 0 & 1 & 1 & 1 & 1 & 1 \end{bmatrix}^T$$
$$\mathbf{x}^3 = \begin{bmatrix} 1 & 0 & 1 & 0 & 0 & 1 & 1 & 1 & 1 \end{bmatrix}^T$$
$$\mathbf{x}^4 = \begin{bmatrix} 1 & 0 & 1 & 0 & 0 & 1 & 1 & 1 & 0 \end{bmatrix}^T$$

It should be noted that sequence s_1 holds only partial information regarding the system state change; we know the ordering of events that forced the system from $\mathbf{x}_0$ to $\mathbf{x}^4$, but we are not able to tell the time instances in which the events actually took place. Adding the time in the sequence gives $s_1 = ((b, t_b), (d, t_d), (f, t_f), (p, t_p))$, where t_b represents the time instance of the occurrence of event b, t_d represents the time instance of the occurrence of event d, and so forth. Having the *timed sequence* defined in this way we can calculate how much time the system spends in a particular state. This kind of information is essential in investigation of system properties that are related to utilization and throughput of the system, due time of events, system transient time, *etc.* An untimed sequence describes only the logical (we might say IF-THEN) behavior of the system.

2.2.1 Automaton

So far we have introduced the basic concept of event-driven systems in an informal way. The tunnel ventilation and the card game examples encompassed a set of events that forced changes of the system states, thus forming a set of sequences. The problem with informal representation is that it is usually difficult to determine all possible sequences that could be generated by the system without some kind the of the system model. One of the most popular modeling tools for DES representation is *automaton* [12–14]. In the following text we give a concise description of the basic notations in automata theory.

Definition 2.2.1 (automaton): An automaton, denoted by A, is defined as a five-tuple

$$A = \{E, X, f, x_0, X_m\}$$

where E is the set of events, X is the set of states, $f : X \times E \to X$ is the *transition function*, x_0 is the initial state and X_m is the set of *marked states*.

In many cases (particularly when one deals with practical implementation of DES) sets E and X have a finite number of elements. The transition function f describes mapping between these two sets in the following way: if there exists an event e that generates transition from state x to state y, then $f(x,e) = y$. If upon the occurrence of event e the system state x does not change we write $f(x,e) = x$. When $f(x,a) = y$ and $f(y,b) = z$ we have

$$f(y,b) = f(f(x,a),b) = f(x,ab) = z \tag{2.10}$$

i.e. the definition of the transition function is generally extended to the *set of sequences*, denoted E^*. An additional property of the transition function should also be mentioned. Given that the set of events that cause transitions from state x is usually a subset of E, $\Gamma(x) \subset E$, it is apparent that the automaton transient function f exists only on the part of its domain (usually in the literature $\Gamma(x)$ is it called the *active event function* and it is a part of the automaton definition). Hence, $f(x,e)$ should not be defined for each event e at each state x.

The set of marked states, X_m, is a subset of X. In general, by using marked states one is able to point out that some states have a special meaning. For example, a marked state could be connected with the notion of an ending or a final state; the state in which the system resides most of the time (from that point of view a marked state can be related to the steady state in time-driven systems).

It should be noted that Definition 2.2.1 covers only the so-called *deterministic automata*, i.e. automata in which the occurrence of one particular event (or a sequence of events) forces the system into a strictly defined state. On the contrary, in nondeterministic automata one event may cause transitions from one state to several states, i.e. the value of the transient function is expressed as a subset of X, $f(x,e) = \{y, z, w\}$.

Having Definition 2.2.1 we can determine an automaton that models the status of jet fans in the tunnel ventilation system from Example 2.1.1, as

$$A_F = \{E_F, X_F, f_F, x_{F0}, X_{Fm}\}$$

where

$E_F = \{\text{ON,OFF}\}$, $X_F = \{0,1\}$, $X_{Fm} = \{1\}$
$f_F(0, \text{ON}) = 1$, $f_F(0, \text{OFF}) = 0$, $f_F(1, \text{OFF}) = 0$, $f_F(1, \text{ON}) = 1$, $x_{F0} = 0$

Observation of automaton A_F exemplifies Definition 2.2.1. Foremost, we see that the transient function f_F is defined on the whole domain since each event from E_F is related with each state from X_F. Secondly, the state $x = 1$ is marked. Its particular importance lies in the fact that it asserts the situation when the jet fans exploit energy, thus, the energy-usage calculation is active as long as the system stays in this state.

For simple automata, as the one describing jet fans, with just a few states and several events, a written form of presentation is suitable. On the other hand, for complex discrete event systems the more convenient way of automaton representation is in graphical form or in the form of a so-called *state transition diagram*, shown in Figure 2.11.

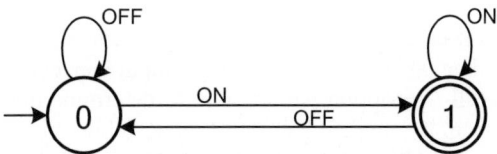

Figure 2.11. State transition diagram of automaton A_F

In mathematical formalism, structures such as a state transition diagram in Figure 2.11, are known as *directed graphs*. The basics of graph theory are covered in later chapters of the book so at this point we shall skip definitions and properties of graphs. For our purpose it is sufficient to note that in the state transition diagram labeled circles represent states and labeled arcs represent events. The initial state is shown as a circle marked with an arrow while a marked state is represented by a double circle. In the literature the state transition diagram is usually referred to as an automaton. We will use the same principle in this book.

The following example demonstrates robotized workcell modeling by using an automaton.

Example 2.2.1 (state transition diagram of a robotized workcell)

We examine the robotized workcell shown in Figure 2.12. Our goal is to design an automaton that models this cell. The machines and the robot are considered failure free, *i.e.* our model does not include breakdowns, malfunctions and other uncontrollable incidents. We assume that both parts have a stochastic arrival time.

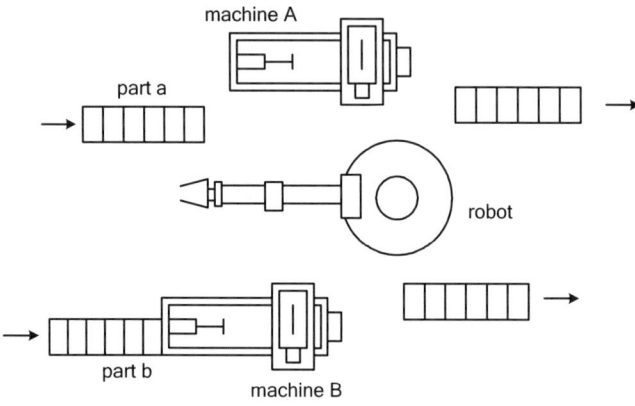

Figure 2.12. Workcell from Example 2.2.1

The cell consists of two machines and one robot. Two types of parts, *a* and *b*, are processed in the following way. Both parts are brought into the cell by input conveyers. Entering the cell, part *a* is picked up by the robot and transported to the machine A. When processing is finished the robot removes the part from the machine and leaves it on the output conveyer. Upon arrival, part *b* is processed in the machine B and then taken by the robot to its output conveyer.

From the workcell description we are able to determine the states and events that are important from the modeling point of view. While the status of machine A (the same is valid for machine B) can be "idle" - I or "work in progress" – W, the situation with the robot is different since it executes three tasks. Hence, its status can be described as "available" – A, "moving part *a* in machine A" – M, "removing part *a* from machine A" – 1 and "removing part *b* from machine B" – 2. For the given specifications the automaton state can be described with three characters, where the first character is related to the robot status, the second character stands for machine A status and the third character for machine B status. Following this notation, the overall cell status "placing part *a* in machine A" while "machine A is idle" and "machine B is working" is written as a state MIW.

Events of interest are those associated with the transitions of the above-defined automaton states. Their notations and descriptions are given in Table 2.2.

Table 2.2. Events in workcell from Example 2.2.1

Event	Description
α	arrival of part a
β	arrival of part b in machine B (processing started)
m	processing of part a in machine A started
f	replacement of part b from machine B started
r	replacement of part b from machine B completed
	replacement of part a from machine A completed
c	replacement of part a from machine A started

Having defined states and events we can start with the determination of the automaton. In a complex DES, automatons for each *component* of the system are built first and then their integration gives a model of the entire system. Here we are using an informal approach – two *part paths* are modeled separately and then put together. First we model only part a path. It is assumed that at the beginning the machines are idle and the robot is available, thus the initial state is AII. Now we should check how events, defined in Table 2.2, influence the given initial state. Upon arrival of part a, event α triggers the transition from state AII into state MII, that is, robot carries the part into machine A while the machines remain idle. Other events, except β, are not related with the initial state (how event β is related with state AII will be discussed later), *i.e.* processing of part a in machine A cannot start (event m) since the part is not placed in the machine, replacement of part a from machine A cannot start (event c) as the part has not been processed yet, and finally, the robot cannot complete replacement (event r) since this task has not started (event f is not considered since it is attached to the part b path). Clearly, the only accessible state from AII, on the part a path, is MII.

Following the same reasoning we can build an automaton state by state. When the part is placed into the machine A, event m generates the transition to the next state, AWI, *i.e.* the robot becomes available, and machine A is processing the part while machine B is still idle. The operational sequence is finished when part processing in machine A is completed (event c, state 1II) and the robot removed the part from the cell that corresponds to event r that releases the robot and returns the automaton in its initial state AII. The model of part a path is shown in Figure 2.13.

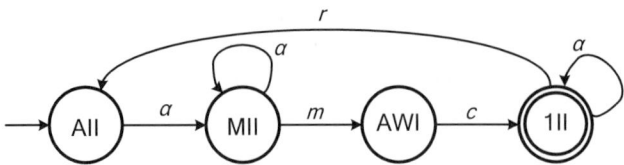

Figure 2.13. An automaton of part *a* path in the workcell from Example 2.2.1

It should be noted that the occurrence of α does not influence states MII and 1II since in both cases the robot is already occupied (the way event α affects state AWI will be discussed at the end of the example).

The automaton that models part *b* path is depicted in Figure 2.14. The initial state of this automaton is the same as for the automaton describing part *a* path. Arrival of part *b* triggers processing in machine B, event β forces the system in AIW. Then, the part is removed from the machine (event *f*, state 2II) and the robot is released (event *r*), which leads the system into the initial state. When that new part *b* arrives while the previous part is still being carried by the robot, state 2II changes into 2IW. Then, the robot is released and the system returns to state AIW. As for the automaton shown in Figure 2.13, some events have no influence on particular states.

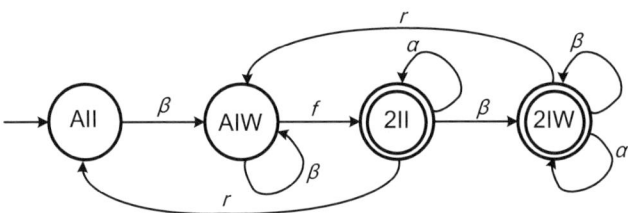

Figure 2.14. An automaton of part *b* path in the workcell from Example 2.2.1

It is evident that the automata in Figures 2.13 and 2.14 do not provide a full description of the workcell. First, some states that are particularly important are missing, and second, events that connect two automata should be added in order to obtain the correct model. By using the same reasoning that has been used for already-determined automata, we can construct the third automaton shown in Figure 2.15.

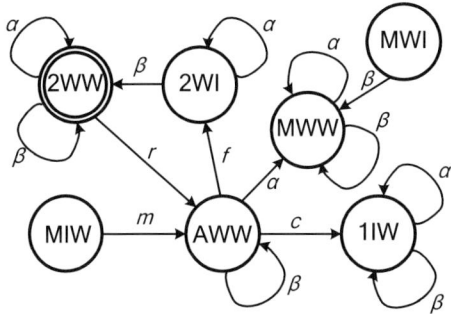

Figure 2.15. Partial automaton of the workcell from Example 2.2.1

Finally, a complete model of the workcell, depicted in Figure 2.16, is obtained by "merging" three automata. Since these automata were determined by an informal approach here we purposely omit formal notions of *parallel composition* and the *product* of two automata, and use the term merge instead. However, in order to verify the final result, the automata of the robot and machines are presented later on, together with a definition of parallel composition. For convenience, a complete model, depicted in Figure 2.16, does not encompass arcs corresponding to events that have no affect on the automaton states.

A survey of the automaton model reveals some interesting properties of the workcell. It can be seen that there exists a state (MWW) with no events that lead the system out of it. This state corresponds to a situation when both machines are processing parts while the robot carries part a. In order to place the part in machine A, the robot should remove the part that has been processed, but this task cannot be done since the robot already holds a part. At the same time machine B is not able to receive new parts since replacement of the part that has been processed requires the robot, which is occupied with an another task. Hence, once the workcell gets in MWW it remains in that state indefinitely; no further events are possible. This means that the automaton blocks without termination of the planned task. This situation is known as a *deadlock* [15] (it should be noted that there exists another form of blocking, called a *livelock*). Deadlock prevention, which is the key concern in the discrete event systems supervisory design, will be discussed and analyzed throughout the book.

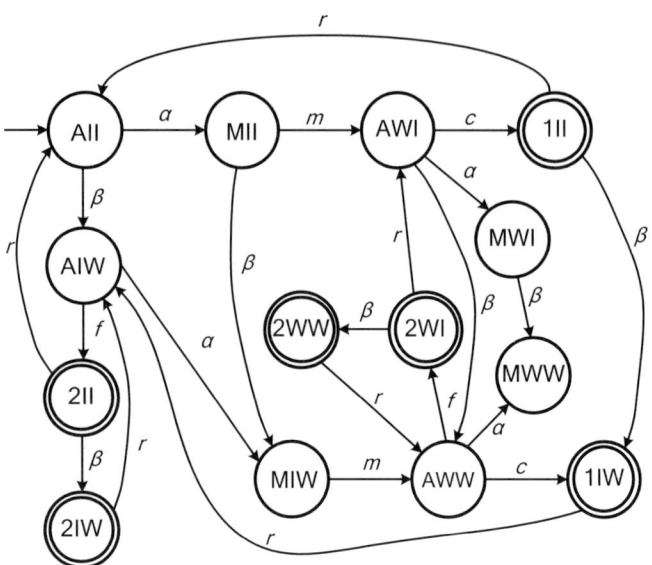

Figure 2.16. Complete automaton of the workcell from Example 2.2.1

In our case, the deadlock situation is in close relation with states AWI and AWW. That is, when the system comes into one of these two states the event α has to be forbidden otherwise the system will be blocked. In the next section we will discuss how we can keep the system described by an automaton out of illegal states.

Before we conclude the example it is worth saying a few words about marked states in the automaton in Figure 2.16. The main task of the workcell is a cyclic repetition of parts processing, hence, no state can be treated as final. However, four states, 2II, 2WI, 2IW, 2WW, 1IW and 1II have been marked. These states are chosen because each time the system gets into one of them, part a or part b eventually leaves the system (recall that robot statuses "1" and "2" stand for tasks related to removal of parts from the machines).

♦

As mentioned in the previous example, a formal description of a joint behavior of a set of automata can be obtained by two operations, a product and a parallel composition. The latter, defined hereafter, is more interesting for our purpose.

Definition 2.2.2 (parallel composition of automata): Given automata A_1 and A_2, their parallel composition is defined as

$$A_1 \parallel A_2 = Ac\left(X_1 \times X_2, E_1 \cup E_2, f((x_1 x_2), e), x_{01} x_{02}, X_{m1} \times X_{m2}\right), \quad (2.11)$$

where Ac is the so-called *accessible operation*, i.e. an operation that deletes all states that are not accessible from the initial state.

A set of states attained by the parallel composition contains all combinations made by states in A_1 and A_2 (the same holds for marked states). This points to the main drawback of automata – each state is represented explicitly. By combining components of real-world systems the number of states can easily explode.

A new set of events, obtained by the parallel composition, is calculated as a union of events in A_1 and A_2. A transient function of the joint automaton is defined as follows:

$$f((x_1,x_2),e) = \begin{cases} (f_1(x_1,e)f_2(x_2,e)) & \text{if } e \in \Gamma_1(x_1) \cap \Gamma_2(x_2) \\ (f_1(x_1,e)x_2) & \text{if } e \in \Gamma_1(x_1) \setminus E_2 \\ (x_1 f_2(x_2,e)) & \text{if } e \in \Gamma_2(x_2) \setminus E_1 \end{cases}$$

In other words, an event e that belongs to both automata can be executed only when the joint automaton arrives in the state that is formed by states that initiate event e in the original automata. Other events can be executed with no restriction.

Automata representations of the robotized workcell components from Example 2.2.1 are shown in Figure 2.17. As may be seen, in accordance with the discussion from the beginning of the example, each machine has two states, I and W, while the automaton representing the robot has four states, A, M, 1 and 2. A set of events in the automata corresponds to those defined in Table 2.2. We demonstrate a parallel composition of automata representing machine A, denoted A_A, and the robot, denoted A_R.

According to the definition a new automaton will have 8 states (4x2): AI, MI, 1I, 2I, AW, MW, 1W and 2W. A set of common events is determined as $E_A \cap E_R = \{c, m\}$. The new states and corresponding events are shown in Figure 2.18.

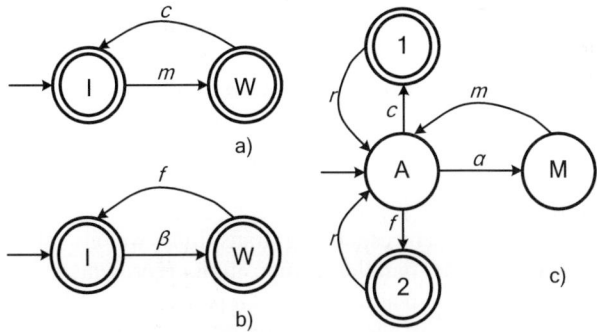

Figure 2.17. Automata representation of the workcell components from Example 2.2.1; (a) machine A, (b) machine B, and (c) robot

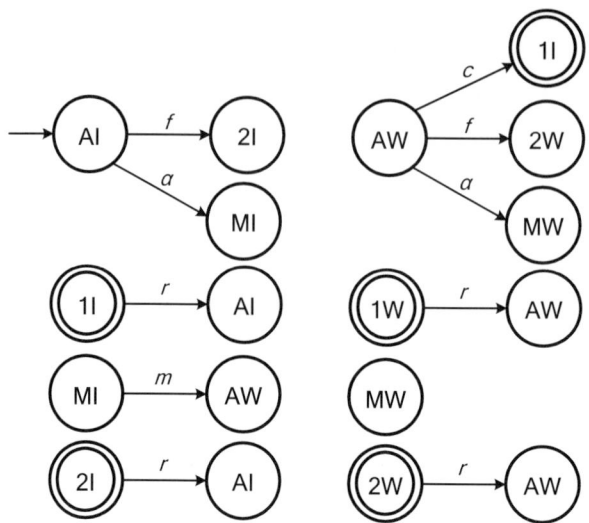

Figure 2.18. States and transitions attained by the parallel composition of the automata (a) and (c) in Figure 2.17

Let us take a closer look at Figure 2.18. A new state AI is formed from the robot state "available" – A and machine state "idle" – I. Both states represent initial states in the original automata, thus, state AI represents the initial state of the joint automaton. From Figure 2.17 we see that $\Gamma(A) = \{c, f, a\}$ and $\Gamma(I) = \{m\}$. Since common events c and m do not belong to both $\Gamma(I)$ and $\Gamma(A)$, their execution is forbidden. The remaining events, f and a, are allowed; the occurrence of f enforces a new state 2I, while event a leads the system into state MI, as shown in the figure. State 1I is composed of states 1 and I, with $\Gamma(1) = \{r\}$ and $\Gamma(I) = \{m\}$. As for the previous state, event m is not allowed, while event r causes a transition to state AI. State 1I is marked because states 1 and I are marked. The next state, MI, illustrates the situation when both automata involved in parallel composition, perform a common event. As $\Gamma(M) = \{m\}$ and $\Gamma(I) = \{m\}$, the condition for execution is satisfied and state MI changes to AW. Further analysis gives the remaining transitions as shown in Figure 2.18. State MW cannot trigger any event since $\Gamma(M) = \{m\}$ and $\Gamma(W) = \{c\}$.

Given new states and corresponding transitions we are able to form an automaton obtained by the parallel composition, represented in Figure 2.19. Similarity with the automaton that models part a path (Figure 2.13) is evident.

The parallel composition of the automaton describing machine B (Figure 2.17 (b)) and the automaton that models joint behavior of the robot and machine A will give a complete model of the workcell. We leave this step to the reader.

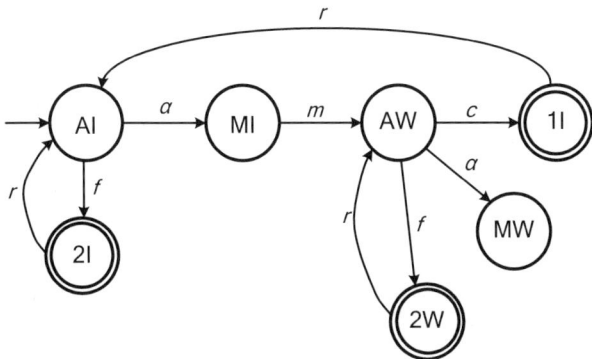

Figure 2.19. Automaton as result of the parallel composition of the automata (**a**) and (**c**) in Figure 2.17

2.2.2 Languages and Supervisory Control of DES

When we introduced the notion of feedback control in time-driven systems we mentioned that usually there exist at least two control objectives; a) the system should be conducted to the desired state and b) this state should be reached in a particular way, *i.e.* the state vector should follow a predefined trajectory. These two objectives are applicable to event-driven systems too. Generally, the goal of supervisory control of DES is to force the system i) to avoid undesirable states and ii) to maintain selected specifications (control policy).

The prospective design of such a supervisor requires two issues to be resolved. First, we have to identify illegal states. This might be difficult, especially if the system is large and has hundreds of states. In the workcell example deadlock states were caused by a structural property of the system and they were recognized when the automaton model of the system was built. On the other hand, in the tunnel ventilation example the forbidden states were imposed by the designer. One way or the other, undesirable states need a formal description in order to be incorporated in the supervisor design and avoided by the controlled system. The second issue, associated with the supervisor design, is specification of system activities once the problem of forbidden states is solved. The question is how to arrange the system states in order to execute the specified tasks? In the workcell example the robot provided services for two machines by handling parts processed by the machines. One possible scenario for the robot is to remove three parts from machine B, then put one part in machine A and then again remove three parts from machine B before it returns to machine A and takes out the part. This job sequence is repeated. Integration of such (cyclic) behavior into the supervisor design needs to be done formally in order to enable analysis of the controlled system.

Let us now recall the ventilation system example. In this example the "control rules" were associated with the system states: switch on the jet fans each time the system arrives at states $\mathbf{x} = [\mathrm{H} \cdot 0]^\mathrm{T}$ and switch off the jet fans when the system leaves states $\mathbf{x} = [\mathrm{H} \cdot 1]^\mathrm{T}$. Also, in the robotized workcell example we pointed out that deadlock was linked with two states, AWI and AWW. The automata theory approaches the supervisory control design from a different perspective. Instead of defining actions (events) that have to be taken when the system gets in a particular state, in the automata theory sequences of events are analyzed. Then, the supervisor's responsibility is to limit the system behavior to those sequences that are admissible or specified by a given control policy. To achieve this, the supervisor i) must "know" the current state of the system, and ii) should be able to prevent the occurrence of specific events. The first requirement is associated with system observability [16, 17] while the second one is related to the system controllability. Although fulfilment of both requests is rarely achieved, observability is easier to handle. Indeed, there are numerous DES observer design techniques that provide the supervisor with information regarding events that cannot be measured directly [18]. The situation with controllability is different. Due to the presence of breakdowns, malfunctions and other irregular incidents, some events are uncontrollable and their occurrence cannot be disabled by the supervisor [19] (it should be noted that there are events that are not related to failures but still cannot be controlled). Usually, a theoretical analysis of systems with uncontrollable events is concerned with determination of the probability that an uncontrollable event will take place. In practice, the best we can do is to employ redundancy in the parts of the system with the highest probability of having a failure.

At this point we return to the notion of the events sequence. As we mentioned earlier in the chapter, events occur asynchronously, one after the other, changing the system state and forming the sequences. In order to be able to trace all sequences that are generated by the system, we introduced the DES modeling tool called an automaton. An automaton, comprised of system events and system states, describes in which way the occurrence of a particular event changes the system state. Starting from the initial state, an automaton A creates a set of untimed sequences $s_1, s_2, \ldots \subset E^*$, written in the form of *strings*. This set of strings, called the *language* of automaton A, is denoted $L(A)$ and defined as

$$L(A) = \left\{ s \in E^* : f(x_0, s) \text{ exists} \right\} \tag{2.12}$$

It can be seen that $L(A) \subseteq E^*$. The concept of languages, generated by an automaton, has a central place in DES supervisory control design and analysis. In the remainder of the section this concept is presented briefly only to provide the reader with a concise insight into the potential that languages offer in DES theory. For further readings one may wish to consult [20–22]

Let us denote the automaton depicted in Figure 2.16 as A_W. Then

$$L(A_W) = \{\alpha, \beta, \alpha m, \alpha \beta, \beta f, \beta \alpha, \alpha mc, \alpha m\alpha, \alpha m\beta, \alpha \beta m, \beta f \beta, \ldots\}$$

It is apparent that strings, belonging to the language generated by automaton A_W, correspond to directed paths in the state transition diagram of A_W (paths are fundamental structural properties of graphs and will be discussed in more details later in the book).

The other language, which is closely related to the notion of deadlock, is associated with marked states. When that last event in a sequence s corresponds to an event that leads the system to the marked state, $f(x_0,s) \in X_m$, we talk about a *marked language* of automaton A, denoted $L_m(A)$ and defined as

$$L_m(A) = \{s \in E^* : f(x_0,s) \in X_m\} \tag{2.13}$$

From Figure 2.16 we have

$$L_m(A_W) = \{\beta f, \alpha mc, \alpha mc\beta, \alpha m\beta f, \alpha m\beta c, \beta \alpha mc, \beta \alpha mf, ...\}$$

For a given string $s = abc$, a is called a *prefix* of s, b is called a *substring* of s and c is called a *suffix* of s. String $s = \varepsilon$ is called an *empty string*. A language $L(A)$ is said to be *prefix-closed* if $L(A) = \overline{L}(A)$, where

$$\overline{L}(A) = \{s \in E^* : \exists c \in E^*, sc \in L(A)\}, \tag{2.14}$$

i.e. $\overline{L}(A)$ contains all prefixes of strings in $L(A)$.

A blocking (deadlock) is related to the prefix-closer. Specifically, an automaton contains a blocking condition if

$$\overline{L_m}(A) \subset L(A). \tag{2.15}$$

According to the usual interpretation, marked states appoint the final stage of the process modeled by the automaton. If blocking occurs, the automaton is not able to get into the marked state; hence, any generated string that ends in a deadlock state cannot be a prefix of a string that ends in the marked state.

To verify relation (2.15) we choose a few strings that belong to $L(A_W)$ and lead automaton A_W in deadlock; $s_1 = \alpha m a \beta$, $s_2 = \alpha m \beta a$, $s_3 = \beta \alpha m a$. From Figure 2.16 we can see that none of these strings is a prefix of strings that end in marked states of A_W, therefore, $\overline{L_m}(A_W) \subset L(A_W)$.

In the automata theory supervisory control is implemented in the form of a function, usually denoted S, which dynamically enables or disables events in a controlled automaton A. Thus, $S(s)$ is a set of all events that are allowed by S after the automaton A has generated string s. As an example, we examine strings $s_1 = \alpha m$ and $s_2 = \beta \alpha m$ generated by A_W. If supervisor S is to prevent a deadlock, then $S(s_1) = \{c, \beta\}$ and $S(s_2) = \{c, f\}$. Further, let us study two more strings, also generated by A_W, $s_3 = \alpha mcram$ and $s_4 = \beta \alpha mcram$. In order to prevent deadlock we must have

$S(s_3) = \{c, \beta\}$ and $S(s_4) = \{c, f\}$. Note that for a given control policy function S could have the same value for different strings, $S(s_1) = S(s_3)$ and $S(s_2) = S(s_4)$.

Before we return to the question posted at the end of Example 2.2.1, that was, how to design a supervisor that prevents a deadlock, we should see in which way the design specifications, which are usually given in a heuristic manner, can be formally specified. The problem is how to represent statements such as "prevent deadlock", "apply last-buffer-first-serve dispatching policy", "task a has a higher priority than task b", and so on, and relate them to the supervisor S.

It is apparent from the earlier discussion that a language generated by an automaton could have a large or even infinite number of strings, hence, making a list of all sequences (strings) that satisfy (or not) required specifications will be not only impractical but in many cases impossible. Since the domain of the control function S is language $L(A)$, it is natural to realize the control function in the form of an automaton, let us denote it as A_S. Once defined, automaton A_S should execute events in parallel with an uncontrolled automaton that is allowed to trigger only events announced by A_S. The issue here is that the determination of supervisor automaton A_S is usually a demanding task that requires practice. For that reason, A_S is not designed directly from the design specification. As a solution, the automata theory offers a choice of standard methods for modeling specifications in the form of an automaton, hereafter denoted A_D. Upon determination of A_D the supervisory automaton is computed as a parallel composition or product of A_D and the automaton that describes the system.

In some cases A_S can be obtained directly from the model of the system; inadmissible states and all events related to them should be simply removed. For example, in the workcell automaton A_W (Figure 2.16) illegal states MWW and MWI that embrace events α and β, can be erased, thus creating supervisor A_S. In this way, each time the system arrives in state AWI or AWW the occurrence of event α will be restricted by A_S.

Discrete event systems are often required to perform some tasks alternately. This specification can be presented in the form of a two-state automaton A_D having transitions that correspond with events that trigger the requested tasks. In our workcell example we can build such an automaton with events α and c, thus preventing a deadlock. The arrival of a new part a will be ignored as long as the previous part is not removed from the workcell (this dispatching policy is known as last-buffer-first-served). Even though part a arrives in the system stochastically and this process cannot be controlled, from the technical point of view that should not be a problem. As the supervisor is implemented in the form of a computer or PLC program, it is not difficult to ignore a signal from the sensor that triggers event α as long as part a is being processed by machine A.

References

[1] Isidori A. Nonlinear Control Systems. London: Springer, 1995.
[2] Slotine JJE, Li W. Applied Nonlinear Control. Englewood Cliffs: Prentice Hall, 1990.
[3] Vidyasagar M. Nonlinear Systems Aalysis, SIAM, 2003.

[4]   Bay JS. Fundamentals of Linear State Space Systems. New York: WCB/McGraw-Hill, 1998.
[5]   Kuo BC. Digital Control Systems. New York: Holt, Rinehart, Winston, 1980.
[6]   Astrom KJ, Wittenmark B. Computer Controlled Systems. Englewood Cliffs: Prentice Hall, 1990.
[7]   Henzinger T, Sastry S. Hybrid systems: Computation and Control. Berlin: Springer-Verlag, 1998.
[8]   Special Issue on Hybrid Systems, Automatica 1999;35;3
[9]   Special Issue on Hybrid Systems, IEEE Trans. Aut. Contr. 1998;43;4
[10] Tornambe A. Discrete-Event Systems Theory. Singapore: World Scientific, 1995.
[11] Kumar R, Garg VK. Modeling and Control of Logical Discrete Event Systems. Boston: Kluwer Academic Publishers, 1995.
[12] Carroll J, Long D. Theory of Finite Automata. Englewood Cliffs: Prentice Hall, 1989.
[13] Hopcroft JE, Ullman JD. Introduction to Automata Theory, Languages and Computation. Reading: Addison-Wesley, 1979.
[14] Wonham WM. Supervisory Control of Discrete Event Systems, Lecture notes, 2005.
[15] Che E, Lafortune S. Dealing with blocking in supervisory control of discrete event systems, IEEE Trans. Aut. Contr. 1991;36;6:724–735.
[16] Lin F, Wonham WM. On observability of discrete event systems, Information sciences 1988;44:173–198.
[17] Cieslak R, Desclaux C, Fawaz A, Varaiya P. Supervisory control of discrete event processes with partial observations, IEEE Trans. Aut. Contr. 1988;33;3:249–260.
[18] Wong KC, Wonham WM. On the computation of observers in discrete event systems, Discrete Event Dynamic Systems 2004;14;1:55–107.
[19] Wonham WM, Ramadge PJ. On the supremal controllable sublanguage of a given language, SIAM J. of Contr. and Optim. 1987;25;3:637–659.
[20] Ito M. Algebraic Theory of Automata & Languages. Singapore: World Scientific, 2004.
[21] Kelly D. Automata and Formal Languages: An Introduction. Englewood Cliffs: Prentice Hall, 1998.
[22] Kozen DC. Automata and Computability. New York: Springer, 1999.

3
Matrix Model and Control of Manufacturing Systems

The widest definition of a manufacturing system (MS) incorporates all the people, facilities and services needed to produce a product or a range of products. From this point of view, the MS design problem is extended beyond the traditional boundaries of machine tool and process selection, together with plant layout and job design. Tasks related to organizational issues and the design of information and control systems represent an increase in the variety of skills required of the MS design experts. The comprehensive nature of the approach (skills required and the amount of work involved) calls for a group of people drawn from related technical and operational functions in the business, which, together with design engineers, provides the set up of a project team.

The manufacturing systems design may be separated into four major steps – analysis, conceptual design, detail design and finally, implementation. Usually, the first step, analysis, deals with issues related to business, *i.e.* market-data collection, analysis of products and processes, analysis of manufacturing strategies, *etc*. The conceptual design is concerned with decisions related to the manufacturing architecture, *i.e.* flowlines, flexible lines, job shops or combinations of these. The architecture mainly depends on the product volume and the product variety. To be competitive in the global market and provide flexible manufacturing in today's high-mix-low-volume manufacturing environment, manufacturing systems have moved away from the old style fixed hardware sequential assembly lines with dedicated workstations. The trend has been toward flexible manufacturing systems (FMS). The flexibility of an FMS can be achieved in several ways:
- machine flexibility – ease of making changes required to produce a given set of part types,
- process flexibility – ability to produce a given set of part types in different ways,
- product flexibility – ability to change over to produce new products economically and quickly,
- routing flexibility – ability to handle breakdowns and continue producing a given set of part types,
- volume flexibility – ability to operate profitably at different production volumes,

- expansion flexibility – ability to expand the system easily and in a modular fashion,
- operation flexibility – ability to interchange ordering of several operations for each part type,
- production flexibility – universe of part types that the manufacturing system can produce.

Once the basic structure of the system is defined, detailed design provides answers to queries regarding the system performance under the initial design. In this stage, for example, calculations indicating where the performance bottlenecks are likely to lie in the system lead to a redesign that will eventually improve the system performance. Then, dynamic simulations of the system under various conditions give information regarding the system robustness, uncertainties, adaptability and sensitivity, to end with the system model. Given the model and the manufacturing policy, the last stage in the detailed design, the *control system* determination, can be carried out.

The control in MSs spreads over all levels of abstraction. The top-level controllers are concerned with decision making on the global market, hence, they have long prediction horizons and large sampling intervals (weekly, monthly, quarterly, *etc.*). Their outputs are usually used as set points for lower-level control loops that manage production lines (workcells) on a shop floor. Design and analysis of these intermediary control loops is the main scope of the book. At the bottom of the MS control structure we have controllers that work in real time and handle machines and tools. These bottom level controllers accept working points from the intermediate level.

Sometimes it is difficult to make a distinction between the three mentioned levels. Furthermore, in some applications there are more than three levels of control [18, 19], especially in the case of decentralized structures [20–23]. Anyway, interaction between various control levels, in a feedback form, is required in order to provide a proper study of the entire system. For example, some events from the bottom level, such as machine malfunctions or completions of tasks, should be supplied to the upper levels to provide an appropriate response of the overall control system.

The agility provided by the capacity of an FMS to be quickly reconfigured to produce new products relies mainly on the extent to which it is possible to efficiently and rapidly reprogram the FMS control system. One of the major components of an FMS control system is a computer-based supervisory controller for monitoring the status of jobs and directing part routing and machine job selection. This supervisor can be seen as an intermediate level of control.

There are many approaches to modeling, simulation and control design for manufacturing systems, including the already-presented automata, Petri nets which will be described in more detail in later chapters, alphabet-based approaches, perturbation methods, control theoretic techniques, expert systems design, and so on. In this chapter we present a *matrix-based* model of FMS that is a part of a detail design of manufacturing systems [1]. This matrix framework is very convenient for computer simulation [2], as well as for a supervisory controller design [3]. It is straightforward to write down the matrix description for a specific manufacturing system since the matrices are given by the bill of material (BOM) [4], Steward's sequencing matrix [5], the resource-requirements matrix, assembly trees, and existing dispatching algorithms. In addition, the matrix-based formulation can be easily modified if there are changes in product requirements or resources available, making the control of the workcell more flexible and re-configurable. We make the following three assumptions that define the sort of discrete-part manufacturing systems:

No pre-emption – once assigned, a resource cannot be removed from a job until it is completed,

Mutual exclusion – a single resource can be used for only one job at a time,

Hold while waiting – a process holds the resources already allocated to it until it has all the resources required to perform a job.

In addition to these assumptions, we assume that there are no machine failures.

This chapter is organized in the following way: first we introduce the system matrices that fully describe an MS; then we use these matrices to determine the system equations that are calculated in and/or algebra. The system equations form recursive matrix model used for simulation and system analysis. In order to be able to investigate dynamic phenomena in an MS, we introduce time into the matrix model. At the end of the chapter, a supervisory controller based on the matrix model is described and a case study is presented.

3.1 System Matrices

Before defining system matrices we introduce basic terms that will be used throughout the chapter and later in the book. Let Π be the set of distinct types of parts produced (or customers served) by an MS. Then each part type $k \in \Pi$ is characterized by a predetermined sequence of job operations $J^k = \{J_1^k, J_2^k, J_3^k, ..., J_{L_k}^k\}$ with each operation employing at least one resource. (Note that some of these job operations may be similar, e.g. J_i^k and J_j^k with $i \neq j$ may both be drilling operations.) We uniquely associate with each job sequence J^k the operations of raw part-in, J_{in}^k, and finished product-out, J_{out}^k. It is assumed, without loss of generality, that each part is fixed on a pallet throughout its processing sequence. Let $R_0 = \{r_0^k\}_{k \in \Pi}$ represent the set of pallets, where r_0^k denotes the pool of multiple pallets devoted to part-type k. Note that the multiplicity of pallets in pool r_0^k gives an upper bound for the number of parts of type k that can be processed concurrently.

Denote the other system resources in addition to the pallets with $R = \{r_i\}_{i=1}^n$, where $r_i \in R$ can represent a pool of multiple resources each capable of performing the same type of job operation. In this notation, $R^k \subset R$ represents the set of resources utilized by job sequence J^k. Note that $R = \underset{k \in \Pi}{\cup} R^k$ and $J = \underset{k \in \Pi}{\cup} J^k$

represent all resources and jobs in a particular FMS. Since the system could be reentrant, a given resource $r \in R^k$ may be utilized for more than one operation $J_i^k \in J^k$ (*sequential sharing*). Also, certain resources may be used in the processing of more than one part-type so that for some $\{l, k\} \in \Pi$, $l \neq k$, $R^l \cap R^k \neq \emptyset$ (*parallel sharing*). Resources that are utilized by more than one operation in either of these two ways are called *shared resources*, while the remaining are called *nonshared resources*. Thus, one can partition the set of system resources as $R = R_s \cup R_{ns}$, with R_s and R_{ns} indicating the sets of shared and nonshared resources, respectively, where $|R_s| = n_s$ and $|R_{ns}| = n_{ns}$, $n_s + n_{ns} = n$. For any $r \in R$ we define the *resource job set* $J(r)$. Obviously, $|J(r)| = 1 (> 1)$ if $r \in R_{ns}$ ($r \in R_s$).

Definition 3.1.1 (resource loop): For each $r \in R$, a set $L(r)$ defined as

$$L(r) = r \cup J(r) \tag{3.1}$$

is called a *resource loop*.

Given a set of jobs and a set of resources that compose a manufacturing system, we can present the system activities in the form of IF-THEN rules. Each rule corresponds to a component of the *logical state vector*, denoted **x**. A job is said to be *activated* (*started*) when all the preconditions (IF part) for its execution are satisfied. When a multitude of jobs requesting the same shared resource are simultaneously activated, a *conflict* is said to have occurred and a decision is needed as to which job the resource should be allocated to. This type of priority assignment in resource allocation constitutes the *problem of dispatching*, which we shall revisit and analyze in Chapter 6.

Now, the formal definitions of system matrices follow.

Definition 3.1.2 (job-sequencing matrix): The *job-sequencing matrix*, $\mathbf{F}_v$, is a matrix that relates the job set and the logical state vector: $\mathbf{F}_v(i,j) = 1$ if job j contributes to construction of the ith component of the logical state vector. Otherwise $\mathbf{F}_v(i,j) = 0$.

Definition 3.1.3 (resource-requirements matrix): The *resource-requirements matrix*, $\mathbf{F}_r$, is a matrix that relates the resource set and the logical state vector: $\mathbf{F}_r(i,j) = 1$ if resource j contributes to construction of the ith component of the logical state vector. Otherwise $\mathbf{F}_r(i,j) = 0$.

These matrices are easy to write down, $\mathbf{F}_v$ is the job-sequencing matrix of Steward (1962) – it is determined from the BOM or assembly tree [6]. Element $\mathbf{F}_v$ (i,j) is equal to 1 if job j is required as an immediate precursor to job i (equivalent in the BOM, if subassembly j is required to produce subassembly i). $\mathbf{F}_r$ is the resource-requirements matrix of Kusiak (1992), which is assigned by the shop floor engineer. It has an element $\mathbf{F}_r\,(i,j)$ equal to 1 if resource j is required for job i. Steward's sequencing matrix $\mathbf{F}_v$ and the resource-requirements matrix $\mathbf{F}_r$ have long been used as heuristic design aids by industrial engineers, with some possibility for limited analysis (as described *e.g.* by Warfield (1973) in the case of $\mathbf{F}_v$ and Kusiak (1992) in the case of $\mathbf{F}_r$). The matrix model elevates these design tools to formal computation elements.

In order to demonstrate development of the matrix model, let us consider the assembly tree depicted in Figure 3.1, which shows the required sequence of actions (jobs) to produce a product. Though the example is a relatively simple one, the technique extends directly to more complicated systems.

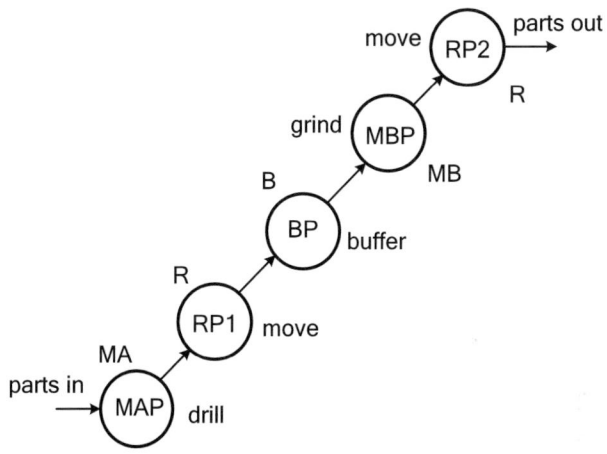

Figure 3.1. Product information for example of the system matrices determination

The job-sequencing matrix can be written directly from Figure 3.1:

$$\mathbf{F}_v = \begin{array}{c} \text{MAP RP1 BP MBP RP2} \\ \begin{bmatrix} 0 & 0 & 0 & 0 & 0 \\ 1 & 0 & 0 & 0 & 0 \\ 0 & 1 & 0 & 0 & 0 \\ 0 & 0 & 1 & 0 & 0 \\ 0 & 0 & 0 & 1 & 0 \\ 0 & 0 & 0 & 0 & 1 \end{bmatrix} \end{array}$$

Resource-requirements information may be given in the form of table or included directly in the product information, as shown in Figure 3.1. From this information one can write down the resource-requirements matrix:

$$\mathbf{F_r} = \begin{bmatrix} \overset{\text{MA}}{1} & \overset{\text{MB}}{0} & \overset{\text{B}}{0} & \overset{\text{R}}{0} \\ 0 & 0 & 0 & 1 \\ 0 & 0 & 1 & 0 \\ 0 & 1 & 0 & 0 \\ 0 & 0 & 0 & 1 \\ 0 & 0 & 0 & 0 \end{bmatrix} \begin{matrix} \text{MAP} \\ \text{RP1} \\ \text{BP} \\ \text{MBP} \\ \text{RP2} \\ {} \end{matrix}$$

Since the first operation of the job sequence does not have any prerequisites among the tasks, all components of the first row of $\mathbf{F_v}$ are equal to 0. The same analogy is applicable to the last row of $\mathbf{F_r}$, that is, the last row corresponds to the parts leaving the system, hence all its components are 0 as no resource is involved in this operation. We shall return to the issue of system inputs and outputs when we define the corresponding matrices.

One of the possible layouts of the workcell that performs a job sequence described by matrices $\mathbf{F_v}$ and $\mathbf{F_r}$, is shown in Figure 3.2.

In the matrix model, matrices $\mathbf{F_v}$ and $\mathbf{F_r}$ belong to the IF part of the rules describing the system. As we mentioned earlier, when all the preconditions for execution of a particular job are satisfied, the job will be started. These consequent parts of the rules are structured by the matrices defined below.

Definition 3.1.4 (job-start matrix): The *job-start matrix*, $\mathbf{S_v}$, is a matrix that relates the logical state vector and the job set: $\mathbf{S_v}(i,j) = 1$ if the jth component of the logical state vector is a prerequisite to start job i. Otherwise $\mathbf{S_v}(i,j) = 0$.

Definition 3.1.5 (resource-release matrix): The *resource-release matrix*, $\mathbf{S_r}$, is a matrix that relates the logical state vector and the resource set: $\mathbf{S_r}(i,j) = 1$ if the jth component of the logical state vector is a prerequisite to start the release of resource i. Otherwise $\mathbf{S_r}(i,j) = 0$.

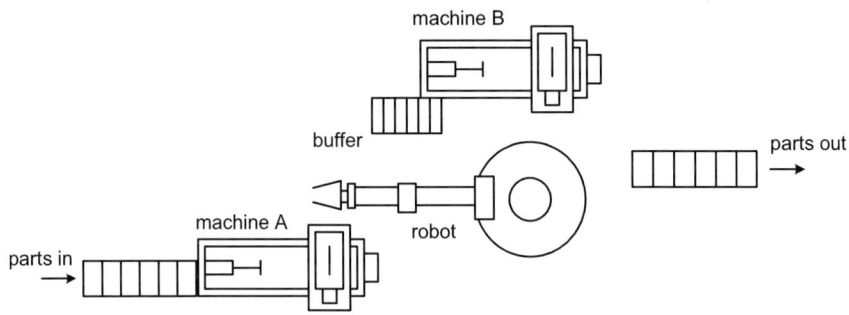

Figure 3.2. The workcell layout for the assembly tree in Figure 3.1

The job-start matrix S_v and the resource-release matrix S_r are new matrices that must be introduced to obtain a complete matrix description of manufacturing systems. In the flowline, matrix S_v has diagonal 1s, while in the job shop, it has multiple ones in the same column corresponding to job-routing decisions.

For the job sequence, depicted in Figure 3.1, matrices S_v and S_r have the following form:

$$S_v = \begin{bmatrix} 1 & 0 & 0 & 0 & 0 & 0 \\ 0 & 1 & 0 & 0 & 0 & 0 \\ 0 & 0 & 1 & 0 & 0 & 0 \\ 0 & 0 & 0 & 1 & 0 & 0 \\ 0 & 0 & 0 & 0 & 1 & 0 \end{bmatrix} \begin{matrix} \text{MAP} \\ \text{RP1} \\ \text{BP} \\ \text{MBP} \\ \text{RP2} \end{matrix}, \quad S_r = \begin{bmatrix} 0 & 1 & 0 & 0 & 0 & 0 \\ 0 & 0 & 0 & 0 & 1 & 0 \\ 0 & 0 & 0 & 1 & 0 & 0 \\ 0 & 0 & 1 & 0 & 0 & 1 \end{bmatrix} \begin{matrix} \text{MA} \\ \text{MB} \\ \text{B} \\ \text{R} \end{matrix}$$

As in the case of matrices F_v and F_r, columns of S_v and S_r, corresponding with inputs and outputs, have all components equal to 0. It is interesting to note that F_v and S_v depend only on job-sequencing information, while all the resource information is contained in F_r and S_r. Furthermore, it is important to keep in mind that the column of matrix F_r corresponding to the robot R, which is a shared resource, has more than one "1". Also, the row of matrix S_r, corresponding to the shared resource, has multiple "1s". We shall return to this issue later when discussing a supervisor design. When an operation requires more than one resource, the corresponding row of F_r has "1" for each resource that participates in the operation.

Raw parts entering and finished products leaving the manufacturing system are described with the following matrices.

Definition 3.1.6 (input matrix): The *input matrix*, F_u, is a matrix that relates the inputs of the system (raw parts entering the system) and the logical state vector: $F_u(i,j) = 1$ if an input j contributes to construction of the ith component of the logical state vector. Otherwise $F_u(i,j) = 0$.

Definition 3.1.7 (output matrix): The *output matrix*, S_y, is a matrix that relates the logical state vector and the outputs of the system (finished products leaving the system): $S_y(i,j) = 1$ if the jth component of the logical state vector is a prerequisite for output i. Otherwise $S_y(i,j) = 0$.

Since the job sequence, shown in Figure 3.1, is a single part processing, input and output matrices have a vector-like form:

$$\mathbf{F}_u = \begin{bmatrix} 1 \\ 0 \\ 0 \\ 0 \\ 0 \\ 0 \end{bmatrix}, \quad \mathbf{S}_y = \begin{bmatrix} 0 & 0 & 0 & 0 & 0 & 1 \end{bmatrix}$$

So far we introduced the system matrices that assemble a set of rules describing the behavior of a manufacturing system. In the next section we present equations that utilize these matrices and provide a mechanism for calculation of the logical state vector in a recursive manner.

3.2 System Equations

As we already pointed out, the matrix model represents a set of rules, so that it is formally a rule base. The previously defined job set J and resource set R are associated with the system matrices and incorporated into the matrix model in the form of vectors. We define a *job vector* $\mathbf{v} : J \to \aleph$ and a *resource vector* $\mathbf{r} : R \to \aleph$ that represent the set of jobs and the set of resources corresponding to their nonzero elements. The set of jobs (resources) represented by $\mathbf{v}$ ($\mathbf{r}$) is called the *support* of $\mathbf{v}$ ($\mathbf{r}$), denoted $sup(\mathbf{v})$ ($sup(\mathbf{r})$); *i.e.* given $\mathbf{v} = [v_1 \ v_2 \ ... \ v_q]^T$, vector element $v_i > 0$ if and only if job $v_i \in sup(\mathbf{v})$. In the same manner, given $\mathbf{r} = [r_1 \ r_2 \ ... \ r_p]^T$, vector element $r_i > 0$ if and only if resource $r_i \in sup(\mathbf{r})$. Usually, index i is replaced with job (resource) notation, hence, r_{MA} stands for the component of resource vector r that corresponds to resource MA.

For example, the workcell shown in Figure 3.2 has a job set $J = \{MAP, RP1, BP, RP2, MBP\}$ and the resource set $R = \{MA, MB, B, R\}$. Then, the vector representation of jobs performed by the robot is $\mathbf{v}_R = [0 \ 1 \ 0 \ 1 \ 0]^T$ and $sup(\mathbf{v}_R)=\{RP1, RP2\}$. A vector that represents shared resources is $\mathbf{r}_s = [0 \ 0 \ 0 \ 1]^T$ with $sup(\mathbf{r}_s)=\{R\}$. The definitions of job and resource vectors imply that the job and resource sets should be ordered.

We proceed further with the determination of system equations by defining a *vector negation*. Given a natural number vector $\mathbf{a} = [a_1 \ a_2 \ ... \ a_n]^T$, its negation $\bar{\mathbf{a}} = [\bar{a}_1 \ \bar{a}_2 \ ... \ \bar{a}_n]^T$ is such that $\bar{a}_i = 0$ if $a_i > 0$, and 1 otherwise. A vector negation is required since state equations and system matrices are Boolean, while job and resource vectors have positive integer components. Consequently, all matrix operations are defined to be in *and/or* algebra, denoted $\triangle$ and $\triangledown$, where multiplication is replaced by AND, and addition is replaced by OR. Hence, for given matrices and vectors

$$\mathbf{A} = \begin{bmatrix} 0 & 1 & 1 \\ 0 & 1 & 0 \end{bmatrix}, \mathbf{B} = \begin{bmatrix} 1 & 1 \\ 1 & 0 \end{bmatrix}, \mathbf{a} = \begin{bmatrix} v_a & v_b & v_c \end{bmatrix}^T, \mathbf{b} = \begin{bmatrix} 3 & 0 \end{bmatrix}^T$$

we have

$$\bar{\mathbf{c}} = \mathbf{A}_\Delta \bar{\mathbf{a}} = \begin{bmatrix} 0 & 1 & 1 \\ 0 & 1 & 0 \end{bmatrix} \Delta \begin{bmatrix} \bar{v}_a \\ \bar{v}_b \\ \bar{v}_c \end{bmatrix} = \begin{bmatrix} (0 \wedge \bar{v}_a) \vee (1 \wedge \bar{v}_b) \vee (1 \wedge \bar{v}_c) \\ (0 \wedge \bar{v}_a) \vee (1 \wedge \bar{v}_b) \vee (0 \wedge \bar{v}_c) \end{bmatrix} = \begin{bmatrix} \bar{v}_b \vee \bar{v}_c \\ \bar{v}_b \end{bmatrix}$$

$$\Rightarrow \mathbf{c} = \begin{bmatrix} v_b \wedge v_c \\ v_b \end{bmatrix}$$

$$\mathbf{d} = \mathbf{A}_\Delta \bar{\mathbf{a}} \nabla \mathbf{B}_\Delta \bar{\mathbf{b}} = \begin{bmatrix} 0 & 1 & 1 \\ 0 & 1 & 0 \end{bmatrix} \Delta \begin{bmatrix} \bar{v}_a \\ \bar{v}_b \\ \bar{v}_c \end{bmatrix} \nabla \begin{bmatrix} 1 & 1 \\ 1 & 0 \end{bmatrix} \Delta \begin{bmatrix} 0 \\ 1 \end{bmatrix} = \begin{bmatrix} \bar{v}_b \vee \bar{v}_c \\ \bar{v}_b \end{bmatrix} \nabla \begin{bmatrix} 1 \\ 0 \end{bmatrix}$$

$$= \begin{bmatrix} \bar{v}_b \vee \bar{v}_c \vee 1 \\ \bar{v}_b \vee 0 \end{bmatrix}$$

where $\wedge$ and $\vee$ are standard symbols for logical AND and OR, respectively. It should be noted that the final step in vector **c** calculation is obtained by DeMorgan's rule.

Having defined all the necessary components, the system equations that outline the matrix model are formalized in the following section.

3.2.1 Logical State-vector Equation

The job vector **v** has two interpretations. As a status output of the workcell, vector **v** denotes *a job-completed vector*; in this role it is denoted as $\mathbf{v}_c$. Hence, $sup(\mathbf{v}_c)$ comprises all operations of the given system that are completed. On the other hand, as an input to the workcell, vector **v** represents *a job-start vector*, denoted as $\mathbf{v}_s$, thus, $sup(\mathbf{v}_s)$ includes all operations of the given system that should be started. The same holds for the resource vector **r**, *i.e.* $sup(\mathbf{r}_c)$ contains all resources that are idle ($\mathbf{r}_c$ is called *an idle-resource vector*) and $sup(\mathbf{r}_s)$ is a set of all resources that should be released ($\mathbf{r}_s$ is called *a resource-release vector*). Then, for given vectors $\mathbf{v}_c$ and $\mathbf{r}_c$, and for specified system matrices, the logical state vector **x** is calculated according to the following equation:

$$\bar{\mathbf{x}} = \mathbf{F}_v \Delta \bar{\mathbf{v}}_c \nabla \mathbf{F}_r \Delta \bar{\mathbf{r}}_c \nabla \mathbf{F}_u \Delta \bar{\mathbf{u}} \quad (3.2)$$

Input vector **u** represents raw parts entering the cell, *i.e.* $sup(\mathbf{u})$ is a set of inputs that have parts ready to be processed. A computed entry of $x_i=1$ in **x** indicates that all conditions required for the rule *i* have been met. As we shall see later, in a closed-loop system controlled by a supervisor, the components of $\mathbf{v}_c$ and $\mathbf{r}_c$ are calculated from the signals measured by sensors and used as a feedback.

It is important to order the jobs correctly in order to obtain lower triangular matrices $\mathbf{F}_v$ and $\mathbf{S}_v$, for then the sequencing of the jobs is causal. A causal ordering is also important as the particular system structure helps to overcome NP-hard complexity problems. When the logical state-vector equation is constructed using the causal ordering of jobs, the system matrix $\mathbf{F}_v$ consists of *diagonal blocks*, one per part path, having a subdiagonal of 1s. If there is an assembly there will be some 1s in $\mathbf{F}_v$ below the diagonal blocks, where 1 in element (i,j) means that job j is the last job in a partial part path and joins rule i in another part path.

Matrices $\mathbf{F}_r$ and $\mathbf{S}_r$ are related as follows: if the ith rule is not the last rule in a partial part path, and there is an entry "1" in position (i,j) of $\mathbf{F}_r$, meaning resource j participates in rule i, then there is an entry "1" in position $(i+1,j)$ of $\mathbf{S}_r^T$, meaning that the resource is released by the next rule. If the ith rule is the last rule on a partial part path, and there is an entry "1" in position (i,j) of $\mathbf{F}_r$, then there is an entry "1" in position (k,j) of $\mathbf{S}_r^T$, meaning that the resource is released by the assembly rule k.

The logical-state vector components should be numbered corresponding to the jobs in rules consequent parts. From the example shown in Figure 3.1, one can read a rule corresponding to the component x_1:

IF *part is ready* **AND** *machine A is ready* **THEN** *rule 1 is TRUE*

In a symbolic form we write

IF PI∈ *sup*(**u**) **AND** MA ∈ *sup*($\mathbf{r}_c$) **THEN** $x_1 = 1$

or shorter $x_1 = u \wedge \text{MA}$.

A complete logical state-vector equation for the considered system has the form

$$\mathbf{x} = \begin{bmatrix} 0 & 0 & 0 & 0 & 0 \\ 1 & 0 & 0 & 0 & 0 \\ 0 & 1 & 0 & 0 & 0 \\ 0 & 0 & 1 & 0 & 0 \\ 0 & 0 & 0 & 1 & 0 \\ 0 & 0 & 0 & 0 & 1 \end{bmatrix} \Delta \overline{\mathbf{v}}_c \mathbf{V} + \begin{bmatrix} 1 & 0 & 0 & 0 \\ 0 & 0 & 0 & 1 \\ 0 & 0 & 1 & 0 \\ 0 & 1 & 0 & 0 \\ 0 & 0 & 0 & 1 \\ 0 & 0 & 0 & 0 \end{bmatrix} \Delta \overline{\mathbf{r}}_c \mathbf{V} + \begin{bmatrix} 1 \\ 0 \\ 0 \\ 0 \\ 0 \\ 0 \end{bmatrix} \Delta \overline{\mathbf{u}}$$

3.2.2 Job-start Equation

The logical state-vector equation may be seen as a transformation of status of jobs and resources into the system state vector. As such, it represents only the prerequisite parts of the rules. The consequent parts of the rules that describe actions taken when a particular component of the state vector attains a logical "1" are described with other three equations. The first one is *a job-start equation* that relates the state vector **x** and the job-start vector $\mathbf{v}_s$:

$$\mathbf{v}_s = \mathbf{S}_v \Delta \mathbf{x} \qquad (3.3)$$

When the system is controlled the components of $\mathbf{v}_s$ stand for requests issued by the supervisor to the system. When all the prerequisites for starting a particular job are satisfied, the corresponding component of the job-start vector is set to "1". For the workcell shown in Figure 3.2 the consequent part of rule 1 is

IF *rule 1 is TRUE* **THEN** *start job in machine A*

In a symbolic form we have

IF $x_1=1$ **THEN** MAP $\in sup(\mathbf{v}_s)$

3.2.3 Resource-release and Product-output Equations

A *resource-release equation* relates the logical state vector $\mathbf{x}$ and the resource-release vector $\mathbf{r}_s$. A resource is released from the task it has been allocated for when the task is completed:

$$\mathbf{r}_s = \mathbf{S}_r \Delta \mathbf{x} \qquad (3.4)$$

From Figure 3.1 one can read that

IF *rule 2 is TRUE* **THEN** *release machine A*

or

IF $x_2=1$ **THEN** MA $\in sup(\mathbf{r}_s)$

For a shared resource there exist at least two rules that release it. In the case of the robot in our example, these rules are

IF *rule 3 is TRUE* **THEN** *release robot R*

IF *rule 6 is TRUE* **THEN** *release robot R*

A product-output equation

$$\mathbf{y} = \mathbf{S}_y \Delta \mathbf{x} \qquad (3.5)$$

describes how the processed products depart from the system. Once the last job on the part path is finished, the corresponding rule is satisfied and the part leaves the system.

3.2.4 Recursive Matrix Model

Generally, the complete task plan could be given by the system matrices $\mathbf{F}_v$, $\mathbf{S}_v$, $\mathbf{F}_r$, $\mathbf{S}_r$, defined above, which are specified by higher-level planners, or, as we show, may be written down in manufacturing systems given the BOM or the assembly tree plus resource-availability information. Additionally, these matrices can easily be extracted from plans generated by typical planning software, including hierarchical planners. Since each matrix has a well-defined function for job sequencing, resource assignment, and resource release, they are straightforward to construct as well as easy to modify in the event of goal changes, resource changes, or failures; that is, they accommodate task planning as well as *task replanning*. The matrix-design technique extends directly to complicated interconnected systems using notions of *block matrix (e.g. subsystem) design*.

In this section we discuss the usage of matrix formulation for computer simulation of manufacturing systems (and other DES). The formal notation of logical rules contains matrices that express the structure of a manufacturing system. As such, these matrices are extremely useful in system analysis and supervisor design. Additionally, when included into system equations (3.2) – (3.5) they provide an apparatus for simulation analysis of the system.

Denoting the discrete event iteration number with k, we can calculate the logical state vector each time an event takes place, *i.e.* a job is completed, resource becomes idle or part enters the system:

$$\bar{\mathbf{x}}(k) = \mathbf{F}_v \Delta \bar{\mathbf{v}}_c(k-1) \triangledown \mathbf{F}_r \Delta \bar{\mathbf{r}}_c(k-1) \triangledown \mathbf{F}_u \Delta \bar{\mathbf{u}}(k-1) \tag{3.6}$$

The equations describing the consequent parts of rules can be rewritten in the same way:

$$\mathbf{v}_s(k) = \mathbf{S}_v \Delta \mathbf{x}(k)$$
$$\mathbf{r}_s(k) = \mathbf{S}_r \Delta \mathbf{x}(k) \tag{3.7}$$
$$\mathbf{y}(k) = \mathbf{S}_y \Delta \mathbf{x}(k)$$

In order to be able to link recursive equations (3.6) and (3.7) we have to relate a job-completed vector $\mathbf{v}_c$ with a job-start vector $\mathbf{v}_s$, and an idle-resource vector $\mathbf{r}_c$ with a resource-release vector $\mathbf{r}_s$. According to its definition, the components of vector $\mathbf{v}_c$ correspond to completed operations, hence, each time a job is completed, the number of parts held by this particular job is increased. At the same time, if a job contributes to a rule(s) that is fulfilled, an already processed part(s) leaves the job and proceeds through the system. In other words

$$\mathbf{v}_c(k) = \mathbf{v}_c(k-1) + \mathbf{v}_s(k) - \mathbf{F}_v^T \mathbf{x}(k) \tag{3.8}$$

The term $\mathbf{F}_v^T \mathbf{x}(k)$ in Equation (3.8) corresponds to parts that have been processed and advance to the next operation. Inclusion of Equation (3.7) in Equation (3.8) gives

$$\mathbf{v}_c(k) = \mathbf{v}_c(k-1) + \mathbf{S}_v \mathbf{x}(k) - \mathbf{F}_v^T \mathbf{x}(k) = \mathbf{v}_c(k-1) + \left[\mathbf{S}_v - \mathbf{F}_v^T\right]\mathbf{x}(k) \quad (3.9)$$

where multiplications and additions are carried out in the standard way.

By following the same reasoning one can find the number of idle resources and the number of finished products in step k as

$$\mathbf{r}_c(k) = \mathbf{r}_c(k-1) + \mathbf{S}_r \mathbf{x}(k) - \mathbf{F}_r^T \mathbf{x}(k) = \mathbf{r}_c(k-1) + \left[\mathbf{S}_r - \mathbf{F}_r^T\right]\mathbf{x}(k)$$
$$\mathbf{y}(k) = \mathbf{y}(k-1) + \mathbf{S}_y \mathbf{x}(k) \quad (3.10)$$

Let us now introduce the *system vector* $\mathbf{m}(k)$ as

$$\mathbf{m}(k) = \begin{bmatrix} \mathbf{u}(k) \\ \mathbf{v}_c(k) \\ \mathbf{r}_c(k) \\ \mathbf{y}(k) \end{bmatrix} \quad (3.11)$$

Then, Equations (3.6) – (3.10) can be written in the following form

$$\bar{\mathbf{x}}(k) = \mathbf{F}_\Delta \bar{\mathbf{m}}(k-1) \;,\quad \mathbf{m}(0) = \mathbf{m}_0$$
$$\mathbf{m}(k) = \mathbf{m}(k-1) + \left[\mathbf{S} - \mathbf{F}^T\right]\mathbf{x}(k) \quad (3.12)$$

with

$$\mathbf{S} = \begin{bmatrix} \mathbf{S}_u \\ \mathbf{S}_v \\ \mathbf{S}_r \\ \mathbf{S}_y \end{bmatrix}, \quad \mathbf{F}^T = \begin{bmatrix} \mathbf{F}_u^T \\ \mathbf{F}_v^T \\ \mathbf{F}_r^T \\ \mathbf{F}_y^T \end{bmatrix}$$

where $\mathbf{S}_u = [\mathbf{0}]$, $\mathbf{F}_y = [\mathbf{0}]$ are null-matrices required for keeping matrix dimensions consistent. If $\mathbf{S}_u \neq [\mathbf{0}]$, then the arrival of parts depends on the system status, *i.e.* factor $\mathbf{S}_u \mathbf{x}(k) \neq \mathbf{0}$ will increase the corresponding component of $\mathbf{u}(k)$, which is in disagreement with the definition of input vector **u** that should be independent and should represent raw parts entering the cell. When $\mathbf{F}_y \neq [\mathbf{0}]$, the part that was considered to have left the system returns to be processed by one of the system operations, which is not allowed. Usually, matrix **S** is called the *activity-start matrix*, and matrix **F** is called the *activity-completion matrix*.

The first equation in Equation (3.12) encompasses logical AND/OR operations, while the second one is calculated by using the standard multiplication and addition, hence Equation (3.12) represents *a hybrid matrix model* of an MS. Even though the hybrid matrix model (3.12) is recursive, it does not capture the system dynamics. The term $\mathbf{Sx}(k)$, representing the start of activities, contributes to the vector **m** components in the same iteration step k, which means that the durations of all tasks in the system are assumed to be equal to 0, *i.e.* activities are completed at the same time as they are started. By tracking $sup[\mathbf{m}(k)]$ we can reconstruct an untimed sequence that describes only logical activities of the system.

The matrix model is very convenient for computer simulation. In the following example we use MATLAB® to simulate the system shown in Figure 3.2 (any other simulation tool could be used as well).

Example 3.2.1 (DES simulation by using the matrix model)

In this example we present results obtained by the simulation of the system shown in Figure 3.2, by using the hybrid matrix model. For convenience, the previously determined system matrices are shown again

$$\mathbf{F}_v = \begin{bmatrix} 0 & 0 & 0 & 0 & 0 \\ 1 & 0 & 0 & 0 & 0 \\ 0 & 1 & 0 & 0 & 0 \\ 0 & 0 & 1 & 0 & 0 \\ 0 & 0 & 0 & 1 & 0 \\ 0 & 0 & 0 & 0 & 1 \end{bmatrix} \quad \mathbf{F}_r = \begin{bmatrix} 1 & 0 & 0 & 0 \\ 0 & 0 & 0 & 1 \\ 0 & 0 & 1 & 0 \\ 0 & 1 & 0 & 0 \\ 0 & 0 & 0 & 1 \\ 0 & 0 & 0 & 0 \end{bmatrix} \quad \mathbf{F}_u = \begin{bmatrix} 1 \\ 0 \\ 0 \\ 0 \\ 0 \\ 0 \end{bmatrix} \quad \mathbf{F}_y = \begin{bmatrix} 0 \\ 0 \\ 0 \\ 0 \\ 0 \\ 0 \end{bmatrix}$$

with columns labeled MAP RP1 BP MBP RP2 for $\mathbf{F}_v$, MA MB B R for $\mathbf{F}_r$, IN for $\mathbf{F}_u$, and OUT for $\mathbf{F}_y$.

$$\mathbf{S}_v = \begin{bmatrix} 1 & 0 & 0 & 0 & 0 & 0 \\ 0 & 1 & 0 & 0 & 0 & 0 \\ 0 & 0 & 1 & 0 & 0 & 0 \\ 0 & 0 & 0 & 1 & 0 & 0 \\ 0 & 0 & 0 & 0 & 1 & 0 \end{bmatrix} \quad \mathbf{S}_r = \begin{bmatrix} 0 & 1 & 0 & 0 & 0 & 0 \\ 0 & 0 & 0 & 0 & 1 & 0 \\ 0 & 0 & 0 & 1 & 0 & 0 \\ 0 & 0 & 1 & 0 & 0 & 1 \end{bmatrix}$$

$$\mathbf{S}_y = [0 \ 0 \ 0 \ 0 \ 0 \ 1] \quad \mathbf{S}_u = [0 \ 0 \ 0 \ 0 \ 0 \ 0]$$

Let us define the system input vector **u** as

Matrix Model and Control of Manufacturing Systems   65

$$\mathbf{u}(k) = \begin{cases} 1 & \text{for } k = 0 \\ 0 & \text{for } k > 0 \end{cases}$$

that is, only one part enters the system at the initial step. If we assume that all resources are idle at the beginning of the simulation, then the initial value of vector **m** is $\mathbf{m}(0) = \mathbf{m}_0 = [1\ 0\ 0\ 0\ 0\ 0\ 1\ 1\ 2\ 1\ 0]^T$. The first component of **m** stands for the system input – it is equal to 1 as defined for $k = 0$. The component that attains the value 2 corresponds to the buffer that has two empty slots at start. Other entries of "1" stand for idle resources. Inclusion of $\mathbf{m}_0$ in Equation (3.12) for $k = 1$, gives the logical state vector $\mathbf{x}(1) = [1\ 0\ 0\ 0\ 0\ 0]^T$, *i.e.* only the first rule, which requires idle machine MA and a part at the input, is satisfied. For given $\mathbf{x}(1)$ we calculate $\mathbf{m}(1) = [0\ 1\ 0\ 0\ 0\ 0\ 1\ 2\ 1\ 0]^T$. The set $sup[\mathbf{m}(1)] = \{MAP, MB, B, R\}$ indicates that task MAP, executed by MA, is finished, while other resources remain idle. Iteratively, for $k = 2$, we get $\mathbf{x}(2) = [0\ 1\ 0\ 0\ 0\ 0]^T$ and $\mathbf{m}(2) = [0\ 0\ 1\ 0\ 0\ 1\ 0\ 2\ 1\ 0]^T$, that is, the robot carries the part to the buffer (job RP1) and machine MA is released.

Simulation results are graphically presented in Figure 3.3. The value of vector **m** can be directly read from graphs. Propagation of the part through the system is clearly seen (left side of Figure 3.3). The task sequence is executed as defined in the assembly tree shown in Figure 3.1. Resource utilization demonstrates that robot R is used twice, exactly as specified in the resource requirements. Since only one part was processed by the system, we can see that only one slot in buffer B has been used. From the graphical representation of the system output (OUT) we conclude that the part leaves the system after 6 iterations.

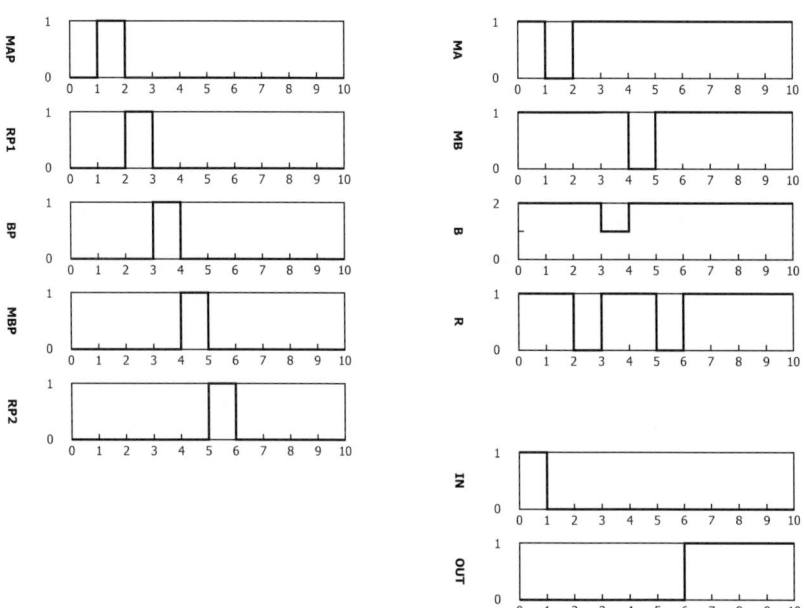

Figure 3.3. Graphical representation of results from Example 3.2.1

Let us now analyze the same system with a different input vector,

$$\mathbf{u}(k) = 1 \quad for \quad k > 0$$

Defining **u**(*k*) in this way we imply that a new part is available for processing in each iteration.

The results obtained by simulation are graphically presented in Figure 3.4. As in the previous case we can see that parts are processed according to the predefined sequence. However, an interesting situation occurs for $k = 5$. At that instant both machines hold parts ready to be transferred further down the line, *i.e.* MAP = 1 and MBP = 1. Since both tasks require robot R, which is idle for $k = 5$, two rules having the robot as a prerequisite are satisfied. This situation is described in the chapter beginning as a conflict. The consequence of a conflict is seen on the graph representing robot R. The value 1 for $k = 5$ becomes -1 for $k = 6$ clearly indicating that two operations simultaneously requested the same resource (since there exists only one resource and two operations, value -1 indicates the lack of resource). From this result it is obvious, as we have already mentioned, that the decision-making supervisor is required in order to provide acceptable system performance.

We conclude this example with the MATLAB® code that has been used for DES simulation based on the matrix model.

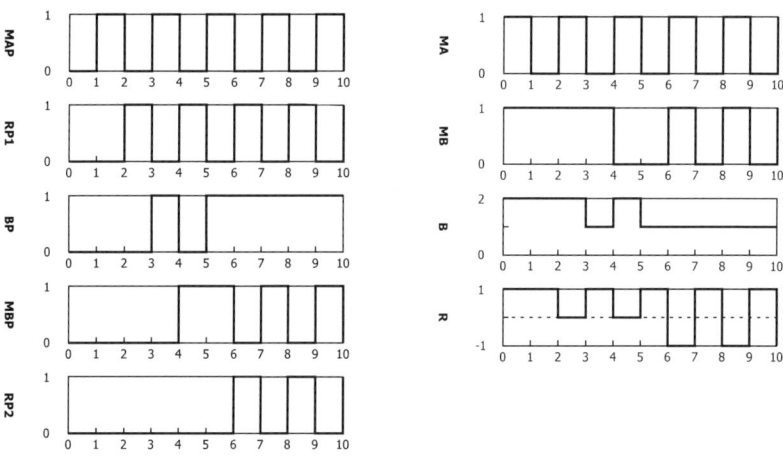

Figure 3.4. Graphical representation of results from Example 3.2.1 for **u**(*k*)=1 for *k*>0

```
clear all;                        % Initial Conditions               function Y = not(X)
% system matrices                 % m(k=0) = [u v r y] = [ ]'
                                  % initial value                    [nx, mx] = size(X);
% job sequencing matrix           m = [ ]';
Fv = [ ];                                                             for i = 1:nx
% resource requirements matrix    % results output format;              for j = 1:mx
Fr = [ ];                         % 3 character component name            if (X(i,j) < 1)
% job start matrix                output(1,:) = [' '];                      Y(i,j) = 1;
Sv = [ ];                         output(2,:) = sprintf(' %2d ', m);      else
% resource release matrix                                                   Y(i,j) = 0;
Sr = [ ];                         % Running the simulation                end
% input matrix                    for i = 3:100;                        end
Fu = [ ];                                                             end
% null matrix                     % calculation of logical state vector
Fy = [ ];                             x = multoa(not(F), not(not(m)));  function R = multoa(X,Y)
% null matrix
Su = [ ];                         % calculation of vector m            [nx, mx] = size(X);
% output matrix                       m = m + (M' * x);                [ny, my] = size(Y);
Sy = [ ];
                                      output(i,:) = sprintf(' %2d ', m); for i = 1:nx
F = [Fu Fv Fr Fy];                                                       for j = 1:my
S = [Su' Sv' Sr' Sy']';           end                                      R(i,j) = ((X(i,1)) | (Y(1,j)));
                                                                           for k = 2:mx
% system matrix                                                              R(i,j) = R(i,j) & ((X(i,k)) | (Y(k,j)));
M = S' - F;                                                                end
                                                                         end
                                                                       end
```

Figure 3.5. MATLAB® code for DES simulation by using the matrix model

♦

3.3 Modeling System Dynamics

It has been shown in the previous example that the model (3.12) describes only logical (static) properties of an MS. Although the prerequisites that are required for an event to start are given by Equation (3.12), we are not able to tell at which particular moment these prerequisites are met, *i.e.* we do not know when the event actually starts. In real applications on actual manufacturing processes, we will be sensing the completion of prerequisite jobs by either using sensors (*e.g.*, proximity, tactile, *etc.*) or via notification from the machines or resources. On the other hand, for the purpose of computer simulation, we must find a way to keep track of the time lapsed in the processing of jobs. To keep track of job time durations, we incorporated the system dynamics into the matrix model in the form of a *lifetime* [7, 16]. That is, a real number d_i, called a lifetime, is associated with each task in an MS. Under the assumption that there are no machine failures, every task that starts will actually finish in a finite time, hence:

$$\begin{aligned} v_{ci}(t) &= v_{si}(t - d_{vi}) \\ r_{ci}(t) &= r_{si}(t - d_{ri}) \end{aligned} \tag{3.13}$$

where d_{vi} and d_{ri} are lifetimes of operation v_i and resource release r_i, respectively. Although we consider the lifetime to be deterministic and known, matrix modeling of the system dynamics allows simulation of MS with stochastic lifetimes as well.

The final goal of an MS modeling and analysis is to prepare the ground for design of an appropriate dispatching supervisor. The nature of this supervisor is determined by its computer-based implementation, usually in the form of a PLC. Since the execution of an algorithm on a PLC is cyclic, the moment at which the supervisor detects completion of an operation does not necessarily coincide with the actual moment in which an operation is finished. Therefore, from the supervisor point of view, the operation lifetime is not d_i but $d_i + \varepsilon_i$ (Figure 3.6). We can rewrite Equation (3.13) as

$$v_{ci}^s(kT_s) = v_{si}(kT_s - d_{vi} - \varepsilon_{vi}) = v_{si}((k - n_{vi})T_s)$$
$$r_{ci}^s(kT_s) = r_{si}(kT_s - d_{ri} - \varepsilon_{ri}) = r_{si}((k - n_{ri})T_s)$$
(3.14)

where $n_i T_s \geq d_i > (n_i - 1)T_s$, T_s is the supervisor sampling (cycle) interval, and n_i is an integer representation of the lifetime expressed in number of sampling intervals. It is apparent that the sampling interval should be small enough to provide an accurate dynamic model.

Introduction of a *shift (delay) operator* q in Equation (3.14) gives

$$v_{ci}^s(q) = q^{-n_{vi}} v_{si}(q)$$
$$r_{ci}^s(q) = q^{-n_{ri}} r_{si}(q)$$
(3.15)

where $y(q) = q^{-n}x(q)$ corresponds with $y(k) = x(k-n)$, i.e. y is delayed n sampling intervals after x. For convenience purpose in the remainder of the book we omit superscript s from $v_{ci}^s(q)$ and $r_{ci}^s(q)$.

By recalling Equation (3.7), Equations (3.15) can be written in the vector form as

$$\mathbf{v}_c(q) = \mathbf{T}_v(q)\mathbf{x}(q)$$
$$\mathbf{r}_c(q) = \mathbf{T}_r(q)\mathbf{x}(q)$$
(3.16)

where $\mathbf{T}_v$ and $\mathbf{T}_r$ are operations and resources release *delay matrices* with elements representing operations lifetimes. Delay matrices are obtained by replacing each entry "1" in $\mathbf{S}_v$ and $\mathbf{S}_r$ with a shift operand representation of the corresponding lifetime.

Due to the existence of shared resources, transformation of the second equation in Equations (3.15) requires additional explanation. Namely, each nonshared resource in $\mathbf{r}$ has its corresponding operation in $\mathbf{v}$ that is responsible for its release. At the same time, a shared resource that is represented by one component in vector $\mathbf{r}$, has several operations in $\mathbf{v}$ it could be released from. As release lifetimes associated with these operations generally differ, the row in $\mathbf{T}_r$ that corresponds to a shared resource could have two or more different entries.

Conversion of Equations (3.16) into recursive form, suitable for simulation, can be done in the same way as in the case of the static recursive model (3.12).

$$\mathbf{v}_c(q) = q^{-1}\mathbf{v}_c(q) + \mathbf{T}_v(q)\mathbf{x}(q) - \mathbf{F}_v^T \mathbf{x}(q) \tag{3.17}$$

$$\mathbf{r}_c(q) = q^{-1}\mathbf{r}_c(q) + \mathbf{T}_r(q)\mathbf{x}(q) - \mathbf{F}_r^T \mathbf{x}(q) \tag{3.18}$$

Finally, the dynamic matrix model of an MS is obtained by including the shift operator q in the logical state-vector equation:

$$\begin{aligned} \overline{\mathbf{x}}(q) &= \mathbf{F}_\Delta q^{-1}\overline{\mathbf{m}}(q) \;, \quad \mathbf{m}(0) = \mathbf{m}_0 \\ \mathbf{m}(q) &= q^{-1}\mathbf{m}(q) + \left[\mathbf{T}(q) - \mathbf{F}^T\right]\mathbf{x}(q) \end{aligned} \tag{3.19}$$

where

$$\mathbf{T}(q) = \begin{bmatrix} \mathbf{S}_u \\ \mathbf{T}_v(q) \\ \mathbf{T}_r(q) \\ \mathbf{S}_y \end{bmatrix}$$

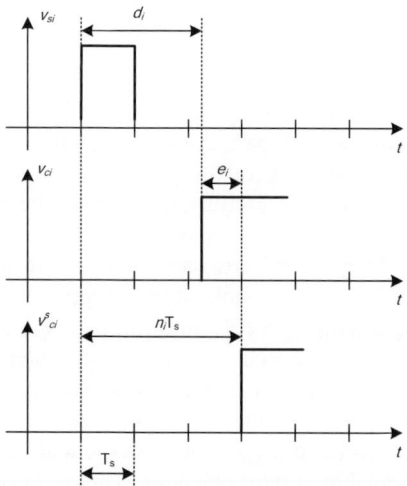

Figure 3.6. Extension of the operation lifetime for the system dynamics modeling

By comparing Equation (3.12) with Equation (3.19) one can notice that the main difference between the two models is in matrix **S** that is replaced with delay matrix **T**(q). The other difference is in vector **m** that comprises the status of jobs and resources. Even though vectors in Equations (3.12) and (3.19) have the same form, **m**(q) represents the state of the system as "seen" by the supervisor. For practical implementations the difference between the actual status of jobs and resources and status expressed in **m**(q) can be ignored in the case of a very small sampling interval.

Before we give an example of system dynamics modeling based on Equation (3.19), there are two issues that have to be further discussed. The simulation of a dynamic model is done such that each element **T**(q)(i,j) of the delay matrix that is not equal to 0 is associated with a *clock*, denoted $C(i,j)$, containing the time passed after the job has been started. All clocks are initially set to zero. When the rule for starting a particular task is satisfied, the corresponding clock is activated. Then, in each sampling interval all active clocks are checked. If some clock is found to be equal to or greater than the corresponding task lifetime, defined as an entry of the delay matrix, the particular task is considered completed. In that case the entry of vector **m** matching this task is incremented. Such realization of model (3.19) is valid as long as there are no resources that can process more than one part at a time. If there exists such a resource, then the simulation algorithm must be modified in a straightforward manner, by expanding the number of clocks for each additional part processed simultaneously by the resource. For example, if **T**(q)(i,j) = q^{-5} stands for some task that lasts 5 sampling intervals and can process 3 parts in the same time, then it is associated with a so-called multipart clock, that is, $C(i,j,1)$, $C(i,j,2)$ and $C(i,j,3)$. The first part entering the task activates $C(i,j,1)$, the second one $C(i,j,2)$ and the third part $C(i,j,3)$. Having its own clock, each part can be tracked separately.

The second issue that needs additional clarification when one considers realization of the dynamic matrix model is related to so-called "hidden" parts. Let us assume that rule x_i, which has job v_i in its prerequisite part and job v_j in its consequent part, is satisfied in the sampling interval k. Further, let processing of the part in v_j follow immediately after processing in v_i. Then, according to Equation (3.19), term $\mathbf{F}^T\mathbf{x}(k)$ removes the part from v_i, i.e. corresponding component of vector **m** is decreased. Processing of the part in v_j starts in the same sampling interval k, but due to the operation lifetime, the part will be completed n_{vj} sampling intervals later, i.e. the component of vector **m** that corresponds with operation v_j will be increased with delay. Therefore, one is not able to tell where the part is if only vector **m**(k) is tracked. For example, it may happen that several parts already entered the system but $sup[\mathbf{m}(k)] = \{\emptyset\}$ since all parts are being processed at that particular sampling interval. However, the results of system performance analysis in the sense of system throughput, resources utilization, *etc.*, are not influenced by the existence of hidden parts. On the other hand, the outcome of the supervisor design that is based on vector **m**(k) as a feedback could be inadequate and could finally generate unacceptable system behavior. This is to be detailed in the next section.

Matrix Model and Control of Manufacturing Systems   71

Example 3.3.1 (DES simulation by using the dynamic matrix model)

Let us consider the system shown in Figure 3.2. The lifetimes of workcell operations are given in Table 3.1. Release of buffer BA, which lasts 2.75 seconds, is the shortest task in the workcell, thus, we choose the simulation sampling interval to be $T_s = 1$ [s]. Extended lifetimes for this sampling interval are specified in the third column of Table 3.1. We see that machine B is the slowest one. For a given job-start matrix $\mathbf{S}_v$ and resource-release matrix $\mathbf{S}_r$ (see Example 3.2.1) we can determine delay matrices $\mathbf{T}_v$ and $\mathbf{T}_r$:

$$\mathbf{T}_v = \begin{bmatrix} q^{-76} & 0 & 0 & 0 & 0 & 0 \\ 0 & q^{-10} & 0 & 0 & 0 & 0 \\ 0 & 0 & q^{-4} & 0 & 0 & 0 \\ 0 & 0 & 0 & q^{-113} & 0 & 0 \\ 0 & 0 & 0 & 0 & q^{-8} & 0 \end{bmatrix} \quad \mathbf{T}_r = \begin{bmatrix} 0 & q^{-15} & 0 & 0 & 0 & 0 \\ 0 & 0 & 0 & 0 & q^{-10} & 0 \\ 0 & 0 & 0 & q^{-3} & 0 & 0 \\ 0 & 0 & q^{-6} & 0 & 0 & q^{-5} \end{bmatrix}$$

Table 3.1. Lifetimes of the workcell tasks

Operation	Lifetime d_i [s]	Extended lifetime n_i
MAP (drill)	76	76
RP1 (move 1)	10	10
BP (buffer)	3.5	4
MBP (grind)	113	113
RP2 (move 2)	7.5	8
release of MA	15	15
release of B	2.75	3
release of MB	10	10
release of R (after RP1)	5.75	6
release of R (after RP2)	4.25	5

There are ten different tasks in the system, and two of them can hold two parts simultaneously, buffer operation BP and buffer release B. Accordingly, the simulation requires eight standard and two multipart clocks. As in the case of the static simulation, we assume that only one part enters the system at the initial step and all resources are idle at the beginning, consequently, $\mathbf{m}(0) = \mathbf{m}_0 = [1\ 0\ 0\ 0\ 0\ 0\ 1\ 1\ 2\ 1\ 0]^T$.

The results obtained by simulation are shown in Figure 3.7. Upon entering the system, the part has been processed in machine A. After 76 sampling intervals (graph MAP) the part is removed from the machine into the buffer, which can be

72 Manufacturing Systems Control Design

clearly seen on graph R – the robot is idle while the part is processed in machine A, then it moves the part (10 sampling intervals) and finally it is released (6 sampling intervals). The part advances through the system and after 211 samples (see graph RP2 that represents the last operation of the system) it leaves the workcell.

In order to get a complete insight into the system dynamic properties we have to simulate a situation with several parts being processed simultaneously. This situation is closer to the real conditions in which the system is fed by parts with predetermined frequency (or stochastically). Given that manufacturing systems are generally designed to work periodically, this kind of simulation provides results that can be used for calculations of production cycles, resources utilizations, bottleneck machines, *etc.* For the moment we shall skip formal definitions of these terms as they are elaborated in more detail in the max-plus algebra section.

Graphical representation of results obtained when a new part is available each time robot R is idle, is given in Figure 3.8.

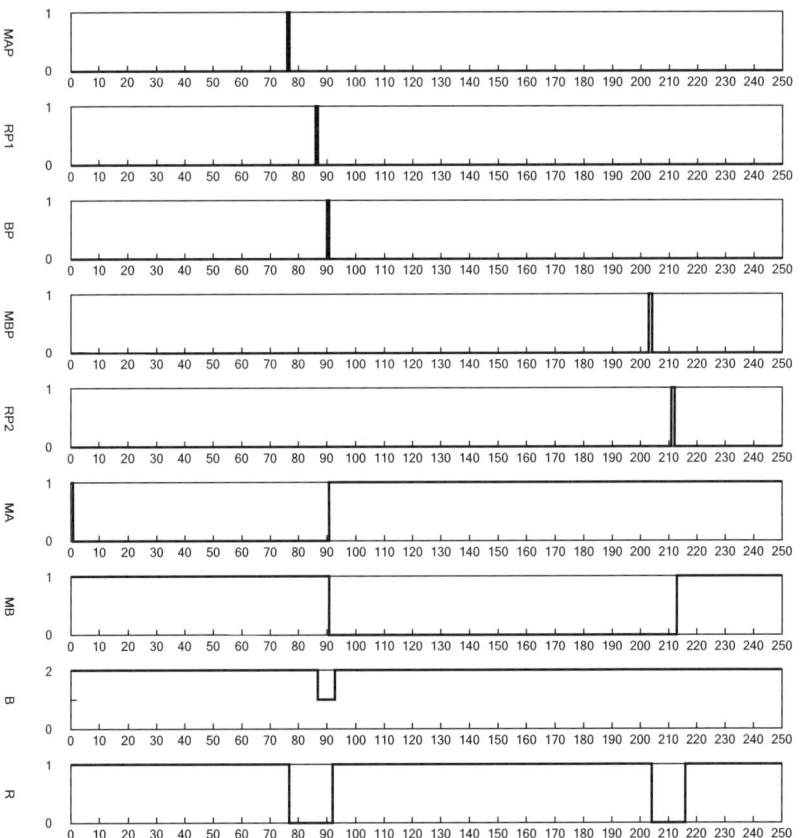

Figure 3.7. Graphical representation of results obtained by the dynamic simulation (one part processed)

Several observations regarding system performance can be made from the attained results. We see that the first part leaves the system after 211 samples, as in the previous simulation when only one part has been passed through the workcell. After that, the time period between departures of two consecutive parts from the system is equal to 123 sampling intervals, which corresponds to the sum of the processing and release lifetimes of machine B (see Table 3.1). Hence, the simulation confirmed, as we expected, that this machine is the system bottleneck since it is the slowest one according to Table 3.1.

The second remark regarding the system behavior is related to the conflict that appeared during the simulation of the static model in Example 3.2.1. From the graph in Figure 3.8 it can be seen that R attains values of 0 and 1, but never –1. This clearly shows that simultaneous requests for the robot R never appeared, *i.e.* there was no conflict. Such a difference between results obtained by simulations of static and dynamic models is common. Even though the structural properties of the system and the static model confirm the existence of conflict, when the system dynamics is included in the matrix model simultaneous requests for shared resource may not occur due to the particular lifetime arrangement.

We conclude this discussion with a note on another interesting phenomenon that is revealed from the results of the dynamic model simulation. From the graphical representation of the first operation in the system, MAP, it is evident that 10 parts have entered the workcell. On the other hand, only 5 parts have arrived at the output. The other 5 parts got trapped in the system; all resources are occupied and none of them can be released since they are all waiting for each other. This condition is known as *circular blocking* and it is equivalent to the already-mentioned deadlock. Analysis of the graphs in Figure 3.8 can clearly show how the system came into deadlock. In sampling interval $k = 806$ machine A just finished processing of the 9th part. At the same time sample buffer B is full (BP = 2 for $k = 806$), machine B is processing the 6th part and robot R is idle. The prerequisites of rule x_2 MAP is completed and robot R is idle, are met, thus, the task in the consequent part, RP1, is started. Since buffer is full, the robot cannot complete RP1. A part that is supposed to leave the buffer and make room for a new one is blocked by the part in machine B that waits to be cleared by robot R that is already holding a part. Resources wait for each other, the system is deadlocked and parts cannot proceed through the line. A similar situation happened with the workcell shown in Figure 2.12.

At the end of the example, let us reorder the job sequence in the workcell by exchanging positions of machines A and B, *i.e.* instead of drill, the first operation in the sequence is grind. The dynamic matrix model is changed correspondingly and the simulation results are shown in Figure 3.9. It can be noticed that deadlock is avoided and the system has cyclic activities. Parts are leaving the workcell with a period of 123 sampling intervals. The operational time of a particular resource can be easily determined from the graphs corresponding to its idleness and activity. For example, graph B clearly shows that the buffer is underutilized as it never accommodates more than 1 part, *i.e.* the system could work correctly with a 1-slot buffer. As expected, the slowest machine is operational 100% of the time (graphs MBP and MB), while the activity periods of the other two resources are approximately 24 % for robot (graphs RP1, RP2 and R) and 74 % for machine A

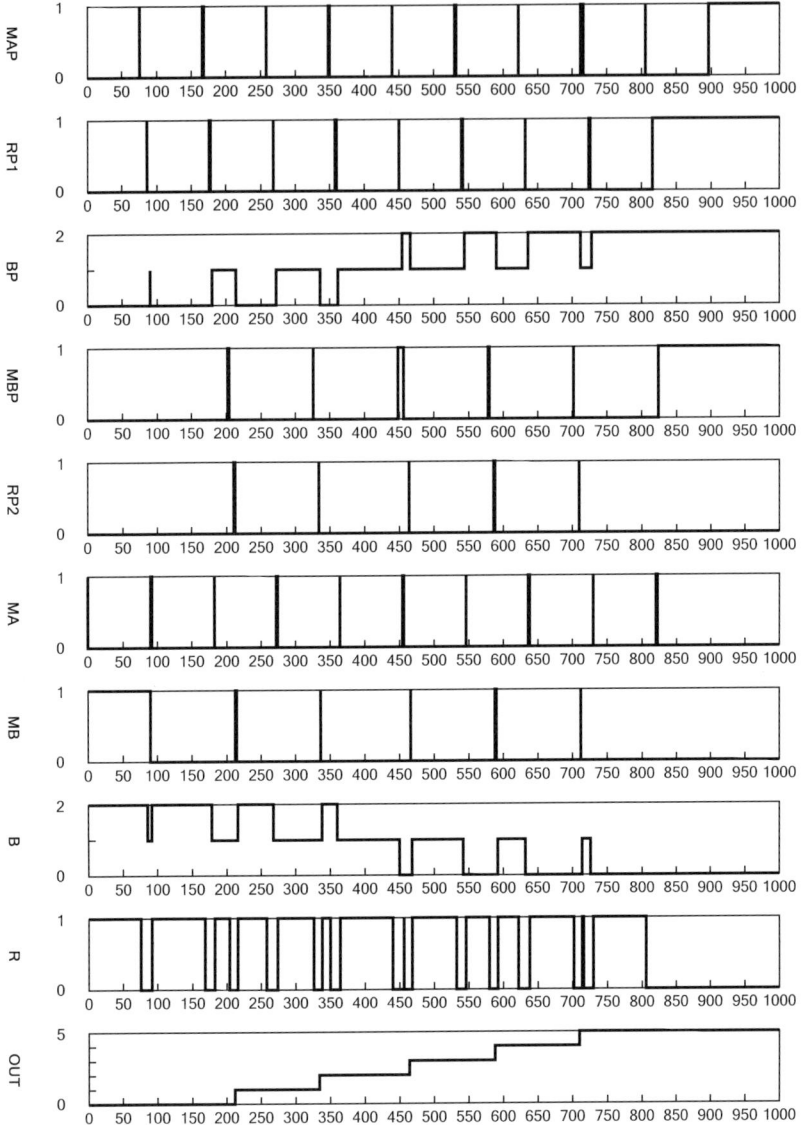

Figure 3.8. Results obtained by simulation based on dynamic matrix model (several parts processed)

(graphs MAP and MA). Comparing these results with the lifetimes in Table 3.1 one can observe that operational times attained from graphs are equal to $(\Sigma d_{oi}+\Sigma d_{ri})/(\text{system cycle})\times 100\%$, where d_{oi} are resource operations lifetimes and d_{ri} resource releases lifetimes.

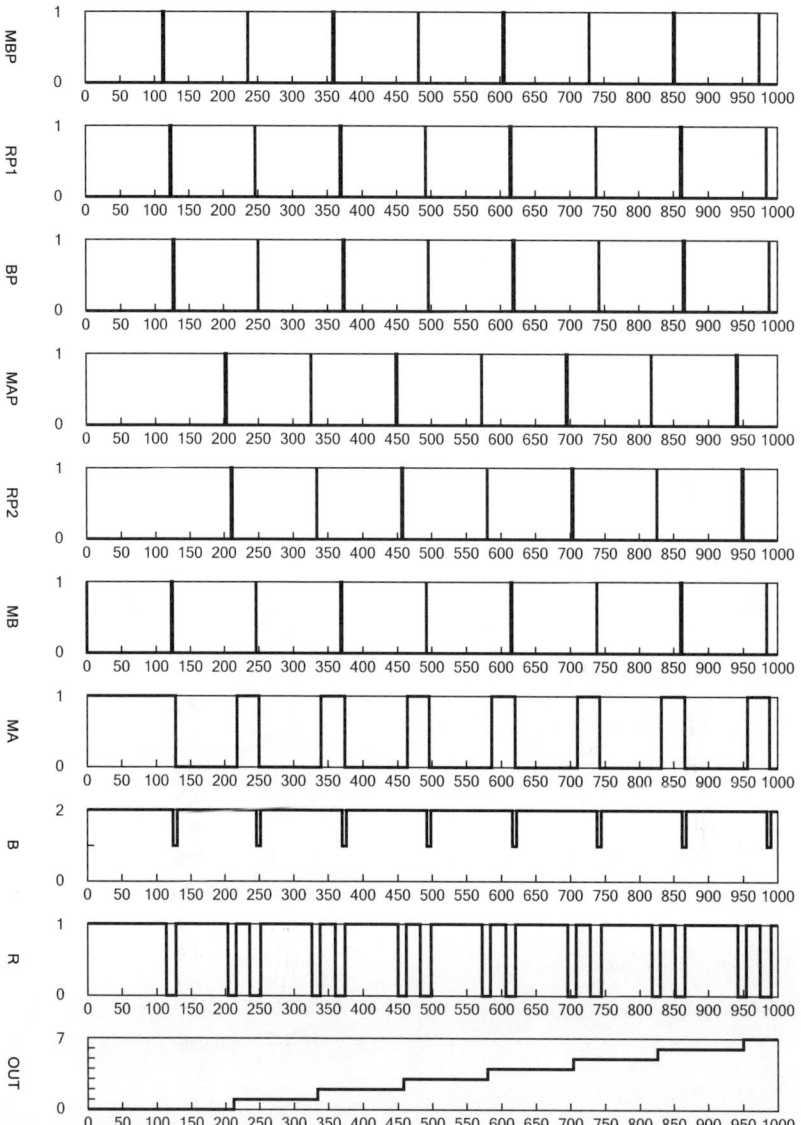

Figure 3.9. Results obtained by simulation based on dynamic matrix model (reordered job sequence)

We conclude this example with the MATLAB® code that has been used for DES simulation based on the dynamic matrix model.

```
% *** MS dynamic simulation ***
% v=[ ]'
% r = [ ]'                                  % Running the simulation
% ud = [ ]'
                                            for i = 2:100;
clear all;
                                              x = multoa(not(F), not(not(m)));
Fv = [ ];
Fr = [ ];                                     Tr_temp = multim(Tr0,x);
Fd= [ ];                                      Tv_temp = multim(Tv0,x);
Tv0= [ ];
Tr0= [ ];                                     for ctn=1:n_tok
Fu = [ ];                                       for k=1:nTv
Fy = [ ];                                         for l=1:mTv
                                                    if(Tv(k,l,ctn)==1)
Sy = [ ];                                             m(k+n_in)=m(k+n_in)+1;
                                                    end
[nTv,mTv]=size(Tv0);
[nTr,mTr]=size(Tr0);                                if(Tv_temp(k,l)>0)
[nFd,mFd]= size(Fd);                                  if(Tv(k,l,ctn)==0)
                                                        Tv(k,l,ctn)=Tv_temp(k,l);
%number of inputs                                       Tv_temp(k,l)=0;
n_in= ... ;                                           end
%number of outputs                                  end
n_out= ... ;
%max. number of jobs per operation (clocks)         if(Tv(k,l,ctn)>0)
n_tok = ... ;                                         Tv(k,l,ctn)=Tv(k,l,ctn)-1;
                                                    end
Tr = zeros(nTr,mTr,n_tok);                        end
Tv = zeros(nTv,mTv,n_tok);                      end
                                              end
Tr_temp = zeros(nTr,mTr);
Tv_temp = zeros(nTv,mTv);                     for ctn=1:n_tok
                                                for k=1:nTr
F = [Fu Fv Fr Fd Fy];                             for l=1:mTr
                                                    if(Tr(k,l,ctn)==1)
% Initial Conditions                                  m(k+n_in+nTv)=m(k+n_in+nTv)+1;
% m(to) = [u v r ud y]'                             end

m = [ ]';                                           if(Tr_temp(k,l)>0)
x = [ ]';                                             if(Tr(k,l,ctn)==0)
                                                        Tr(k,l,ctn)=Tr_temp(k,l);
output(1,:) = [' ... op1 op2 ... r1 r2 ... '];        Tr_temp(k,l)=0;
                                                      end
                                                    end

function R = multim(T0,x,T)                         if(Tr(k,l,ctn)>0)
                                                      Tr(k,l,ctn)=Tr(k,l,ctn)-1;
[nT, mT, oT] = size(T);                             end
                                                  end
R=zeros(nT,mT,oT);                              end
for i = 1:nT                                  end
  for j = 1:mT
    for k=1:oT                                m = m - (F' * x);
      if (T(i,j,k)==0)
        R(i,j,k) = T0(i,j) * x(j,1);          output(i,:) = sprintf(' %2d ', m);
        break
      end                                   end
    end
  end                                       % Displaying the results
end                                         output
```

Figure 3.10. MATLAB® code for DES simulation by using the dynamic matrix model

3.4. Matrix Controller

In the preceding example a simple case demonstrated how modification of the job sequence could entirely change the system behavior. In many cases reordering of jobs is not allowed since product shape and quality depend on production sequence that is firmly defined and should be strictly followed. The problem is that when described as a BOM, or in some other engineering form, the job sequence does not disclose potential difficulties that might develop when the structure of an MS, which executes this particular sequence, is determined. In the previous section we showed how two of these potential difficulties, conflict and deadlock, can be exposed by using static and dynamic simulations of an MS. Based on the matrix model, these simulations provided a complete insight into the system performance.

The other advantage of the matrix model is its convenience when it comes to the integration of the supervisory controller into the already-defined system model. In this section we describe a matrix controller as a part of a closed-loop manufacturing control system, whose foundation is set on already-defined system matrices and system equations. At the beginning, let us recall the main objective of the supervisory control of DES. As we stated in Chapter 2, the controller should force the system to a) avoid undesirable states and b) maintain selected specifications (control policy). In many cases, a) and b) are achieved simultaneously, that is, implementation of a particular control policy at the same time prevents the system from getting into adverse states. For example, such a control policy is "last-buffer-first-served" that is known to avoid deadlock in most cases. On the other hand, "first-buffer-first-served" dispatching usually ends in system deadlock. As a result, its realization requires additional consideration to provide an algorithm that concurrently prevents deadlock.

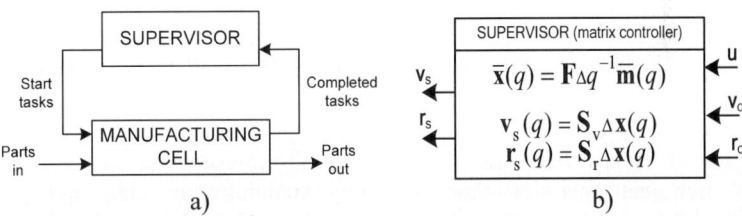

Figure 3.11. A closed-loop manufacturing control system (**a**), and internal structure of the supervisor (**b**)

A supervisor based on the matrix model basically checks the conditions required for performing the next jobs in the MS by utilizing the logical state-vector equation (3.2). This equation is in some ways similar to the differential equation (2.4) in linear system theory. Based on these conditions, stored in the logical state vector $\mathbf{x}$, the job-start equation (3.3) computes which jobs are activated and may be started, and the resource-release equation (3.4) computes which resources should be released (due to completed jobs). These equations are analogous to the output equation in (2.4). Then, as already mentioned, the controller sends *commands* to the MS, namely, vector $\mathbf{v}_s$, whose "1" entries denote which jobs are to be started, and vector $\mathbf{r}_s$, whose "1" entries denote which resources are to be released.

Completed tasks, which outline feedback vector **m**, are given by the system sensors. Structured in this way, a supervisor and a corresponding manufacturing cell represent a closed loop discrete event control system shown in Figure 3.11a.

However, this controller still does not implement any dispatching policy. It only executes rules that describe the required job sequence, as shown in Figure 3.11b. For systems that do not encompass shared resources this structure suffices. Nevertheless, in the case of systems with shared resources simultaneous requests for two or more concurrent tasks could be issued. To resolve this situation a supervisor has to select which jobs to initiate, *i.e.* it has to make a decision regarding the priority. This is needed since the resource-requirements matrix $\mathbf{F}_r$ has several 1s in the same column. In this situation, as has been shown in Example 3.2.1, the component of vector **m** corresponding to a shared resource attains a value of –1, which is not allowed. In order to solve a potential conflict and turn the controlled system to a "decision-free" structure (*cf.* Cofer and Garg 1992), it is therefore necessary to add an extra *dispatching control input*. The high entry selects which of the jobs will be preferred.

The easiest way to prevent conflicts and uniquely define the system activities is to employ this new input into the logical state-vector equation. Given that all prerequisites of a particular rule are met, additional conditions in the form of a vector, denoted $\mathbf{u}_d$, can attain the value 0 and block the rule. In this way the supervisor is able to forbid execution of any controllable task in the system. Vector $\mathbf{u}_d$ is called a *dispatching vector* (or *conflict-resolution* vector) and is generally determined as a function of feedback signals comprised in vector **m**,

$$\mathbf{u}_d(q) = h(\mathbf{m}(q)) \quad , \mathbf{u}_d(0) = \mathbf{u}_{d0} \tag{3.20}$$

where h is a control function. Depending on the way one selects the control function to generate $\mathbf{u}_d$, different dispatching strategies can be selected. These strategies fall mainly into two categories: Buffer and Part/Machine [8, 9]. Examples of the buffer category are: first-buffer-first-serve, last-buffer-first-serve, shortest nonfull queue, shortest remaining capacity, and shortest queue next. Examples of the part/machine category are: shortest imminent operation time, largest imminent operation time, shortest remaining processing time, largest remaining processing time, machine with least work and least slack time, *etc*. Although determination of an appropriate h is important, the objective of this section is not an elaboration on how the control function depends on a particular dispatching policy or how to prove the existence of a control function for a particular strategy. For our purpose it is sufficient to say that in some cases the control function attains a simple form of matrix multiplication, while in the case of large manufacturing systems with demanding policies it could be very complex or its implementation might even be questionable.

As far as the resolution of shared-resource conflict in an MS is concerned Equation (3.20) can provide suitable results. On the other hand, if the dispatching policy requires information regarding the exact arrangement of processed parts in the system, control based only on vector **m** may cause improper system performance due to the existence of hidden parts (as previously explained). An elegant way to overcome this problem is to introduce an additional vector, $\mathbf{m}^s(q)$ that is calculated by the supervisor according to the following relation

$$\mathbf{m}^s(q) = q^{-1}\mathbf{m}^s(q) + \left[\mathbf{S} - \mathbf{F}^T\right]\mathbf{x}(q), \quad \mathbf{m}^s(0) = \mathbf{m}_0^s \qquad (3.21)$$

One can notice that Equation (3.21) has the same structure as the second equation in Equation (3.12), which describes the evaluation of the system vector in the static matrix model. For this reason, vector $\mathbf{m}^s$ can be seen as prediction of feedback vector $\mathbf{m}$. Components of $\mathbf{m}^s$ are increased by the term $\mathbf{Sx}(q)$ immediately upon fulfillment of corresponding rules, while data obtained from the system, contained in $\mathbf{m}$, are delayed due to operations lifetimes. Having both vectors available, the supervisor design is not restricted only to signals gathered from sensors, thus,

$$\mathbf{u}_d(q) = h\big(\mathbf{m}(q), \mathbf{m}^s(q)\big) \qquad (3.22)$$

Once defined, the dispatching vector is integrated in the logical state-vector equation by a *dispatching matrix* (or *conflict-resolution matrix*) $\mathbf{F}_d$ in the following way:

$$\bar{\mathbf{x}}(q) = \mathbf{F}_\Delta q^{-1}\bar{\mathbf{m}}(q) \triangledown \mathbf{F}_d \Delta \bar{\mathbf{u}}_d(q) \qquad (3.23)$$

Equations (3.21) – (3.23) outline a new internal structure of the supervisor, depicted in Figure 3.12.

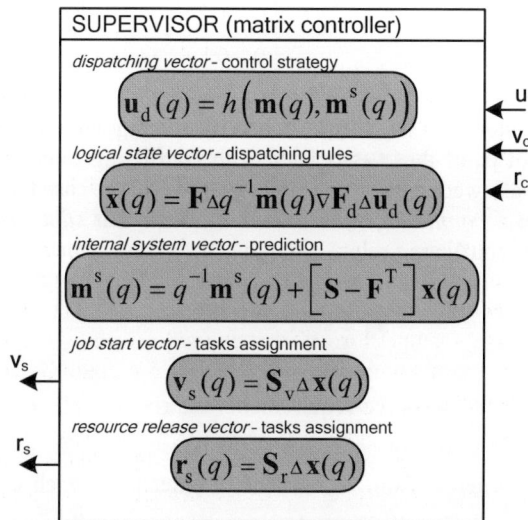

Figure 3.12. An internal structure of the supervisor based on the matrix controller

In each sampling interval recursive equations are executed as shown in Figure 3.12, from the top to the bottom. First, the matrix controller calculates a dispatching vector based on current data from the system and internal system vector $\mathbf{m}^s$. Then, the logical state vector is determined, a new value of $\mathbf{m}^s$ is evaluated and task assignments are issued.

A newly introduced component of the supervisor, dispatching matrix, needs further explanation. As for the system matrices, $\mathbf{F}_d$ has logical elements 0 and 1. Its structure and components depend on the applied dispatching policy. Given that the system comprises shared resources, the primary concern in the supervisory design is conflict resolution. Hence, the first step in determination of $\mathbf{F}_d$ is allocation of conflicting rules, which are related to columns of the resource-allocation matrix $\mathbf{F}_r$ containing more than one entry "1" (as we already stated, resources corresponding to those columns are shared resources). As a first thought we could say that for each "1" on these columns, a new column is constructed in $\mathbf{F}_d$ having only one entry "1" in the corresponding position for each "1" in $\mathbf{F}_r$. Established in this way, $\mathbf{F}_d$ would provide that each shared resource column in $\mathbf{F}_r$ is associated with as many components of the dispatching vector as it has entries of "1". In Example 3.2.1 the last column of $\mathbf{F}_r$ that corresponds to shared resource R, has two 1s. As a result, matrix $\mathbf{F}_d$ would have two columns, the first column with "1" in the 2nd position and the second column with "1" in the 5th position, while all other elements should be equal to 0:

$$\mathbf{F}_d = \begin{bmatrix} 0 & 0 \\ 1 & 0 \\ 0 & 0 \\ 0 & 0 \\ 0 & 1 \\ 0 & 0 \end{bmatrix}$$

Consequently, vector $\mathbf{u}_d = [u_{d1}\ u_{d2}]^T$. Since conflicting rules are concurrent, only one component of $\mathbf{u}_d$ of those associated with conflicting rules that belong to the same resource is allowed to have its value equal to 1 in the case of conflict.

Let us define a *conflicting-rules vector* $\mathbf{x}_d$, such that $sup(\mathbf{x}_d)=\{x_i, x_j, x_k, ...\}$, where x_i, x_j, x_k are conflicting rules. Binary vector $\mathbf{x}_d$ can be determined from $\mathbf{F}_r$ as

$$\hat{\mathbf{x}}_d = \hat{\mathbf{F}}_r \Delta \bar{\mathbf{r}}_s \tag{3.24}$$

where $\hat{\mathbf{F}}_r$ is a reduced resource-requirements matrix, *i.e.* all rows corresponding with rules that have an output operation in the consequent part are erased from the matrix. In order to get $\mathbf{x}_d$ from $\hat{\mathbf{x}}_d$ one has to enter 0 for each component of the conflicting rules vector that matches a row removed from the resource-requirements matrix. Recalling $\mathbf{F}_r$ from Example 3.2.1, Equation (3.24) gives

Matrix Model and Control of Manufacturing Systems   81

$$\hat{\mathbf{F}}_r \Delta \bar{\mathbf{r}}_s = \begin{bmatrix} \overset{MA}{1} & \overset{MB}{0} & \overset{B}{0} & \overset{R}{0} \\ 0 & 0 & 0 & 1 \\ 0 & 0 & 1 & 0 \\ 0 & 1 & 0 & 0 \\ 0 & 0 & 0 & 1 \\ 0 & 0 & 0 & 0 \end{bmatrix} \Delta \begin{bmatrix} 1 \\ 1 \\ 1 \\ 1 \\ 0 \end{bmatrix} = \begin{bmatrix} 1 \\ 0 \\ 1 \\ 1 \\ 1 \\ 0 \end{bmatrix} = \hat{\bar{\mathbf{x}}}_d \rightarrow \mathbf{x}_d = \begin{bmatrix} 0 \\ 1 \\ 0 \\ 0 \\ 1 \\ 0 \end{bmatrix}$$

Calculation of $sup(\mathbf{x}_d) = \{x_2, x_5\}$ confirms the already-determined conflicting rules that correspond with shared resource R.

For a given $\mathbf{x}_d$, one can determine the (i,j)th component of the dispatching matrix by using the following relation:

$$f_d(i, j) = \begin{cases} 1 & \text{if } x_d(i)=1 \text{ and } j=\sum_{k=1}^{i} x_d(k) \\ 0 & \text{otherwise} \end{cases} \qquad (3.25)$$

There is an observation regarding determination of the dispatching matrix as described above. If two (or more) shared resources contribute to one rule, then two (or more) columns in $\mathbf{F}_r$ would have "1" at the row that corresponds to this particular rule. In that case, according to our discussion, two (or more) components of the dispatching vector should be associated with this rule. Therefore, matrix $\mathbf{F}_d$ is supposed to obtain the form as shown below (it is assumed that the system has two shared resources R1 and R2),

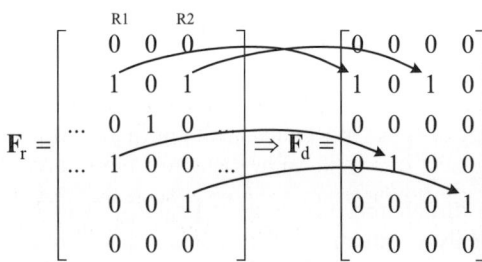

However, Equation (3.24) gives $\mathbf{x}_d = [0\ 1\ 0\ 1\ 1\ 0]^T$ and then by applying Equation (3.25), one obtains

$$\mathbf{F}_d = \begin{bmatrix} 0 & 0 & 0 \\ 1 & 0 & 0 \\ 0 & 0 & 0 \\ 0 & 1 & 0 \\ 0 & 0 & 1 \\ 0 & 0 & 0 \end{bmatrix}$$

i.e., a conflicting rule that involves two shared resources is linked to only one component of the dispatching vector. Determination of that component involves coordination of conflict-resolution strategies of all resources that participate in the considered rule.

Example 3.4.1 (DES simulation with conflict resolution – closed-loop dynamic matrix model)

In Example 3.3.1 simulation of the dynamic model of the workcell depicted in Figure 3.2 showed that an uncontrolled system gets in deadlock. Herein we are concerned with determination of the control function that would prevent conflict and avoid deadlock. As our intention is to illustrate a closed loop-manufacturing system with a simple dispatching strategy, no formal methods will be used in the example.

Conflicting rules vector $\mathbf{x}_d$ and dispatching matrix $\mathbf{F}_d$ are already determined and have the form

$$\mathbf{x}_d = \begin{bmatrix} 0 \\ 1 \\ 0 \\ 0 \\ 1 \\ 0 \end{bmatrix}, \quad \mathbf{F}_d = \begin{bmatrix} 0 & 0 \\ 1 & 0 \\ 0 & 0 \\ 0 & 0 \\ 0 & 1 \\ 0 & 0 \end{bmatrix}$$

Components of the dispatching vector $\mathbf{u}_d = [u_{d1} \ u_{d2}]^T$ will be calculated in two steps. First, we analyze situations that could cause conflict. As previously elaborated, this condition is met when the robot is idle and both machines have parts to be removed, which can be clearly seen from the matrix model. We define the control function in the form of a rule:

IF $sup(\mathbf{v}_c) \cap \{MAP, MBP\} = \{MAP, MBP\}$
THEN $\mathbf{u}_d = [0 \ 1]^T$
ELSE $\mathbf{u}_d = [1 \ 1]^T$

Since the operation MBP is the last operation in the sequence, this strategy prefers pulling the parts from the system.

The second step in the design is related to evaluation of circumstances that could lead to the deadlock. As we already discussed, in an uncontrolled system a deadlock occurred because parts have been pushed into the workcell by the robot. From the graphs shown in Figure 3.8 it can be seen that once the buffer is full and machine MB is processing the part, additional entry of parts should be blocked. This can be employed by extending the above control rule:

IF $sup(\mathbf{v}_c) \cap \{MAP, MBP\} = \{MAP, MBP\}$ OR $B \in sup(\mathbf{m}_c)$
THEN $\mathbf{u}_d = [0\ 1]^T$
ELSE $\mathbf{u}_d = [1\ 1]^T$

The results obtained by simulation of the workcell controlled by the supervisor are graphically shown in Figure 3.13. System is stable with no conflict. It is evident from the graphs that control signal u_{d1} is equal to 0 as long as the buffer is full, which blocks operation RP1 and prevents incoming of new parts into the workcell.

◆

As we mentioned earlier, in many cases the control function may be realized by simple matrix operations. This is especially suitable when MS is represented by the matrix model. The simplest form of the dispatching policy is defined as

$$\mathbf{u}_d = \mathbf{S}_d \Delta \mathbf{x} \qquad (3.26)$$

where $\mathbf{S}_d$ is a *dispatching vector release matrix*. In Equation (3.26) vector $\mathbf{u}_d$ is directly related to the logical state vector $\mathbf{x}$. Execution of a particular rule and entry "1" in the corresponding element of $\mathbf{S}_d$ will increase the value of the associated component of vector $\mathbf{u}_d$. In general, the structure of $\mathbf{S}_d$ depends on matrix $\mathbf{F}_d$, the job ordering and dispatching strategy.

One convenient method to determine the dispatching vector release matrix is the *reordering* of rows of matrix $\mathbf{F}_d^T$. Usage of $\mathbf{F}_d$ is intuitively understandable since the dispatching matrix defines the way conflict-resolution vector $\mathbf{u}_d$ is connected to the system, therefore, its transpose implies that rules that are already known as conflicting and that encompass the dispatching vector in their prerequisite parts, also have vector $\mathbf{u}_d$ components in their consequent parts. Hence, no additional calculation of rules that release the dispatching vector is required. However, the process of reordering of rows must be made with care, otherwise the method could end up in system deadlock.

The main idea behind row rearrangement is related to the job sequencing performed by the shared resource. Let us consider the robotized workcell shown in Figure 2.13. This system, having the robot as a shared resource, can be described with seven IF-THEN rules. Three of them related to operations performed by the robot are involved in the conflict. Given that $\mathbf{x}_d = [1\ 0\ 1\ 0\ 0\ 1\ 0]^T$, the corresponding transposes of dispatching matrix and dispatching vector are defined as

$$\mathbf{F}_d^T = \begin{bmatrix} 1 & 0 & 0 & 0 & 0 & 0 \\ 0 & 0 & 1 & 0 & 0 & 0 \\ 0 & 0 & 0 & 0 & 1 & 0 \end{bmatrix}, \; \mathbf{u}_d = \begin{bmatrix} u_{d1} \\ u_{d2} \\ u_{d3} \end{bmatrix}$$

If we denote operations executed by the robot as RP1 (placing part *a* in machine A), RP2 (removing part *a* from machine A) and RP3 (removing part *b* from machine B), then, for example, a possible repeatable sequence could be $s_1 =$ (RP1, RP3, RP2). From $\mathbf{F}_d$ and $\mathbf{u}_d$, defined above, we see that rule x_1 is controlled by u_{d1}, rule x_3 is controlled by u_{d2} and rule x_6 is controlled by u_{d3}. Let execution of operation RP1 be related to fulfillment of x_1, RP2 to x_3 and RP3 to x_6. Then, in order to realize sequence s_1, matrix $\mathbf{S}_d$ and the initial value of the dispatching vector should be

$$\mathbf{S}_d = \begin{bmatrix} 0 & 0 & 0 & 0 & 0 & 1 & 0 \\ 1 & 0 & 0 & 0 & 0 & 0 & 0 \\ 0 & 0 & 1 & 0 & 0 & 0 & 0 \end{bmatrix}, \; \mathbf{u}_{d0} = \begin{bmatrix} 1 \\ 0 \\ 0 \end{bmatrix}$$

For a given matrix $\mathbf{S}_d$ and according to Equation (3.26), execution of rule x_1 releases u_{d2}, rule x_3 releases u_{d3}, while execution of x_6 releases u_{d1}. Implemented in this way, the control strategy prevents conflict and accomplishes the required sequence.

Rearrangement of rows can be easily done by matrix operation:

$$\mathbf{S}_d = \mathbf{\Phi}_\Delta \mathbf{F}_d^T \tag{3.27}$$

where $\mathbf{\Phi}$ is a transformation matrix defined in the following way: when dispatching vector component u_{di} is released by the rule that is controlled by the component u_{dj}, then $\mathbf{\Phi}(i,j) = 1$, otherwise it is 0. In our case

$$\mathbf{\Phi} = \begin{array}{c} \phantom{\begin{bmatrix}} u_{d1}\; u_{d2}\; u_{d3} \phantom{\end{bmatrix}} \\ \begin{bmatrix} 0 & 0 & 1 \\ 1 & 0 & 0 \\ 0 & 1 & 0 \end{bmatrix} \begin{array}{c} u_{d1} \\ u_{d2} \\ u_{d3} \end{array} \end{array}$$

Due to its simplicity, the dispatching strategy (3.26), with matrix $\mathbf{S}_d$ determined according to Equation (3.27), is very restrictive. In general, it allows only one part to enter the part path, which leads to poor resources utilization and low system throughput. This situation can be demonstrated if we return to the system shown in Figure 2.12. Assuming that processing and setup times of machine A are much shorter than those of machine B, the dispatching strategy determined above will force the robot and machine A to remain inactive, although they might have enough time to process several parts while waiting for machine B.

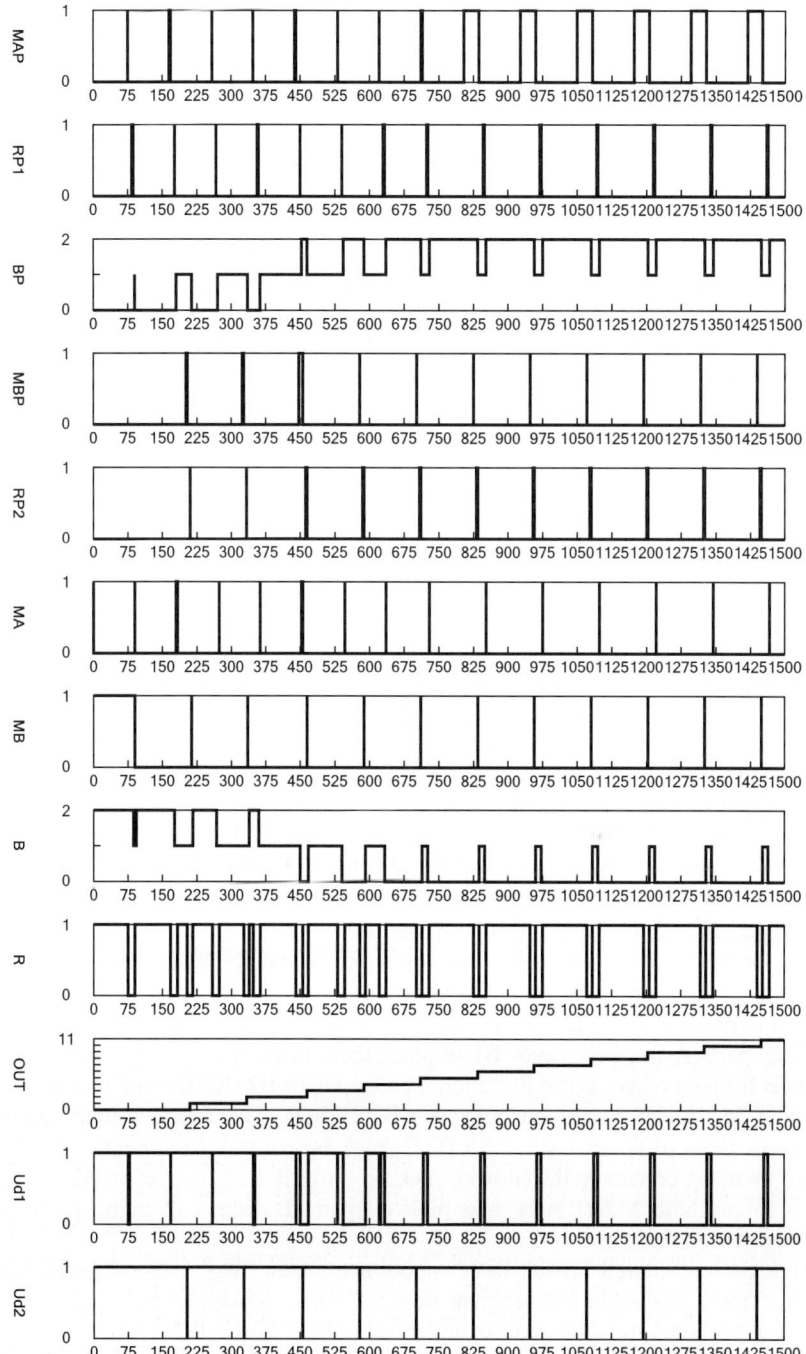

Figure 3.13. Graphical representation of results for the closed-loop system (Example 3.4.1)

The other remark that should be made is the set up of the dispatching vector initial value. Incorrect assignment of $\mathbf{u}_{d0}$ can block components of the logical state vector so that the system cannot start with activities. In our example, this condition happens when $\mathbf{u}_{d0} = [0 \ 1 \ 0]^T$. In this case the first rule that corresponds to placement of a part in machine A and the sixth rule that executes replacement of a part from machine B are blocked. Therefore, no other activities can start after the processing of a part in machine B is finished.

3.5. A Case Study: Implemetation of the Matrix Controller

This section presents the implementation of the matrix controller for supervision of an intelligent material handling (IMH) cell [17]. Then, we show that the actual implementation and the simulated system give commensurate results. The IMH cell belongs to the class of so-called *multipart re-entrant flowline* (MRF) systems, which are described in Chapter 5. The objective of this section is to show the versatility of the system developed with this matrix formulation. The supervisor based on the matrix framework permits implementation of different methodologies for conflict resolution, as well as optimization of the resource assignment and part throughput. The given technical information includes the matrix controller development in LabVIEW®.

3.5.1 Intelligent Material Handling (IMH) Workcell Description

The IMH cell is composed of three robots, three conveyors, ten sensors and two simulated machines. Different configuration of re-entrant flowline problems can be accomplished with this structure. The image and the part flowline for a specific layout of the IMH cell are depicted in Figures 1.1 and 3.14.

For this specific layout the robot defined as R1 (a CRS robot) can perform four different tasks, $|J(R1)|=4$. Two tasks (R1u1 and R1u2) are related to picking up part-types A and B from the input-parts area, which are to be placed on the conveyor denoted B1. The other two tasks (R1u3 and R1u4) are associated with picking up final products A and B from conveyor B3 and placing them in the output-parts area. A PUMA robot, R2, performs three different tasks, $|J(R2)|=3$: pick up parts A from conveyor B1 to place them in machine M1 (R2u1), pick up parts B from conveyor B1 to place them on conveyor B2 (R2u2), and pick up parts A from M1 to be placed on conveyor B2 (R2u3). The Adept robot, R3, also performs three different tasks, $|J(R3)|=3$: pick up parts A from conveyor B2, to place them on conveyor B3 (R3u1), pick up parts B from conveyor B2 to place them in machine M2 (R3u2), and pick up parts B from M2 to be placed on conveyor B3 (R3u3).

For the considered layout, three robots manipulate two different parts, while two of them manipulate re-entrant flow of parts. Machines M1 and M2 are simulated by activating valve-air cylinders controlled from a PC.

Due to the existence of shared resources this configuration of the IMH cell presents a dispatching problem. Both phenomena, conflict and deadlock, may occur in the case of an inappropriate dispatching strategy. Since up to now deadlock prevention and avoidance have not been discussed, we concentrate on determination of the control policy that provides a conflict resolution. It is shown, without additional elaboration and formal proof that the obtained strategy is deadlock free (we return to this issue in Chapter 6).

The matrix model can be directly written down from Figure 3.14, which shows both job sequencing and resource assignment. From Figure 3.14 one can find that the system is described with 20 rules. The job sets that correspond with job sequences for two part paths and the set of resources are defined as follows:

- part A path

J^1={R1u1,B1AS,R2u1,M1P,R2u3,B2AS,R3u1,B3AS,R1u3}

- part B path

J^2={R1u2,B1BS,R2u2,B2BS,R3u2,M2P,R3u3,B3BS,R1u4}

- set of resources

R={B1AA,B1BA,M1A,B2BA,B2AA,M2A,B3AA,B3BA,R1A,R2A,R3A}
with a set of shared resources R_s={R1A, R2A, R3A}.

The description of jobs performed by nonshared resources is given in Table 3.2.

Table 3.2. Description of jobs in IMH cell

Notation	Description
B1AS	transporting part A on conveyor B1
M1P	processing part A in machine M1
B2AS	transporting part A on conveyor B2
B3AS	transporting part A on conveyor B3
B1BS	transporting part B on conveyor B1
B2BS	transporting part B on conveyor B1
M2P	processing part B in machine M2
B3BS	transporting part B on conveyor B1

88   Manufacturing Systems Control Design

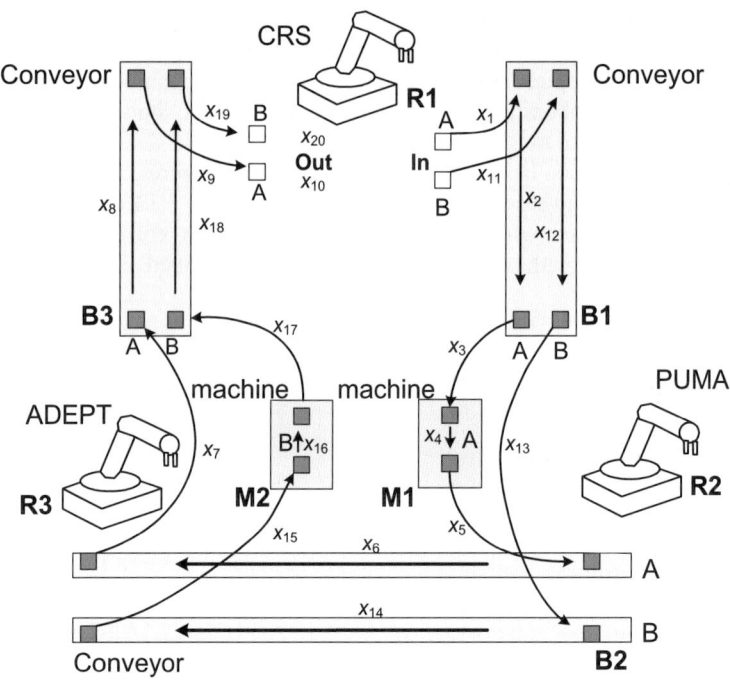

Figure 3.14. A layout with the parts paths of the IMH-cell used in a case study

The nomenclature used in the IMH is as follows: "RXuY" means job "Y" is executed by robot "X", "BxyS" means that product type "y" is transported by conveyor "x", "MxP" stands for machine "x" is busy, "BxyA" means that conveyor "x" is available for product type "y", "MxA" denotes machine "x" is available, "RxA" stands for robot "x" is idle. Note that instead of having three different resources for conveyors B1, B2 and B3, six different resources are used. This is because of the two different materials paths on each conveyor. For example, conveyor B1 has paths B1A and B1B, which are denoted as B1AA and B1BA when they are available, and denoted as B1AS and B1BS when they are carrying material.

Given the system layout and the system description, one can determine the system matrices, herein shown "graphicaly" with black and white rectangles, indicating "1" and "0", respectively.

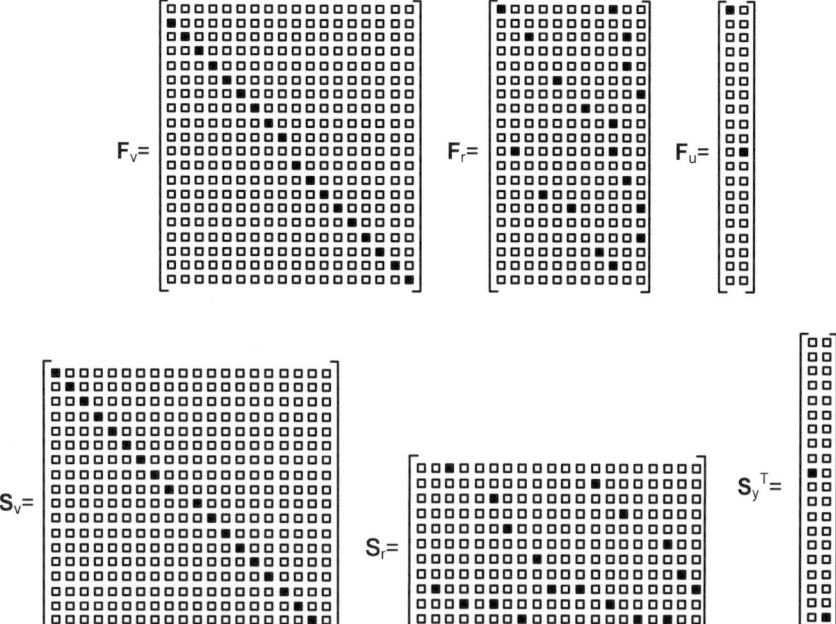

3.5.2 IMH Workcell Dispatching Strategy

The last three columns of $\mathbf{F}_r$ correspond to the shared resources R1A, R2A and R3A. From the number of 1s in those columns we see that R1A is involved in four conflicting rules, while each of the remaining robots, R2A and R3A, contribute in three, which finally gives ten conflicting rules. According to the definition, $\mathbf{F}_d$ is constructed by creating a new column for each "1" appearing in $\mathbf{F}_r$ for the shared resources, hence, the dispatching matrix will have 10 columns. By using Equations (3.24) and (3.25) we obtain:

It should be noted that columns of $\mathbf{F}_d$ have been rearranged in order to group components of the dispatching vector that belong to the same shared resource. Specifically, R1 is controlled with u_{d1}, u_{d2}, u_{d3} and u_{d4}, R2 with u_{d5}, u_{d6} and u_{d7}, and R3 with u_{d8}, u_{d9} and u_{d10}.

The conflict resolution used in the IMH cell for the layout shown in Figure 3.14 is an augmented version of the last-buffer-first-serve dispatching strategy, we call it ALBFS, modified for multipath systems. Herein we demonstrate an additional way of realization of the control function h by using a so-called *temporary system vector*, $\mathbf{m}^t$. At the beginning of each sampling interval all components of the dispatching vector attain the value "1". Then, the logical state vector $\mathbf{x}$ is calculated according to Equation (3.23) and the obtained value is included in

$$\mathbf{m}^t(q) = q^{-1}\mathbf{m}(q) + \left[\mathbf{S} - \mathbf{F}^T\right]\mathbf{x}(q)$$

i.e. temporary system vector is attained by allowing execution of all conflicting rules ($u_{di}=1$, for $i=1,10$) for current data from sensors comprised in the system vector $\mathbf{m}$. When some of the shared resources (robots in our system) are requested by more than one operation, the corresponding component of $\mathbf{m}^t$ would have a negative value, thus pointing out the occurrence of conflict. If that happens, the ALBFS dispatching strategy blocks some of the conflicting rules and the logical state vector is recalculated, this time with no conflict. This new vector is used by the supervisor for determination of task assignments (calculation of vectors $\mathbf{v}_s$ and $\mathbf{r}_s$). The procedure repeats in each sampling interval.

ALBFS policy, implemented upon calculation of vector $\mathbf{m}^t$, is given as the set of rules (recall that component of vector $\mathbf{y}$ corresponding with job (resource) Z is denoted y_Z):

IF $m^t_{R1A} < 0$ (resource R1A requested more than once) THEN
    IF $m^t_{R1U4} > 0$ (job R1U4 requested R1A) THEN
        $u_{dR1U1} = 0$ AND $u_{dR1U2} = 0$ AND $u_{dR1U3} = 0$ AND $u_{dR1U4} = 1$
    ELSE IF $m^t_{R1U3} > 0$ THEN
        $u_{dR1U1} = 0$ AND $u_{dR1U2} = 0$ AND $u_{dR1U3} = 1$ AND $u_{dR1U4} = 0$
    ELSE IF $m^t_{R1U1} > 0$ THEN
        $u_{dR1U1} = 1$ AND $u_{dR1U2} = 0$ AND $u_{dR1U3} = 0$ AND $u_{dR1U4} = 0$
    ELSE
        $u_{dR1U1} = 0$ AND $u_{dR1U2} = 1$ AND $u_{dR1U3} = 0$ AND $u_{dR1U4} = 0$

IF $m^t_{R2A} < 0$ THEN
    IF $m^t_{R2U3} > 0$ THEN
        $u_{dR2U1} = 0$ AND $u_{dR2U2} = 0$ AND $u_{dR2U3} = 1$
    ELSE IF $m^t_{R2U1} > 0$ THEN
        $u_{dR2U1} = 1$ AND $u_{dR2U2} = 0$ AND $u_{dR2U3} = 0$
    ELSE
        $u_{dR2U1} = 0$ AND $u_{dR2U2} = 1$ AND $u_{dR2U3} = 0$

IF $m^t_{R3A} < 0$ THEN
    IF $m^t_{R3U3} > 0$ THEN
        $u_{dR3U1} = 0$ AND $u_{dR3U2} = 0$ AND $u_{dR3U3} = 1$
    ELSE IF $m^t_{R3U2} > 0$ THEN
        $u_{dR3U1} = 0$ AND $u_{dR3U2} = 1$ AND $u_{dR3U3} = 0$

ELSE
$$u_{dR3U1} = 1 \text{ AND } u_{dR3U2} = 0 \text{ AND } u_{dR3U3} = 0$$

As we already mentioned, deadlock avoidance is an inherent property of pulling strategies such as ALBFS. Therefore, in the determination of the rules stated above only conflict was considered. If we analyze the rules then we can see that between the two final jobs needed to manufacture products A and B, R1U3 and R1U4, the supervisor is designed to prefer products B.

3.5.3 Implementation of the Matrix Controller on the IMH Workcell

The matrix controller is implemented on a PC in a LabVIEW® graphical programming environment. In LabVIEW®, one can sequence and control different processes at the same time. The processes we are interested in are operations implemented in manufacturing process, like execution of a robots' trajectories, machining jobs and transferring parts using conveyors. The matrix controller runs on the PC that has three serial ports for communication with three robots. It also has a digital acquisition card that receives digital signals from capacitive proximity sensors. The same card is used for sending digital signals to activate machine jobs.

In Figure 3.15 one can find three levels of intelligent control depicted in [10, 11]. The first level is *organization*, which is the highest level of intelligence and in our case it is presented as the matrix-based controller structure. The main purpose of our implementation is to present the advantages and great potential of the organization level realized in the form of the matrix-based controller shown in Figure 3.12. The second level is the *coordination level*. This level contains a set of independent modules that are composed by robot programming sentences encrypted in VAL-like commands [12, 13]. These program modules define the jobs to be done by the robots (*i.e.* the sequence of VAL commands needed to command robots to perform pick and place tasks). Then, once the task or job is selected by the organization level, the coordination level sequences the steps needed for each of these jobs. In our case, the IMH cell's coordination level sends commands sequentially to the appropriate robot to accomplish the desired task (Figure 3.16).

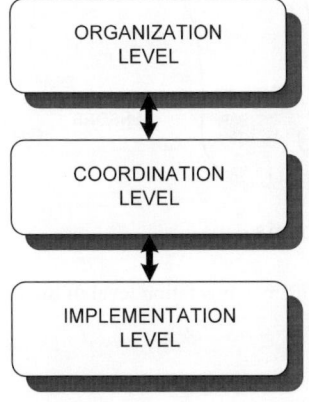

Figure 3.15. Three levels of intelligent control

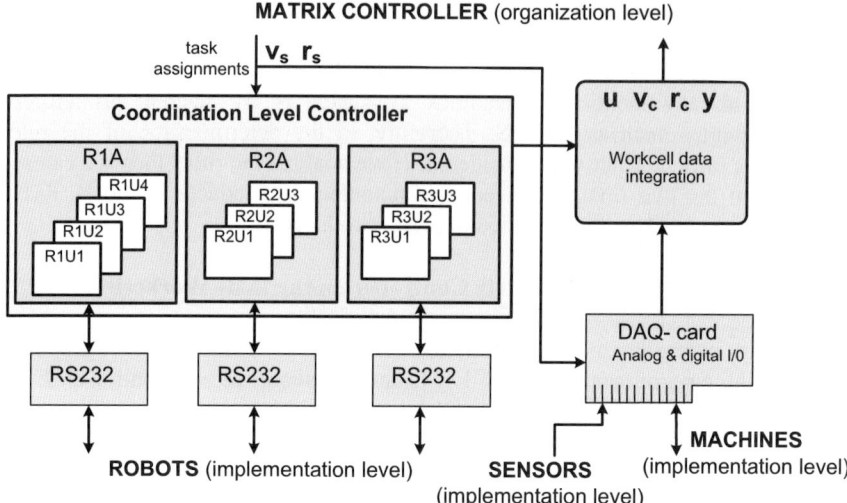

Figure 3.16. Coordination level of the IMH workcell

The last level of the system is the *implementation level* (Figure 3.17), which is accomplished by the robot drivers and controllers. When the robot controller receives a VAL command via the serial port, it performs low-level control calculations and strategies such as interpolation, proportional derivative (PD) control, proportional integrate derivative (PID) control, fuzzy logic control, neural-network control or any other low-level control strategy to manipulate the robotic arms.

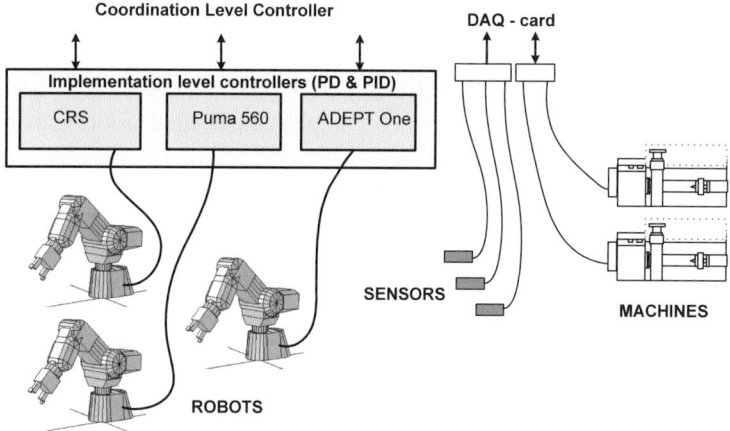

Figure 3.17. Implementation level of the IMH workcell

3.5.4 The Matrix Controller in LabVIEW Graphical Environment

The purpose of this section is to explain the development of the matrix controller by using LabVIEW® [14, 15]. The key equations of the matrix formulation, described in previous sections, are graphically represented in LabVIEW® (Figure 3.18).

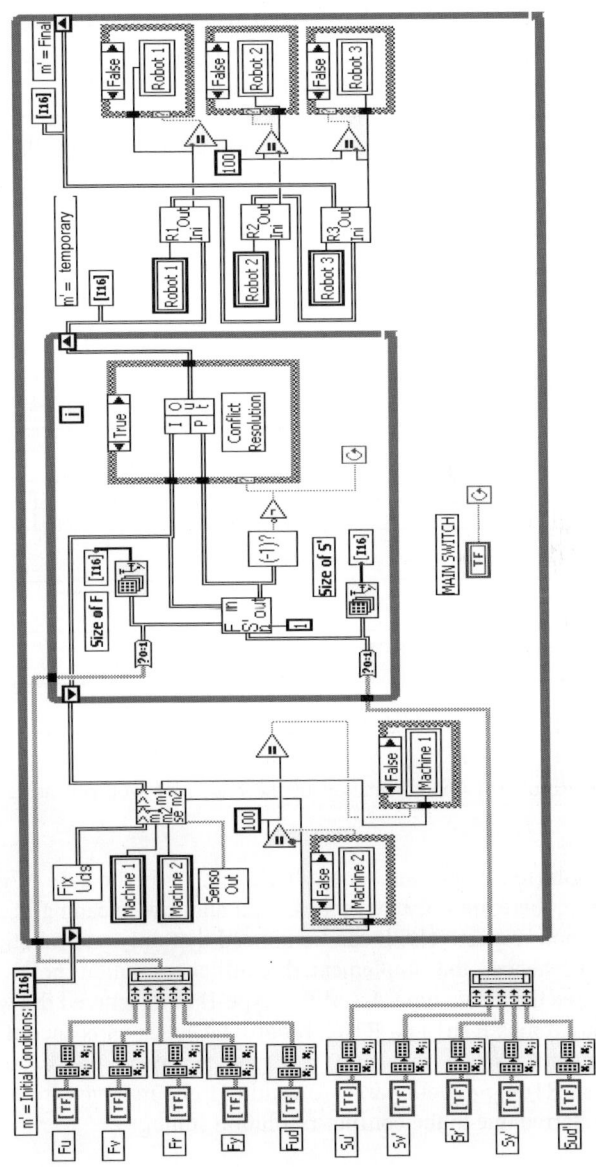

Figure 3.18. The matrix controller in LabVIEW® graphical environment

The entire diagram is used as a single LabVIEW® block (or function) representing the WHILE "Main loop" that is executed in each sampling interval. Inputs in the block are system matrices (shown on the top) and sensors signals. LabVIEW® block "Fix Uds" is positioned inside the Main loop. This block sets all the components of conflict-resolution vector to 1 at the beginning of the cycle, as we described in the previous paragraph. If conflict is detected on any of the robots, the "Conflict Resolution" block deactivates rules according to ALBFS strategy. Inside the main loop, an internal loop is used to calculate the logical state vector **x** by applying Equation (3.23). Function MULTOA(X,Y), already implemented for MATLAB® simulation, is used for that purpose.

The performance of the IMH workcell is shown in Figure 3.19. The results have been obtained in real time directly from the matrix controller implemented in LabVIEW®.

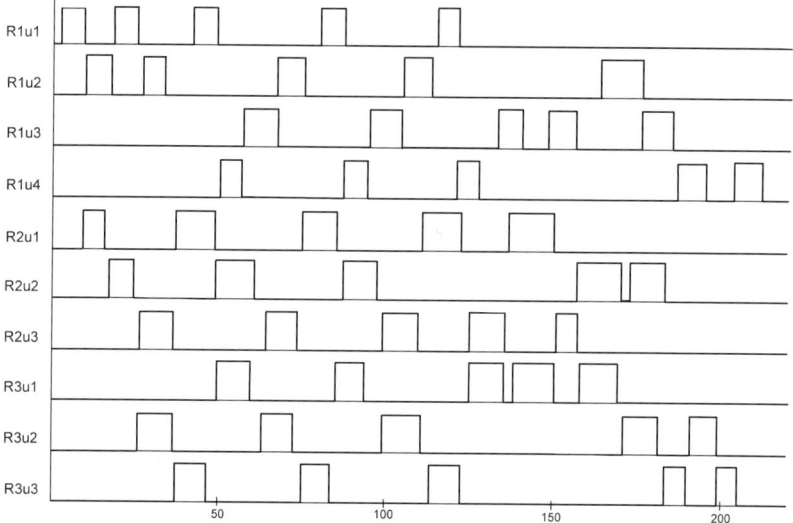

Figure 3.19. Results of implementation of the real IMH workcell using ALBFS conflict resolution

Each graph (line) represents one robotic job. As in the case of graphs obtained by simulation, there are only two states, high and low, meaning that a job is being executed or not, respectively. It can be noticed that only one robotic job goes high at any time, hence, the implemented conflict-resolution policy achieved the requested objective. Five type-A and five type-B parts entered the workcell. As we can see from graphs R1u1 and R1u2, robot R1A loads parts according to the A-B-A-B... sequence. When the third part-type A enters the system, robot R1A executes task R1u4, *i.e.* both parts are waiting to be removed from the system and part B is preferred due to the conflict-resolution strategy.

3.6. Excersises

1. For the system described in Example 4.3.1 do the following:
   a) determine the matrix model,
   b) simulate the matrix model by using MATLAB® code given in Figure 3.5,
   c) determine the dispatching matrix F_d, the dispatching vector release matrix S_d and dispatching vector u_d that will execute the task sequence a) loading M_1 – unloading M_2 – loading M_3,
   d) simulate the matrix model with obtained supervisor by using MATLAB® code given in Figure 3.5 (extend vectors and matrices used in the code in order to include F_d, S_d and u_d),
   e) determine the dynamic matrix model of the system,
   f) simulate the dynamic matrix model by using MATLAB® code given in Figure 3.10.

References

[1] Pastravanu OC, Gürel A, Lewis FL, Huang HH. Rule-Based Controller Design Algorithm For Discrete Event Manufacturing Systems, Proceedings of the American Control Conference 1994;1:299-305.

[2] Tacconi DA, Lewis FL. A New Matrix Model for Discrete Event Systems: Application to Simulation, IEEE Contr. Sys. Mag. 1997;17;5:62-71.

[3] Lewis FL, Huang HH. Control System Design for Flexible Manufacturing Systems, in Flexible Manufacturing Systems: Recent Developments, Elsevier, 1994.

[4] Noori H, Radford R. Production and Operations Management. New York: McGraw-Hill, 1995.

[5] Steward DV. Systems Analysis and Management: Structure, Strategy, and Design. New York: Petrocelli Books, 1981.

[6] Elsayed EA, Boucher TO. Analysis and Control of Production Systems (2nd Ed.). Englewood Cliffs: Prentice-Hall, 1994.

[7] Smolic-Rocak N, Bogdan S, Kovacic Z, Reichenbach T, Birgmajer B. Dynamic modeling and Simulation of FMS by using VRML, CD Proceedings of 15th IFAC World Congress 2002.

[8] Panwalker SS, Iskander W. A survey of scheduling rules, Operations Research 1977;26;1:45-61.

[9] Lewis FL, Huang HH, Jagannathan S. A Systems Approach to Discrete Event Controller Design for Manufacturing Systems Control, Proceedings of the American Control Conference 1993;2:1525-1531.

[10] Saridis G.N. Architectures of Intelligent Controls, in Intelligent Control Systems. New York: IEEE Press, 1995.

[11] Antsaklis PJ, Passino KM. An Introduction to Intelligent and Autonomous Control Systems. Norwell: Kluwer, 1992.

[12] Shimano B. VAL: A Versatile Robot Programming and Control System, Proceedings of the IEEE Computer Society's Third International Computer Software & Applications Conference 1979;3:878-883.

[13] Larson TM. Robotic Control Language, Advances in Instrumentation 1983;38;1:665-675.

[14] Mireles J, Lewis FL. Intelligent Material Handling: Development and Implementation of a Matrix-Based Discrete Event Controller, IEEE Trans. Ind. Electr. 2001;48;6.
[15] Mireles J, Lewis FL, Gurel A. Deadlock Avoidance for Manufacturing Multipart Reentrant Flow Lines Using a Matrix-Based Discrete Event Controller, Int. J. Production Research 2002;40;13:3139-3166.
[16] Bogdan S, Lewis FL, Gurel A, Kovacic Z. Timed matrix-based model of flexible manufacturing systems, Proceedings of the IEEE International Symposium on Industrial Electronics 1999;3:1373-1378.
[17] Mireles J, Lewis FL, Gurel A, Bogdan S. Deadlock Avoidance Algorithms and Implementation , a Matrix Based Approach, in Deadlock Resolution in Computer-Integrated Systems, Marcel Dekker, 2005.
[18] Bauer A, Bowden R, Browne J, Duggan J, Lyons G. Shop Floor Control Systems - From Design to Implementation. London: Chapman & Hall, 1991.
[19] Gullander P. On Reference Architectures forDevelopment of Flexible Cell Control Systems, PhD thesis, Gotenborg University, 1999.
[20] Leitão P, Quintas A. A Manufacturing Cell Controller Architecture, Proceedings of Flexible Automation and Intelligent Manufacturing Conference 1997:483-493.
[21] Maturana F, Norrie DH. Multi-Agent Mediator Architecture for Distributed Manufacturing, J. Intell. Manufact. 1996;7:257-270.
[22] Heikkilä T, Kollingbaum M, Valckenaers P, Bluemink GJ. manAge: An agent architecture for manufacturing control, Proceedings of the 2nd International Workshop on Intelligent Manufacturing Systems 1999:127-136.
[23] Tönshoff HK, Seilonen I, Teunis G, Leitão P. A Mediator-based approach for decentralised production planning, scheduling and monitoring, Proceedings of CIRP International Seminar on Intelligent Computation in Manufacturing Engineering 2000: 89-95.

4

Matrix Methods for Manufacturing Systems Analysis

Since its first practical implementation in the 18th century [1], when Euler proved that it is impossible to visit all the bridges in Köingsberg and then to return to a starting point by passing each bridge only once, graph theory has been successfully applied for solving various problems. From computer networks, today's world information highways - to transportation systems, whose rapid growth requires increased safety and reliability, methods developed by graph theory offer a convenient way to analyze data associated with planning, organization and other related phenomena. Graph theory can easily answer questions such as: what is the communication lines bandwidth required for successful transmission of a particular amount of information between two places on the network, how many trains are needed in order to make a particular timetable feasible, which is the optimal way between two cities where required energy is concerned.

When we talk about manufacturing systems, the first thing that comes to mind is a set of machines processing raw materials in order to make a product. Located on the factory floor according to a specified layout, machines can be understood as points that exchange both materials (parts) and information following a certain plan. One of the most suitable ways to represent this scheme of material or information flow is by using graphs.

In this chapter we describe the basic concepts of graphs. First, we introduce basic graph definitions followed by matrix representations of the graphs. At the end of the section an illustrative example of a manufacturing system modeled by a graph is given. The second section of the chapter is concerned with string composition. String composition is a method for analysis of graph properties based on a particular string-manipulation algorithm. In that section we present string operators and their properties, concluding with an example of the shortest-path determination in an AGV system. The last part of the chapter is devoted to max-plus algebra, which is an extremely useful tool for analysis of a special class of manufacturing systems. We give only the basics of max-plus since deeper insight into its theory would require much more space and time. Furthermore, max-plus theory covers only a particular group of discrete event systems while the DES class we are interested in has a wider application. We show how the max-plus equation is derived from the matrix description of the system. Since the theory is still being developed we hope that a way to extend it to a broader class of manufacturing systems will soon appear. However, it is important to understand the max-plus concept in order to comprehend problems related to the inclusion of operational times in the system analysis. We close the max-plus section with an example. At the end of the chapter problems for exercise are given.

4.1 Basic Definitions of Graphs

First we need to define a graph [2].

Definition 4.1.1 (graph): A *graph* is a structure formed by a set of *nodes V* and a set of *arcs E*. Arcs in *E* represent pairs of nodes in *V*, $G = (V, E)$.

In the mathematical literature nodes are called *vertices*, while arcs are called *edges*. These two names are the origins of symbols *V* and *E*. In a graph, nodes represent places or locations while arcs represent connections between these places. The word place should be taken conditionally when manufacturing systems are considered as there are two ways of representing it. Specifically, in graph representation of the system, a node may represent the *occurrence of some event*, while arcs may be used to show *relations between events* – which event(s) are prerequisite(s) for the occurrence of a particular event. On the other hand, a node may represent *the system state* while in that case arcs represent *events* that lead to this particular state.

As shown in Figure 4.1, the node is graphically represented by a circle and the symbol for an arc is a line drawn between two nodes. In the graph shown in the figure a set of nodes is $V = \{a, b, c, d\}$ while a set of arcs is $E = \{(a,b), (b,c), (b,d), (c,c), (c,d)\}$. It should be noted that set *E* can also be defined as $E = \{(b,a), (c,b), (c,c), (d,b), (d,c)\}$.

Arc (*c,c*), which is different from all the other arcs shown in Figure 4.1 begins and ends in the same node. This type of arc is called *a loop* (or *a self-loop*).

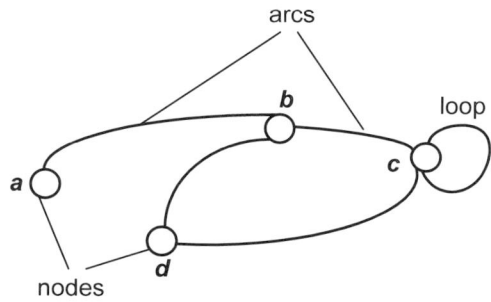

Figure 4.1. A graph

We say that two nodes, n_1 and n_2, are adjacent if there is an arc between them. In that case we call the arc incident to both nodes n_1 and n_2. The degree of node n_1 is equal to the number of arcs incident to it. A graph is called regular of degree r if all nodes in the graph have the same degree equal to r.

A graph is called a multigraph if it contains more than one arc between two nodes or if there are self-loops.

Let us now consider the nodes of the graph shown in Figure 4.1 as street intersections and arcs as streets between these intersections. The question is: can we drive a car from node a to node d? From Figure 4.1 we know only that a is connected with d but we are not able to tell if we can actually get from a to d since streets connecting these two intersections may be one-way streets. To answer to this question we have to add one more property to arcs: direction. We showed that set E for the graph in Figure 4.1 can be defined in two different ways, which means that in an *undirected* graph set E is a *not ordered* set of pairs of nodes. Now, we can go on and define a *directed graph*.

Definition 4.1.2 (directed graph): A *directed graph* $G = (V, E)$ is a graph with ordered set E, *i.e.* pairs of nodes in a directed graph are ordered.

A directed graph is shown in Figure 4.2. The first node, n_1, in the ordered pair (n_1,n_2) is called the *origin* and node n_2 is called the *destination*. In the graphical representation the direction from n_1 to n_2 is shown as an arrow.

We describe the graph in Figure 4.2 as $V = \{a, b, c, d\}$, $E = \{(a,b), (b,a), (b,c), (c,c), (c,d), (d,b)\}$. Continuing our analogy of Figure 4.2, the street connecting a and b may be driven on in both directions while the street that links intersections b and c is a one-way street, *i.e.* it can be passed only from b to c.

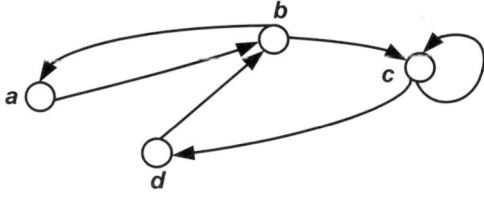

Figure 4.2. A directed graph

Now, having defined directions in the graph, we may answer the question: traveling from a to d is possible by passing through intersections b and c.

The answer to the first question raises another: how far is a from d or how much time do we need to pass along the established route if we drive with a predetermined speed? The answer requires the inclusion of a *weight* property to the notion of arc, *i.e.* a numerical value is associated with each arc in a graph specifying length, time or cost of the arc (weights can be associated with nodes as well). Graphs with weighted arcs (nodes) are called *weighted graphs* (Figure 4.3).

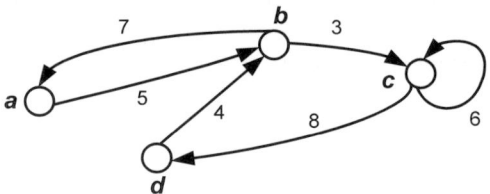

Figure 4.3. A weighted directed graph

The weights shown in Figure 4.3 may represent the average time needed to travel along the corresponding arc. According to the figure, going from *c* to *d* requires 8 time units, while a trip from *c* to *c* would require 6 time units. It should be noted that weights of arcs connecting the same nodes are not necessarily equal. In our example, arcs (*a,b*) and (*b,a*) have different weights. The difference in times required for passing these arcs could be caused by a different number of street lanes for example. The direction from *a* to *b* may have more lanes thus providing conditions for faster traffic, which makes the traveling time shorter than for the trip from *b* to *a*.

Our question, related to the distance between *a* and *d*, may now be answered. From Figure 4.3 we find that traveling from *a* to *d* would take 16 time units.

The ordering of the set *E*, *i.e.* the introduction of directions in a graph, as well as setting weights to arcs (nodes) has many consequences. As we throughout this book deal with directed graphs, in the text that follows we define terms and structures that are needed for the investigation of basic directed graph-properties.

Definition 4.1.3 (upstream, downstream node): In a directed graph, $G = (V, E)$, a node n_1 is called the *upstream node* to node n_2 if there exists an arc $(n_1, n_2) \in E$. In that case, node n_2 is called the *downstream node* to node n_1.

An upstream node is sometimes called a *predecessor* and a downstream node is called a *successor*. When there is more than one node upstream of node *n*, we define a set called a *preset* of *n* that contains all such nodes. Downstream nodes of node *n* belong to the set referred to as the *postset* of *n*. The importance of preset and postset concepts will be shown later in the chapter related to Petri nets.

In the graph from Figure 4.3 the preset of node *b* is {*a*, *d*} while the postset of *b* is {*a*, *c*}.

Definition 4.1.4 (path): Having a directed graph $G = (V, E)$, a *path* is a sequence of nodes $(n_1, n_2, n_3, ..., n_j)$ such that n_i is upstream of n_{i+1} for $i = 1, 2, ... j-1$.

We may also speak of a path as a sequence of arcs that connects a sequence of nodes belonging to the path.

Definition 4.1.5 (path weight): Having a directed weighted graph, $G = (V, E)$, and a path, $\sigma = (n_1, n_2, n_3, ..., n_j)$, we define the *weight of path* σ as a sum of the weights of arcs of which it is composed

$$\sigma_w = \sum_{i=1}^{j} w_i$$

There is, however, a difference between path weight and *path length*. Sometimes these two terms are confused in the literature, especially when it comes to transportation systems. In graphs that represent these systems, weights associated with arcs usually stand for distances between nodes showing kilometers or miles. Summing weights of arcs along a path gives the path weight that actually represents length. Because of this, path weight can be misinterpreted as path length. The reason why these two expressions have to be distinguished will be given later. Now, let us define the concept of path length.

Definition 4.1.6 (path length): Having a directed weighted graph $G = (V, E)$, and a path $\sigma = (n_1, n_2, n_3, ..., n_j)$, we define the *length of path* σ as the number of arcs of which the path is composed. We denote path length as σ_ℓ.

We can recognize several paths from the graph in Figure 4.3; $\sigma_1=(a, b, c)$, $\sigma_2=(a, b, c, d)$, $\sigma_3=(b, c, d, b)$. The lengths and weights of these paths are as follows: $\sigma_{1\ell} = 2$, $\sigma_{2\ell} = 3$, $\sigma_{3\ell} = 3$, $\sigma_{1w} = 8$, $\sigma_{2w} = 16$, $\sigma_{3w} = 17$.

Path σ_3 has an interesting property; the *initial* and the *final* node of this path are the same. This kind of path is called a *circle (cycle)*. As will be seen later in the book, circles are very important structures in the analysis of discrete event dynamic systems. At this point, without further explanation, we define the notion of a *maximum cycle mean*. First, the *mean weight of a path* is characterized.

Definition 4.1.7 (mean weight of a path): The *mean weight of a path* σ in a directed weighted graph $G = (V, E)$, is defined as

$$\overline{\sigma}_w = \frac{\sigma_w}{\sigma_\ell}$$

When this path σ is a cycle, the mean weight of the path is called the *cycle mean*.

Definition 4.1.8 (maximum cycle mean and critical circuit): The *maximum cycle mean* of directed weighted graph $G = (V, E)$, is defined as

$$\lambda = \max_{c}(\overline{\sigma}_w)$$

where c ranges over the set of circuits of G. The circuit that corresponds with λ is called a *critical circuit*.

The concept of reachability is closely related with a notion of a path.

Definition 4.1.9 (reachability): Having a directed graph, $G = (V, E)$, and nodes n_i, $n_f \in V$, we say that node n_f is *reachable* from node n_i if there exists a path such that $\sigma = (n_i, n_{i+1}, n_{i+2}, ..., n_f)$, i.e. n_i is the initial node and n_f is the final node of the path.

In our example, node b is reachable from node c and node c is reachable from node b. Actually, each node in the graph shown in Figure 4.3 is reachable from any other node. This type of graph is called a *strongly connected graph*.

In order to be able to manipulate with graphs, to analyze their properties and to make conclusions regarding the systems modeled by graphs, we need to introduce some kind of graph representation. Pure graphical interpretation of a graph is easy to handle and can provide valuable information when the number of nodes is small. As the number of nodes increases, the graphical interpretation becomes impossible to comprehend.

A graph representation is a very important issue especially when it comes to computer memory and computational times. In the following text we show several possible graph representations suitable for programming, with a special emphasis on matrices that can be related to graphs in one way or another. Later, we use these matrices to find graph properties that are of special interest for manufacturing systems analysis and design.

Generally, when we want to prepare a graph representation structured in a way suitable for computer programming, we may choose one of two basic concepts: arc-structured or node-structured data [3]. Each of them has its own benefits and drawbacks.

In node-structured data we use an array of length N, where N is the number of nodes. An entry i, corresponding with node n_i, is a set (a list) of nodes that are destination nodes of arcs starting in node i, together with weights of arcs. Table 4.1 shows the node-structured data representation of the graph shown in Figure 4.3.

Table 4.1. Node-structured data representation of the graph

Entry i (node)	Destination	Weight
1 (a)	b	5
2 (b)	a	7
	c	3
3 (c)	c	6
	d	8
4 (d)	b	4

This structure offers several benefits – finding nodes adjacent to a particular node is simple and fast and so is adding a node (or an arc) to the structure. A problem arises if a node (or an arc) has to be deleted from the structure. Furthermore, testing whether two nodes are adjacent may be time consuming.

The other approach to graph representation is arc-structured data. In this approach we keep a list of arcs by maintaining the origin and destination nodes of the corresponding arc together with arc weight. Table 4.2 shows the arc-structured data representation of the graph shown in Figure 4.3.

Table 4.2. Arc-structured data representation of the graph

Entry i (arc)	Origin	Destination	Weight
1	a	b	5
2	b	a	7
3	b	c	3
4	c	c	6
5	c	d	8
6	d	b	4

Arc-structured data is space efficient. As in the previous case, including a new node or an arc in the structure is easy. The only drawbacks are the time-consuming search for arcs incident to a particular node and determining which two nodes are adjacent.

The structures representing graphs can be more complicated than the one we described, depending on the data that have to be included in the graph description. Getting a structure suitable for graph analysis is not always straightforward. In an example concerning AGV path planning, which is presented in Section 4.2, the structures that describe the graph contain details such as circular and straight path segment points, the vehicle orientation with respect to segment direction and even the vehicle actions upon arrival in a particular node. By combining the given facts and by extracting information from these structures, bottom-up design finally ends with data suitable for computer graph analysis so we can, for example, predict node reachability or plan the shortest path.

4.1.1 Matrix Representation of the Graph

Even though arc-structured and node-structured representations of the graph meet computer programming requirements such as space efficiency due to memory constraints and fast computation of iterative algorithms, they lack the rigorous mathematical characterization that makes them inappropriate for theoretical analysis of graphs. The most convenient way to investigate the composition of graphs or to treat a graph as a structure that presents the dynamical behavior of a system, is by using the matrix representation of the graph [4]. By representing a graph as a matrix we can define mathematical operators that can be used for studying various properties of systems modeled by graphs.

One of the matrices that describe the structure of a graph is the *adjacency matrix*. This matrix shows relations between nodes.

Definition 4.1.10 (adjacency matrix): Having a directed graph, $G = (V,E)$, an *adjacency matrix* **G** is defined as a matrix with the number of rows and columns equal to the number of nodes in G, with element g_{ij} equal to 1 if node n_j is upstream of node n_i and to 0 otherwise.

According to the definition, an entry (i, j) corresponds with an arc from node n_j to node n_i. Although this notation may seem strange at first, it is very convenient for manipulations with matrices and vectors. By maintaining this form we can write matrix equations in the standard way.

For undirected graphs matrix **G** is symmetrical, as by definition element $g_{ij} = g_{ji} = 1$ when nodes n_i and n_j are adjacent and $g_{ij} = g_{ji} = 0$ otherwise. This difference between adjacent matrices of undirected (Figure 4.1) and directed (Figure 4.2) graphs is shown below; $\mathbf{G_{ud}}$ represents an undirected graph and $\mathbf{G_d}$ represents a directed graph. The fact that the diagonal element is equal to 1 is evidence for a loop in a graph.

$$\mathbf{G_{ud}} = \begin{array}{c} \\ a \\ b \\ c \\ d \end{array} \begin{array}{c} \begin{array}{cccc} a & b & c & d \end{array} \\ \left[\begin{array}{cccc} 0 & 1 & 0 & 0 \\ 1 & 0 & 1 & 1 \\ 0 & 1 & 1 & 1 \\ 0 & 1 & 1 & 0 \end{array} \right] \end{array} \qquad \mathbf{G_d} = \begin{array}{c} \\ a \\ b \\ c \\ d \end{array} \begin{array}{c} \begin{array}{cccc} a & b & c & d \end{array} \\ \left[\begin{array}{cccc} 0 & 1 & 0 & 0 \\ 1 & 0 & 0 & 1 \\ 0 & 1 & 1 & 0 \\ 0 & 0 & 1 & 0 \end{array} \right] \end{array}$$

The adjacency matrix of a directed graph can be used to identify more than connections between nearby nodes: it can also show links between nodes that are far from each other. Let us assume that $g_{ij}=1$, i.e. there is an arc from j to i, and $g_{jk}=1$, i.e. there is an arc from k to j. Then, it is obvious that there is a path from k to i containing two arcs. Now, let us assume that all other entries of the ith row and kth column in the adjacency matrix are equal to 0. Then, the multiplication of the row and the column will give $g_{ik}=g_{ij}\cdot g_{jk}=1$. When we have more entries of 1, for example, $g_{im}=1$ and $g_{mk}=1$, then multiplication will give $g_{ik}= g_{ij}\cdot g_{jk} + g_{im}\cdot g_{mk}=2$, i.e. the result shows that there are two paths from k to i, each containing two arcs. We see that by multiplying adjacency matrix **G** one can tell whether two nodes are connected and, when they are, how many possible paths lie between them. Powers of adjacency matrix are calculated by standard matrix multiplication:

$$\mathbf{G}^r = \mathbf{G}^{r-1} \cdot \mathbf{G} \qquad (4.1)$$

i.e. an entry of $\mathbf{G}^r$ is found as:

$$g_{ij}^r = \sum_k g_{ik}^{r-1} \cdot g_{kj}, \quad i,j,k = 1,2,...,n \qquad (4.2)$$

where n is the number of nodes of the corresponding graph. If $g_{ij}^r = m > 0$ then there are m different ways to get from node n_j to n_i by passing r arcs.

If we calculate $\mathbf{G}^2$ for the graph shown in Figure 4.2 we obtain

$$\mathbf{G}^2 = \begin{bmatrix} 0 & 1 & 0 & 0 \\ 1 & 0 & 0 & 1 \\ 0 & 1 & 1 & 0 \\ 0 & 0 & 1 & 0 \end{bmatrix} \cdot \begin{bmatrix} 0 & 1 & 0 & 0 \\ 1 & 0 & 0 & 1 \\ 0 & 1 & 1 & 0 \\ 0 & 0 & 1 & 0 \end{bmatrix} = \begin{bmatrix} 1 & 0 & 0 & 1 \\ 0 & 1 & 1 & 0 \\ 1 & 1 & 1 & 0 \\ 0 & 1 & 1 & 0 \end{bmatrix}$$

The result demonstrates that, for example, there is one path from c to b ($g_{23}^2 = 1$) that passes over two $(r = 2)$ arcs (see Figure 4.2), while there is no way to get from d to b by passing two arcs. Diagonal elements g_{11}^2, g_{22}^2 and g_{33}^2 equal to 1 show second-order ($r = 2$) circles. Further multiplication will give paths passing 3, 4, ..., *etc.* arcs. For example, computation of $\mathbf{G}^4$ gives $g_{23}^4 = 2$: there are two paths from c to b that pass over 4 arcs – *path*1={c,d,b,a,b}, *path*2={c,c,c,d,b}.

Although results obtained from adjacency matrix multiplication show not only the existence of the route from one node to the other but also how many routes there are in the graph and how many arcs have to be taken for a particular route, these results do not tell us *how* we can "travel" between nodes or what would be the cost of the "trip". Later in this section we show how the adjacency matrix must be broadened in order to comprise more detailed graph representation. In Section 4.2 we also present a procedure for the determination of paths between nodes in a graph.

Now, let us define a matrix that associates nodes with arcs. This matrix is called the *incidence matrix*.

Definition 4.1.11 (incidence matrix): Having a directed graph, $G = (V, E)$, an *incidence matrix* **W** is defined as a matrix with the number of rows equal to the number of nodes and the number of columns equal to the number of arcs, with elements defined as follows: if there exists an arc (n_i, n_j), $i \neq j$, represented in **W** with column l, then $w_{il} = 1$, $w_{jl} = -1$ and the other elements of column l are equal to 0. For an arc (n_i, n_i) represented in **W** with column l, $w_{il} = 0$.

According to the definition, an incidence matrix has elements –1, 1, and 0. Entry of –1 (1) indicates that the corresponding node is the destination (origin) of an arc represented by the consequent column. Since both the destination and origin of a self-loop is the same node, in a column representing a self-loop arc all entries are 0. The incidence matrix $\mathbf{W_g}$ of the graph shown in Figure 4.2 is given below.

$$\mathbf{W_g} = \begin{array}{c} a \\ b \\ c \\ d \end{array} \begin{bmatrix} 1 & -1 & 0 & 0 & 0 & 0 \\ -1 & 1 & 1 & 0 & 0 & -1 \\ 0 & 0 & -1 & 0 & 1 & 0 \\ 0 & 0 & 0 & 0 & -1 & 1 \end{bmatrix}$$

Note that we can infer the existence of loops in a graph from the incidence matrix (there are columns with all elements equal to 0), but it is not apparent which node contains a loop. Although column 4 has all entries equal to 0, we are not able to tell which node is involved in the loop. If we draw a graph represented by the incidence matrix $\mathbf{W_g}$, the loop may close around any of the four nodes (Figure 4.4).

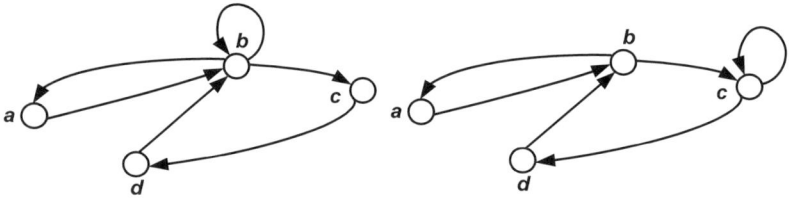

Figure 4.4. Directed graphs with the same incidence matrices and different loops

Even though both matrices, adjacent and incident, represent a graph structure that can be exploited for a survey of a variety of graph properties, they need to be further extended in order to provide adequate information essential in analysis of system dynamics (in the case of manufacturing systems) or of some other features (determination of distances in the case of transportation systems). In an unweighted multigraph, for example, each entry of the adjacency matrix may represent the number of arcs between nodes. On the other hand, as the adjacency matrix of the weighted directed graph has entry 1 for each arc in the graph, we can get a more detailed picture of the graph if we just replace these 1s with weights of corresponding arcs. Concurrently, each 0 from the adjacency matrix has to be replaced by an element, denoted ε, that would stand for a nonexisting arc. In this new matrix, symbol e is used for zero-weight arcs.

Definition 4.1.12 (weighted adjacency matrix): Having a weighted directed graph, $G = (V, E)$, a *weighted adjacency matrix* $\mathbf{A}$ is defined as a matrix with the number of rows and columns equal to the number of nodes in G, with elements defined as follows: if there exists an arc (n_i, n_j) then a_{ji} is equal to its weight, otherwise $a_{ji} = \varepsilon$. For a zero-weight arc (n_i, n_j) entry $a_{ji} = e$.

The elements ε and e will be discussed in more detail in the section dedicated to max-plus algebra.

For the graph shown in Figure 4.3 the weighted adjacency matrix $\mathbf{A}$ is given as:

$$\mathbf{A} = \begin{array}{c} \\ a \\ b \\ c \\ d \end{array} \begin{array}{cccc} a & b & c & d \end{array} \\ \left[\begin{array}{cccc} \varepsilon & 7 & \varepsilon & \varepsilon \\ 5 & \varepsilon & \varepsilon & 4 \\ \varepsilon & 3 & 6 & \varepsilon \\ \varepsilon & \varepsilon & 8 & \varepsilon \end{array} \right]$$

So far we have used analogies from transportation systems to explain basic graph concepts. In the following example we show how graphs can be used for modeling manufacturing systems, which is our main interest. We conclude this section with this example.

Example 4.1.1 (a graph representation of a manufacturing line)

Let us consider a manufacturing line with two machines, M_1 and M_2, shown in Figure 4.5. Our objective is to model this system by a graph. For that purpose, we have to first identify the operations in the system and their order. Next, we need to specify which observation would be represented by nodes – as we stated earlier, we may use nodes to represent events or system states. In this example, nodes represent events.

The figure illustrates that parts visit both machines: after being processed in machine M_1 they proceed to machine M_2 and then leave the system. Therefore, two operations may be identified; operation MP_1 on machine M_1 and operation MP_2 on machine M_2. Each machine can process only one part at a time.

Events that can be characterized as interesting for system analysis are:

- e_1 = part is present at the beginning of the line,
- e_2 = start of operation MP_1,
- e_3 = end of operation MP_1,
- e_4 = start of operation MP_2,
- e_5 = end of operation MP_2,
- e_6 = part leaves the system.

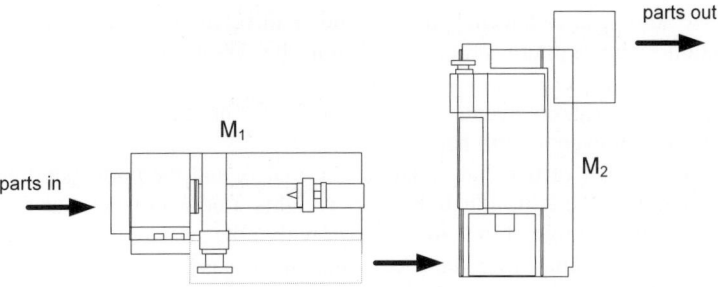

Figure 4.5. A manufacturing line containing two machines

108   Manufacturing Systems Control Design

The type of analysis performed and the system properties will determine which events to model or which physical phenomena will be taken as system states. In our example we could have also selected other events (e = tool in machine M_1 breaks), but these events are not interesting for our study or we may consider that the probability of occurrence of these events is zero (machine tool is believed to be unbreakable).

Since nodes represent events in the example, a graph model of this system would have 6 nodes. Now, let us see how events influence one another (which corresponds with the determination of arcs). It is clear that operation MP_1 cannot start (event e_2) if there is no part at the beginning of the line (event e_1). Furthermore, if machine M_1 is already executing operation MP_1, event e_1 cannot take place since the machine cannot process two parts at a time. These two facts define arcs (e_1, e_2) and (e_3, e_2). The existence of arc (e_2, e_3) is obvious – operation MP_1 cannot end if it had not been started. Similarly, we can define arcs (e_3, e_4), (e_5, e_4) and (e_4, e_5). The final step in the line is the departure of the part, which is represented by arc (e_5, e_6). Thus, having defined relations between nodes, we can go on to write down the adjacency matrix of the system and draw its graph.

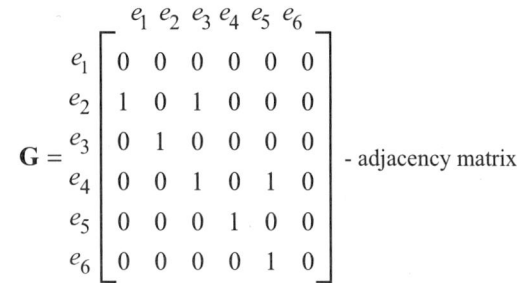

Figure 4.6. Graph representation of the system shown in Figure 4.5

As we know, an adjacency matrix and a directed graph give only limited information about system properties. From the graph in Figure 4.6 we can understand that, for example, event e_3 has an impact on events e_4 and e_2, *i.e.* the occurrence of e_3 triggers events e_4 and e_2, but we cannot say *when*, after e_3 takes place, event e_2 or event e_4 will happen.

We have shown earlier that an adjacency matrix and the corresponding graph can be extended by the introduction of arc weights. Let us broaden our discussion by assigning an operational time to each task within the system. Operation MP_1 is a task, the machine setup after the processing of the part is also a task and so is traveling of the part from M_1 to M_2. Tasks "consume" the time between events, and can therefore be associated with arcs. For events and arcs defined in our example, we can identify the following tasks (operational times are given in parentheses):

- (e_1, e_2) – part enters the machine M_1, (t_U),
- (e_2, e_3) – operation MP_1, (t_{MP1}),
- (e_3, e_2) – setup of machine M_1, (t_{M1}),
- (e_3, e_4) – part travels from machine M_1 to machine M_2, (t_T),
- (e_4, e_5) – operation MP_2, (t_{MP2}),
- (e_5, e_4) – setup of machine M_2, (t_{M2}),
- (e_5, e_6) – part departs the system, (t_Y).

The weighted adjacency matrix obtains the following form:

$$\mathbf{A} = \begin{array}{c} \\ e_1 \\ e_2 \\ e_3 \\ e_4 \\ e_5 \\ e_6 \end{array} \begin{array}{c} \begin{array}{cccccc} e_1 & e_2 & e_3 & e_4 & e_5 & e_6 \end{array} \\ \left[\begin{array}{cccccc} \varepsilon & \varepsilon & \varepsilon & \varepsilon & \varepsilon & \varepsilon \\ t_U & \varepsilon & t_{M1} & \varepsilon & \varepsilon & \varepsilon \\ \varepsilon & t_{MP1} & \varepsilon & \varepsilon & \varepsilon & \varepsilon \\ \varepsilon & \varepsilon & t_T & \varepsilon & t_{M2} & \varepsilon \\ \varepsilon & \varepsilon & \varepsilon & t_{MP2} & \varepsilon & \varepsilon \\ \varepsilon & \varepsilon & \varepsilon & \varepsilon & t_Y & \varepsilon \end{array} \right] \end{array}$$

Figure 4.7 shows the weighted graph representation of the system shown in Figure 4.5.

Figure 4.7. Weighted graph representation of the system shown in Figure 4.5

From matrix **A,** or the graph, among other things, we find that event e_3 occurs t_{MP1} time units after event e_2. Also, we know that the part leaves the system (event e_6) $t_{MP2} + t_Y$ time units after machine M_2 starts its processing (event e_4). Therefore, having defined the weighted adjacency matrix that incorporates operational times, we can study the dynamic properties of the system: machine cycles, machine utilization, system throughput, etc. In Chapter 3 we showed how these data can be revealed from the system model by using matrix operations.

How the system is modeled will depend on the designer's priorities – the designer will highlight events relevant to his/her requirements. Therefore, in the text that follows we show a different model of the same system.

Let us assume that information relevant to model building are events related to the start of operations, parts incoming and parts leaving the system. Thus we identify events e_1, e_2, e_4 and e_6. The occurrence of e_1 activates e_2 after t_U time units (assuming M_1 is ready). Since e_3 and e_5 are not considered, the operational times of tasks connecting these events with other events have to be somehow incorporated in the model. From the graph shown in Figure 4.7 we see that machine M_1 starts processing the next part $t_{MP1} + t_{M1}$ time units after the previous part entered the machine. The same holds for machine M_2 (operational times $t_{MP2} + t_{M2}$). Moreover, the part processed in M_1 enters M_2 after $t_{MP1} + t_T$ time units.

The graph model of the system is shown in Figure 4.8. Operational times are: $t_1 = t_{MP1} + t_{M1}$, $t_2 = t_{MP2} + t_{M2}$, $t_3 = t_{MP1} + t_T$ and $t_4 = t_{MP2} + t_Y$. The new weighted adjacency matrix has the form:

$$\mathbf{A} = \begin{array}{c} \\ e_1 \\ e_2 \\ e_4 \\ e_6 \end{array} \begin{bmatrix} \overset{e_1}{\varepsilon} & \overset{e_2}{\varepsilon} & \overset{e_4}{\varepsilon} & \overset{e_6}{\varepsilon} \\ t_U & t_1 & \varepsilon & \varepsilon \\ \varepsilon & t_3 & t_2 & \varepsilon \\ \varepsilon & \varepsilon & t_4 & \varepsilon \end{bmatrix}$$

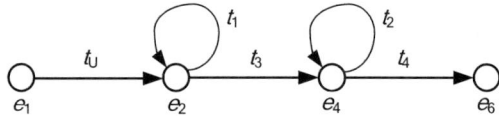

Figure 4.8. Alternative weighted graph of the system shown in Figure 4.5

There are obvious differences between the two models. The second model is reduced and has two loops (diagonal elements t_1 and t_2 in the new matrix **A**). While in the first model we know exactly when the processing of parts in both machines is finished, in the second model these events cannot be tracked directly (actually, we intentionally removed them from consideration).

♦

4.2 String Composition

In the previous section we showed how an adjacency matrix can be used to ascertain whether there exists a path between two particular nodes. It was mentioned that information regarding the existence of a path between nodes does not give any additional data that would answer how one can travel from one node to the other. In order to solve the problem of path finding in a graph, some other form of matrix should be used for graph description.

In this section we describe in detail the string-composition algorithm introduced in [5]. We also extend the notation proposed in [6], where implementation of the string composition to manufacturing systems analysis and design was explored. At the end of the section we give an example of string composition implemented to the problem related to path planning in an AGV system.

String composition is only one of many methods concerned with the shortest path problem. The most popular and well known method is Dijkstra's algorithm [7, 11]. It finds the shortest path from a single origin to all destinations by examining the length of each outgoing arc of a selected node. Every node in a graph is visited only once. As a result, the algorithm gives the distances but actual paths are not known directly. Only the predecessor of the corresponding node is given. Another popular method is the Bellman–Ford (B–F) algorithm. In this algorithm, nodes can be visited more than once and all arcs are checked in each iteration. A more efficient variation of the B–F algorithm, called shortest label first (SLF), is proposed by Bertsekas in [8]. The results of these algorithms are the same as for Dijkstra's: distances and predecessors.

A very popular algorithm that gives all-nodes shortest paths is the Floyd–Warshall algorithm proposed in [9, 10]. The input to the algorithm is an $n \times n$ weighted adjacency matrix **A**, with weights associated with distances. The final result is a matrix whose (i,j) element represents the shortest distance between nodes i and j. To get the actual path from i to j, the algorithm should be changed in order to track calculations in each iteration. The Floyd–Warshall algorithm is similar to the shortest-path computation in the use of max-plus algebra, which we shall describe in the following section.

Two definitions, required for the rest of the section, follow.

Definition 4.2.1 (word): A *word* is a sequence of alphabetical and/or numerical characters. A single character is a word.

Definition 4.2.2 (string): A *string* is a sequence of words having the symbol "-" between two consecutive words.

A few examples of words and strings are:
- words: w; abcd; e4T68u; resource12,
- strings: abcd-e4T68u; resource12-w-r4568-w-abcd

We introduce the following string operations; *multiplication (series composition)* denoted with the multiplication symbol "•", and *addition (parallel composition)* denoted with the standard addition symbol "+".

A string S ending with word A is denoted as S_S–A, where S_S is a substring of S, i.e. S_S is a sequence of words in S followed by word A.

A string S beginning with word A is denoted as A–S_S, where S_S is a substring of S, i.e. S_S is a sequence of words in S that follows word A.

Let $S_1 = S_{S1}$–A and $S_2 = A$–S_{S2} be two strings. Then, multiplications of S_1 with S_2 from the right and the left are defined as follows:

$$S_1 \bullet S_2 = S_{S1} - A \bullet A - S_{S2} = S_{S1} - A - S_{S2} \qquad (4.3)$$

$$S_2 \bullet S_1 = A - S_{S2} \bullet S_{S1} - A = 0 \tag{4.4}$$

where "0" stands for an empty string.

From Equations (4.3) and (4.4) we see that the multiplication of two strings forms a nonempty string if the string that is multiplied from the right ends with the word that is the beginning of the right multiplicand. The result of the multiplication is a string that is composed of two substrings connected with a common word. When the left multiplicand does not end with the word that is the first one of the right multiplicand, the result is an empty string. As the results of left and right multiplications are different, string multiplication is not commutative.

Having strings S_1, S_2, and S_3, the following holds:

$$S_1 + S_2 = S_2 + S_1, \quad (S_1 + S_2) + S_3 = S_1 + (S_2 + S_3)$$

$$(S_1 + S_2) \bullet S_3 = (S_1 \bullet S_3) + (S_2 \bullet S_3)$$
$$S_3 \bullet (S_1 + S_2) = (S_3 \bullet S_1) + (S_3 \bullet S_2)$$

$$S_1 \bullet 0 = 0$$

$$S_1 + 0 = S_1$$

String addition is commutative and associative with an empty string as a zero element. Now we extend the given operations to a particular type of matrix called a *string matrix*. Each string matrix is associated with a graph and may be obtained from its adjacency matrix.

Definition 4.2.3 (string matrix): A *string matrix* **S**, associated with graph $G = (V, E)$ and its adjacency matrix **G**, is an $n \times n$ matrix with string entry s_{ij} obtained as follows: for each g_{ij} that has entry 1, $s_{ij} = A_i - A_j$, where A_i is a word-identifying node n_i and A_j is a word-identifying node n_j. If $g_{ij} = 0$, $s_{ij} = 0$, i.e. if there are no arcs between nodes, the entry is an empty string.

It is clear that a string matrix can also be determined directly from the graph. For the directed graph shown in Figure 4.2, the adjacency matrix has this form:

$$\mathbf{G_d} = \begin{array}{c} \\ a \\ b \\ c \\ d \end{array} \begin{array}{c} a\ b\ c\ d \\ \left[\begin{array}{cccc} 0 & 1 & 0 & 0 \\ 1 & 0 & 0 & 1 \\ 0 & 1 & 1 & 0 \\ 0 & 0 & 1 & 0 \end{array}\right] \end{array}$$

According to Definition 4.2.3, a string matrix **S** associated with adjacency matrix $\mathbf{G_d}$ is given below:

$$\mathbf{S} = \left[\begin{array}{cccc} 0 & a-b & 0 & 0 \\ b-a & 0 & 0 & b-d \\ 0 & c-b & c-c & 0 \\ 0 & 0 & d-c & 0 \end{array}\right]$$

From matrix **S** we may read that there is, for example, a connection between nodes b and c (string element $s_{32} = c-b$).

Even though the string matrix determined as described in Definition 4.2.3 may seem a little confusing (the connection between nodes i and j is represented as string $j-i$), for manufacturing systems analysis this form of the matrix is very convenient. We shall see later that string composition is mainly used for determination of circular waits among resources. The form described in Definition 4.2.3 gives us wait relations directly, thus providing the conditions for a straightforward determination of circular waits. If one imagines that arcs represent part flow in an MS and that nodes represent resources, then in order to proceed from node i to node j, a part should be first processed in i, which means that j waits for i to finish its task. In other words, the connection between i and j is represented with the string $j-i$.

A more intuitive form of the string matrix may be obtained if the adjacency matrix is first transposed and then Definition 4.2.3 is used. The other way is to determine the string matrix directly from the graph. If there is an arc from node i to node j then the entry that corresponds with the row representing node i and the column representing node j is equal to $i-j$. This method can be used if one deals with problems related to path determination (the shortest, the fastest, the cheapest, etc.).

Transposition of $\mathbf{G_d}$ from our example gives:

$$\mathbf{G_d^T} = \begin{array}{c} \\ a \\ b \\ c \\ d \end{array} \begin{array}{c} a\ b\ c\ d \\ \left[\begin{array}{cccc} 0 & 1 & 0 & 0 \\ 1 & 0 & 1 & 0 \\ 0 & 0 & 1 & 1 \\ 0 & 1 & 0 & 0 \end{array}\right] \end{array}$$

The string matrix for $\mathbf{G_d}^T$ has the form:

$$\mathbf{S} = \begin{bmatrix} 0 & a-b & 0 & 0 \\ b-a & 0 & b-c & 0 \\ 0 & 0 & c-c & c-d \\ 0 & d-b & 0 & 0 \end{bmatrix}$$

From the string matrix $\mathbf{S}$ we may find that there is a connection between nodes b and c (string element $s_{23} = b\text{--}c$), only this time we may read this information directly, in a more natural way.

Determined one way or the other, the string matrix gives information regarding connections between nodes. We only have to make sure to follow the conventions, chosen at the beginning throughout the entire procedure of string composition.

A problem may arise if there is more than one arc between two nodes (multigraph). This situation is common in AGV systems as layout designers usually plan alternative routes that should be used when the main route is occupied by another vehicle or if taking the main route is forbidden for some reason. Figure 4.9a shows a graph that has two arcs between nodes c and d. By adding a *virtual node* e, as shown in Figure 4.9b, the alternative route is split into two arcs; $c\text{--}e$ and $e\text{--}d$. Even though the inclusion of virtual nodes increases the dimension of the string matrix, this way of solving the alterative-route problem is quite simple and straightforward.

We have shown in the previous section how the multiplication of the adjacency matrix gives the number of paths between two nodes. We also found how many nodes should be visited to get from node to node, but we were not able to tell how to travel from one node to another. We will now use string composition to find a solution to this problem.

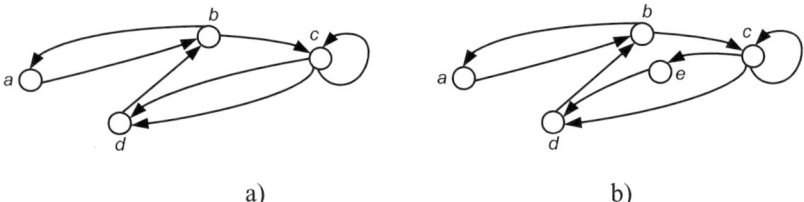

a)   b)

Figure 4.9. Splitting of alternative route by virtual-node inclusion

Let us extend string composition to matrices. Powers of string matrix $\mathbf{S}$ are calculated as follows:

$$\mathbf{S}^r = \mathbf{S}^{r-1} \bullet \mathbf{S} \qquad (4.5)$$

i.e. an entry of $\mathbf{S}^r$ is found as:

$$s_{ij}^r = \sum_k s_{ik}^{r-1} \cdot s_{kj} , \quad i, j, k = 1, 2, ..., n \tag{4.6}$$

where n is the number of nodes in the corresponding graph. In Equation (4.6) standard multiplication should be replaced with series string composition, while standard addition should be replaced with parallel string composition.

For the graph shown in Figure 4.2 we have:

$$\mathbf{S}^2 = \mathbf{S} \bullet \mathbf{S} = \begin{bmatrix} 0 & a-b & 0 & 0 \\ b-a & 0 & b-c & 0 \\ 0 & 0 & c-c & c-d \\ 0 & d-b & 0 & 0 \end{bmatrix} \bullet \begin{bmatrix} 0 & a-b & 0 & 0 \\ b-a & 0 & b-c & 0 \\ 0 & 0 & c-c & c-d \\ 0 & d-b & 0 & 0 \end{bmatrix}$$

$$= \begin{bmatrix} a-b-a & 0 & a-b-c & 0 \\ 0 & b-a-b & b-c-c & b-c-d \\ 0 & c-d-b & c-c-c & c-c-d \\ d-b-a & 0 & d-b-c & 0 \end{bmatrix}$$

We see that there are three diagonal elements that are not null string: diagonal strings a–b–a, b–a–b and c–c–c. This result shows that there are two second-order circles in the graph. Also, all existing second-order paths (containing two arcs, i.e. all paths with $\sigma_1 = 2$) are represented with corresponding strings. Third-order paths ($\sigma_1 = 3$) can be found as:

$$\mathbf{S}^3 = \mathbf{S}^2 \bullet \mathbf{S}$$

$$= \begin{bmatrix} 0 & a-b-a-b & a-b-c-c & a-b-c-d \\ b-a-b-a & b-c-d-b & b-a-b-c+b-c-c-c & b-c-c-d \\ c-d-b-a & c-c-d-b & c-d-b-c+c-c-c-c & c-c-c-d \\ 0 & d-b-a-b & d-b-c-c & d-b-c-d \end{bmatrix}$$

The string $s_{23}^3 = b$–a–b–$c + b$–c–c–c demonstrates that there are two third-order paths between nodes b and c, while $s_{33}^3 = c$–d–b–$c + c$–c–c–c shows that there are two third order circles that start and end with node c.

How far one may go with string matrix composition depends on the specific problem. After n multiplications of the string matrix all paths in the graph will be revealed. Graphs with a large number of nodes require many multiplications, thus, finding all paths can be a time-consuming task that may need huge computational power. There are many ways to solve this problem, depending on the final objective of string composition.

For example, the given results show that the circle exposed by composition repeats in several string matrix elements (b–c–d–b, c–d–b–c and d–b–c–d in $\mathbf{S}^3$

represent the same circle). As our aim is only to find circles in the graph, equation (4.5) can be redefined. Duplicate values use computation time needlessly and do not give any new information. As proposed in [6], the matrix composition can be changed in the following way:

$$s_{ij}^r = \sum_k s_{ik}^{r-1} \cdot s_{kj} \ , \ i,j = 1,2,...,n \ , \ k \geq i+1 \tag{4.7}$$

thus eliminating duplicated circles and restricting the required calculations. A closer look at Equation (4.7) makes it clear that by forcing k to be greater than i we do not check if already passed nodes belong to currently calculated circles. If a node that corresponds with the current row (i) belongs to some circle that includes a passed node ($< i$), then this circle is already determined and there is no need to check previously passed nodes – only nodes above the current row ($> i$) should be checked.

We can even further exploit the fact that the string composition objective is the determination of circles. If a graph has n nodes then the nth composition should give the circle that includes all nodes in the graph (if one exists), i.e. when performing the nth composition we have to calculate only the first diagonal term. As the (n–1)th composition exposes circle(s) that comprise(s) n–1 nodes, only the first two diagonal terms should be calculated. We may proceed in the same manner. Finally, we conclude that only (n–r+1) diagonal terms must be calculated and checked for possible circles. By keeping the original string matrix **S** unchanged we need to determine (n–r) rows and (n–r+1) diagonal terms of **S**^{r} (the rth composition) in order to calculate the r ordered circles in the corresponding graph.

When matrix composition is used for some other purpose, a string matrix can be structured so that the calculation of string composition will not require large computational capacity. In practice, problems of path determination usually have some restrictions that can help in rearrangement of the corresponding string matrix. For example, in many situations nodes are divided into three groups: origins, destinations and bypasses. Only traveling from origin to destination and *vice versa* is allowed. Nodes that lie on the paths between origins and destinations belong to a bypass group. Having groups of nodes we may structure a string matrix so that the first k rows represent origins, followed by l rows that correspond to bypasses, and the final m rows that should stand for destinations. Assumptions regarding allowed travel routes reduce the number of required calculations since we can skip compositions of a row and a column that stand for an origin (destination) as trips from origin to origin (destination to destination) are forbidden. Moreover, prior to the rth composition, all diagonal elements of **S**$^{r-1}$ can be set to 0 since they represent circles. In the following example we show how the given system restrictions define the string matrix structure and restrict string composition.

Example 4.2.1 (an AGV shortest-path determination by using the string composition)

An AGV system layout is shown in Figure 4.10. Since formation of the string matrix is based on data that are structured in a particular way, prior to determination of the system string matrix we have to describe these structures.

The layout is composed of segments and nodes. A segment is an object defined by its properties. It can be circular or straight: circular segments are defined by three and straight segments by two Cartesian points (in world, *i.e.* shop floor, coordinates). We differentiate two types of segments based on traveling direction: unidirectional and bidirectional. Each segment has a weight factor that can be associated with some physical property (maximum allowed speed, segment length, *etc.*). Segments form paths.

A node is an object defined as the point on a segment. Each node has a set of parameters related to a vehicle – actions that should be performed by the vehicle and positions of forks (approach speed, fork orientation, fork pick-up elevation, departure speed, *etc.*). Nodes are grouped as origins, destinations and bypasses.

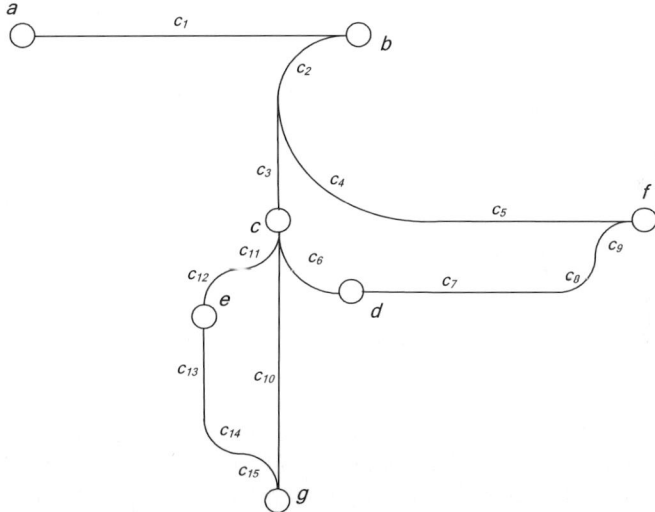

Figure 4.10. An AGV system layout composed of nodes and segments

The layout has 7 nodes and 15 segments:

- set of nodes

$N = N_o \cup N_b \cup N_d = \{a\} \cup \{b, c, d, e\} \cup \{f, g\}$

- set of segments

$C = C_s \cup C_c = \{c_1, c_3, c_5, c_7, c_{10}, c_{13}\} \cup \{c_2, c_4, c_6, c_8, c_9, c_{11}, c_{12}, c_{14}, c_{15}\}$

118   Manufacturing Systems Control Design

where N_o, N_b and N_d are sets of origin, bypass and destination nodes, respectively, while C_s and C_c are sets of straight and circular segments, respectively. All segments, except c_4, c_5 and c_{10}, are bidirectional. We denoted segments as c_i since the letter s is used for string-matrix elements.

As may be seen from Figure 4.10 that some paths are formed of only one segment, while others comprise two and more segments:

$(a, b) \rightarrow \{c_1\}$ - bidirectional,
$(b, c) \rightarrow \{c_2, c_3\}$ - bidirectional,
$(b, f) \rightarrow \{c_2, c_4, c_5\}$ - unidirectional,
$(c, d) \rightarrow \{c_6\}$ - bidirectional,
$(c, e) \rightarrow \{c_{11}, c_{12}\}$ - bidirectional,
$(c, g) \rightarrow \{c_{10}\}$ - unidirectional,
$(d, f) \rightarrow \{c_7, c_8, c_9\}$ - bidirectional,
$(e, g) \rightarrow \{c_{13}, c_{14}, c_{15}\}$ - bidirectional.

Based on the layout data we can form the system string matrix:

$$\mathbf{S} = \begin{bmatrix} 0 & a-b & 0 & 0 & 0 & 0 & 0 \\ b-a & 0 & b-c & 0 & 0 & b-f & 0 \\ 0 & c-b & 0 & c-d & c-e & 0 & c-g \\ 0 & 0 & d-c & 0 & 0 & d-f & 0 \\ 0 & 0 & e-c & 0 & 0 & 0 & e-g \\ 0 & 0 & 0 & f-d & 0 & 0 & 0 \\ 0 & 0 & 0 & 0 & g-e & 0 & 0 \end{bmatrix}$$

Calculation of $\mathbf{S}^2$ gives:

$$\mathbf{S}^2 = \begin{bmatrix} 0 & 0 & a-b-c & 0 & 0 & a-b-f & 0 \\ 0 & 0 & 0 & b-c-d+b-f-d & b-c-e & 0 & b-c-g \\ c-b-a & 0 & 0 & 0 & c-g-e & c-b-f+c-d-f & c-e-g \\ 0 & d-c-b & 0 & 0 & d-c-e & 0 & d-c-f \\ 0 & e-c-b & 0 & e-c-d & 0 & 0 & e-c-g \\ 0 & 0 & f-d-c & 0 & 0 & 0 & 0 \\ 0 & 0 & g-e-c & 0 & 0 & 0 & 0 \end{bmatrix}$$

Note that all diagonal elements are set to 0 as circular paths are not of interest when it comes to shortest-path determination. From $\mathbf{S}^2$ we may see that there are two second-order paths from b to d and from c to f. Further string compositions give paths of 3rd, 4th, *etc.* order. The calculation of these paths we leave to the reader for exercise.

Our goal is to find the shortest paths from all origins to all destinations and *vice versa*. In our example these paths are between node a and nodes f and g. In other words, we have to check string elements s_{16}, s_{61}, s_{17}, and s_{71} after each composition. The results are:

$$s_{16}^2 = a\text{–}b\text{–}f$$
$$s_{16}^4 = a\text{–}b\text{–}c\text{–}d\text{–}f$$
$$s_{17}^3 = a\text{–}b\text{–}c\text{–}g$$
$$s_{17}^4 = a\text{–}b\text{–}c\text{–}e\text{–}g$$
$$s_{17}^5 = a\text{–}b\text{–}f\text{–}d\text{–}c\text{–}g$$
$$s_{17}^6 = a\text{–}b\text{–}f\text{–}d\text{–}c\text{–}e\text{–}$$
$$s_{61}^4 = f\text{–}d\text{–}c\text{–}b\text{–}a$$
$$s_{71}^4 = g\text{–}e\text{–}c\text{–}b\text{–}a.$$

There are two alternative paths from a to f and four possible paths from a to g. As there are segments that are not bidirectional, there is only one path from f to a and from g to a. If we suppose that segment weight $\sigma_w(c_i)$ stands for length, which can be determined easily from input data structure (points that define segments), then it is easy to find the shortest path among the given alternatives:

$$\sigma_w(a, f) = \min\{\sigma_w(s_{16}^2), \sigma_w(s_{16}^4)\}$$

$$\sigma_w(a, g) = \min\{\sigma_w(s_{17}^3), \sigma_w(s_{17}^4), \sigma_w(s_{17}^5), \sigma_w(s_{17}^6)\}$$

where

$$\sigma_w(s_{16}^2) = \sigma_w(c_1) + \sigma_w(c_2) + \sigma_w(c_4) + \sigma_w(c_5)$$
$$\sigma_w(s_{16}^4) = \sigma_w(c_1) + \sigma_w(c_2) + \sigma_w(c_3) + \sigma_w(c_6) + \sigma_w(c_7) + \sigma_w(c_8)$$
$$+ \sigma_w(c_9)$$
$$\sigma_w(s_{17}^3) = \sigma_w(c_1) + \sigma_w(c_2) + \sigma_w(c_3) + \sigma_w(c_{10})$$
$$\sigma_w(s_{17}^4) = \sigma_w(c_1) + \sigma_w(c_2) + \sigma_w(c_3) + \sigma_w(c_{11}) + \sigma_w(c_{12})$$
$$\sigma_w(c_{13}) + \sigma_w(c_{14}) + \sigma_w(c_{15})$$
$$\sigma_w(s_{17}^5) = \sigma_w(c_1) + \sigma_w(c_2) + \sigma_w(c_4) + \sigma_w(c_5)$$
$$+ \sigma_w(c_9) + \sigma_w(c_8) + \sigma_w(c_7) + \sigma_w(c_6) + \sigma_w(c_{10})$$

$$\sigma_w(s_{17}^6) = \sigma_w(c_1) + \sigma_w(c_2) + \sigma_w(c_4) + \sigma_w(c_5) + \sigma_w(c_9)$$
$$+ \sigma_w(c_8) + \sigma_w(c_7) + \sigma_w(c_6) + \sigma_w(c_{11}) + \sigma_w(c_{12})$$
$$+ \sigma_w(c_{13}) + \sigma_w(c_{14}) + \sigma_w(c_{15})$$

Using the described procedure, one may find not only the shortest path between nodes but also the optimal one (the shortest path is optimal if distance is considered). If segment weight is associated with some criterion, then the determined path will be optimal for that particular criterion.

The result of origin–destination path finding is a set of all possible routes in a system that an AGV can pass through. This provides an option for the calculation of shortest paths between all the nodes in the system. It is very important to have this possibility since there are situations when a vehicle moves out of a segment it is currently passing. That may happen if there is a loss of communication between the vehicle and the supervisor or if some problems with navigation occur (loss of visual contact between laser source and mirrors). In addition, if a vehicle is manually controlled it can be switched to autonomous mode at any position on the shop floor. In this case, a navigation system provides information regarding current vehicle position and then the vehicle autonomously moves towards the closest segment. Once the vehicle is on the segment, the supervisor, having all possible routes, sends the vehicle information regarding the path it should take to get to the desired node.

The other reason why all system paths should be at the supervisor's disposition is AGV dispatching. When a vehicle approaches the bypass node it sends a request to the supervisor for the next segment allocation. When the requested segment is occupied, the supervisor allocates an alternative segment (if there is one) to the vehicle. Without knowing all the possible routes beforehand the supervisor would not be able to dispatch the vehicles according to a desired strategy.

♦

4.3 Max-plus Algebra

In general, there are two main approaches to analysis and modeling of discrete event dynamic systems (DEDS). When the designer is investigating only the ordering of events that may occur in DEDS, his/her main concern will be the system *logical behavior*. On the other hand, if the system is studied in order to examine time instants at which a particular event took place, then the *temporal behavior* of DEDS should be analyzed and modeled.

The algebra, called *max-plus*, is one of the mathematical frameworks suitable for the latter case. Although max-plus theory is very convenient when synchronization phenomenon in DEDS is considered, it is not able to handle "nonlinear" problems such as concurrency (in recent years there have been some results that extend the theory to nonlinear cases [15, 16] and systems with so-called *switching functions* [12]). The limitation directed by concurrency constricts the use of max-plus algebra to a special class of DEDS. This class of systems is called *event graphs*.

In this section we give only the basic definitions and properties of max-plus algebra. For those who want to learn more about the topic, very good resources are [13, 14, 17]. At the end of the section an illustrative example for the manufacturing system modeled as a marked graph (the order of operations is known beforehand) is given.

The maximization and addition in max-plus theory are defined over the extended set of real numbers.

Definition 4.3.1 (extended set of real numbers): A set $\Re_\varepsilon$ is a set of real numbers that includes element ε, $\Re_\varepsilon = \Re \cup \{\varepsilon\}$, where the numerical value of $\varepsilon = -\infty$.

Definition 4.3.2 (maximization in max-plus): Maximization over $\Re_\varepsilon$, represented by $\oplus$, is defined as

$$x \oplus y = \max(x, y) \tag{4.8}$$

Definition 4.3.3 (addition in max-plus): Addition over $\Re_\varepsilon$, represented by $\otimes$, is defined as

$$x \otimes y = x + y \tag{4.9}$$

Having defined the basic operations in max-plus, we can identify neutral elements of the algebra. Element ε is the neutral element with respect to maximization,

$$x \oplus \varepsilon = \varepsilon \oplus x = \max(x, \varepsilon) = \max(\varepsilon, x) = x$$

while e is the neutral element with respect to addition,

$$x \otimes e = e \otimes x = x$$

The numerical value of e equals 0. It should be noted that ε is an absorbing element of $\otimes$,

$$x \otimes \varepsilon = \varepsilon \otimes x = \varepsilon$$

The operations $\oplus$ and $\otimes$ can be extended to matrices. This is very important since there is a unique relation between a graph and its weighted adjacency matrix. Implementation of max-plus algebra to matrices allows detailed analysis of graphs, thus providing thorough insight into systems modeled with graphs.

If we calculate $\mathbf{C} = \mathbf{A} \oplus \mathbf{B}$, then

$$c_{ij} = a_{ij} \oplus b_{ij} = \max(a_{ij}, b_{ij}) \tag{4.10}$$

It is clear that maximization is defined only for matrices of the same size as it is performed element-by-element.

Entries of matrix **C**, obtained by max-plus matrix addition, $\mathbf{C} = \mathbf{A} \otimes \mathbf{B}$, are calculated as

$$c_{ij} = \bigoplus_{k} a_{ik} \otimes b_{kj} = \max_{k}(a_{ik} + b_{kj}) \tag{4.11}$$

Matrix addition in max-plus theory is defined only if the number of columns of **A** is equal to the number of rows of **B**. It should be noted that when **A** and **B** are square matrices of the same order, then $\mathbf{A} \otimes \mathbf{B} \neq \mathbf{B} \otimes \mathbf{A}$. The identity matrix **E** in max-plus algebra has diagonal elements equal to e and other entries equal to ε.

Maximization and addition of matrices in max-plus represents a *parallel* and a *series composition*, respectively. Matrix compositions can be comprehended if one imagines a set of nodes that symbolize cities [18]. As cities are connected by roads and rails, two graphs may be used to separately describe two possible ways of traveling (Figure 4.11). Let **A** and **B** be weighted adjacency matrices of these two graphs; a_{ij} and b_{ij} representing road and rail distances between cities j and i. The parallel composition (maximization) of **A** and **B** gives matrix **C** with element c_{ij} equal either to a_{ij} or b_{ij} depending on which distance, road or rail, is longer. In other words, matrix **C** has entries that match one of the two alternative routes between cities j and i. Even though the original graphs offered the possibility of traveling from j to i either by road or by rail, in the graph attained by their parallel composition only one route remains feasible. According to Figure 4.11, node b can be approached from node a by road C2 or by rail T1. Since C2 $\oplus$ T1 = 5 $\oplus$ 6 = 6 = T1, on the graph shown in Figure 4.12, only rail T1 is left.

On the other hand, the series composition of **A** and **B** tells us whether it is possible to travel from j to i using both means of transport, starting with train and then switching to a car. If this trip is feasible, then c_{ij} receives a value that is equal to the maximum distance of all possible routes obtained by adding road distance a_{ik} to rail distance b_{kj}, where k is an intermediate node (the node where the passenger changes a train for a car). From Figure 4.12 we see that this kind of trip is possible if one wants to get from node d to node b. First rail T5 should be taken followed by road C2. The distance traveled is C2 + T5 = 5 + 3 = 8.

In the previous section the shortest-path problem was solved by using string composition. Max-plus also offers a solution to that problem by changing the maximum operation with the minimum operation. In this case ε should attain the value of $+\infty$, while all other properties of max-plus, defined above, still hold. Within this new framework we can find the shortest distances that consist of r arcs from the following expression:

$$\mathbf{A}^r = \mathbf{A}^{r-1} \otimes \mathbf{A} \qquad (4.12)$$

where A is an $n \times n$ weighted adjacency matrix and $r < n$. Entries in $\mathbf{A}^r$ are calculated as

$$a_{ij}^r = \underset{k}{\oplus} a_{ik}^{r-1} \otimes a_{kj} = \min_k (a_{ik}^{r-1} + a_{kj}) \qquad (4.13)$$

Just as in the case of string composition, diagonal elements of $\mathbf{A}^r$ correspond to circles, i.e. a'_{ii} stands for the shortest path, containing r arcs, from node i to itself.

As the shortest path between two nodes may contain more than one arc, in order to get the correct result we have to compare matrix $\mathbf{A}$ with all matrices $\mathbf{A}^r$ calculated for $r = 2, 3, \ldots, n-1$:

$$a_{ij}^{\min} = \min(a_{ij}, a_{ij}^2, \ldots, a_{ij}^{n-1}) . \qquad (4.14)$$

Applying Equation (4.10) to Equation (4.14) it follows that

$$\mathbf{A}^{\min} = \mathbf{A} \oplus \mathbf{A}^2 \oplus \ldots \oplus \mathbf{A}^{n-1} \qquad (4.15)$$

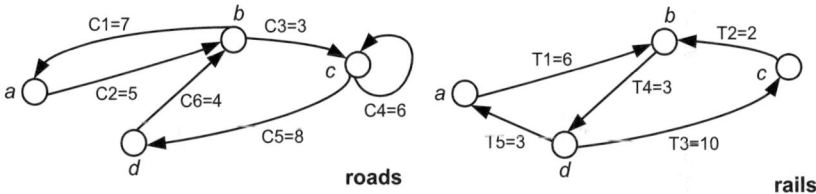

Figure 4.11. Two weighted graphs representing road and rail connections

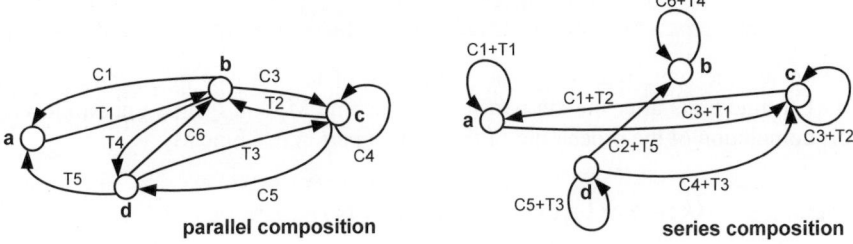

Figure 4.12. Parallel and series compositions of graphs shown in Figure 4.11

Equations (4.12)–(4.15) are applicable to the max operation also by substituting min with max and $+\infty$ with $-\infty$.

4.3.1 DEDS Model in Max-plus Algebra

Let us now return to our main objective that is the determination of dynamical properties of discrete event systems. In order to focus on the subject and to show why max-plus is suitable for DEDS analysis, let us consider the discrete event system analyzed in Example 4.1.1. Remember, Figure 4.13 shows the graph representation of the manufacturing system we examined. Note that notations have been changed in order to be consistent with max-plus nomenclature – a time instant at which event i occurs for the kth time is denoted as $x_i(k)$. To simplify the problem we will concentrate on event x_1 – the start of processing of the part in machine M_1. We have to find time instants at which event x_1 occurs.

From the graph we see that the processing of the part cannot start if the part has not entered the system and machine M_1 has not finished the processing of the previous part. Additionally, x_1 is shifted t_U time units after event u and t_1 time units after the occasion of x_1. If we assume that u and x_1 occur simultaneously at time instant t then the next occasion of x_1 will take place after both tasks that start with events u and x_1 are finished, i.e. x_1 will occur at time instant $\max(t+t_U, t+t_1)$.

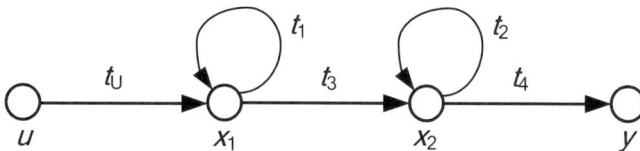

Figure 4.13. Graph representation of the system in Example 4.1.1

The manufacturing system example reveals that two operations, *maximization* and *addition*, play an essential role in the investigation of dynamic properties of discrete event systems. This makes max-plus algebra the first choice for the study of the dynamics of event graphs.

From our discussion we can define the equation for the time instant in which machine M_1 starts the processing of the kth part;

$$x_1(k) = \max\left(x_1(k-1)+t_1, u(k)+t_U\right)$$

Following the same reasoning, the processing of the kth part in machine M_2 can start t_3 time units after the kth part leaves machine M_1 and t_2 time units following the completion of the processing of the $(k-1)$th part in machine M_2:

$$x_2(k) = \max\left(x_1(k)+t_3, x_2(k-1)+t_2\right)$$
$$= \max\left(x_1(k-1)+t_1+t_3, u(k)+t_U+t_3, x_2(k-1)+t_2\right)$$

For the output we can write

$$y(k) = x_2(k) + t_4$$

By using Equation (4.8) and (4.9) these expressions become:

$$x_1(k) = x_1(k-1) \otimes t_1 \oplus u(k) \otimes t_U$$
$$x_2(k) = x_1(k-1) \otimes (t_1 + t_3) \oplus x_2(k-1) \otimes t_2 \oplus u(k) \otimes (t_U + t_3)$$
$$y(k) = x_2(k) \otimes t_4$$

where $\otimes$ has priority over $\oplus$. The upper equations can be affirmed in max-plus matrix formulation:

$$\mathbf{x}(k) = \begin{bmatrix} t_1 & \varepsilon \\ t_1 + t_3 & t_2 \end{bmatrix} \otimes \mathbf{x}(k-1) \oplus \begin{bmatrix} t_U \\ t_U + t_3 \end{bmatrix} \otimes \mathbf{u}(k)$$
$$\mathbf{y}(k) = \begin{bmatrix} \varepsilon & t_4 \end{bmatrix} \otimes \mathbf{x}(k)$$

The obtained model completely describes the system dynamics. It has a well-known *linear state space* form:

$$\mathbf{x}(k) = \mathbf{A} \otimes \mathbf{x}(k-1) \oplus \mathbf{B} \otimes \mathbf{u}(k), \; \mathbf{x}(0) = \mathbf{x}_0$$
$$\mathbf{y}(k) = \mathbf{C} \otimes \mathbf{x}(k) \tag{4.16}$$

Note that matrix **A** in Equation (4.16) is not necessarily an adjacency matrix of the graph that represents the system described by this equation.

Once a discrete event dynamic system is modeled by Equation (4.16), a whole range of various phenomena may be investigated [19]. For example, in the case of manufacturing systems, it is now possible to determine the slowest (bottleneck) part of the system. Furthermore, by studying the influence of operational times, the designer is able to decide in which part of the system an extra resource should be integrated in order to maximize the system throughput and/or improve resources utilization. Additionally, the propagation of disturbances through the system can be explored, in order to answer how much time the system needs to return to the steady state.

Questions concerning the cyclic activity of the system are especially interesting. As we mentioned, cyclic behavior of a discrete event system is treated as a stable state, whereas in time-driven systems this manifestation is considered as marginally stable. In fact, many discrete event systems, particularly manufacturing systems, are designed to start working periodically after a short transient time. In the text that follows we are concerned with the properties of the model (4.16) that lead to periodic activities.

Let us again think of the manufacturing system shown in Figure 4.5 and represented by the graph in Figure 4.13. Suppose that the first processing part is deposited into the system at a time instant 0, *i.e.* $u(1) = 0$. From the max-plus model of the system we calculate that $x_1(1) = t_U$, $x_2(1) = t_U + t_3$ and $y(1) = t_U + t_3 + t_4$. The part has propagated through the system and left it at time instant $t_U + t_3 + t_4$. Now suppose that parts arrive into the system each T_{in} time units starting at $t = 0$, i.e. $u(k) = (k-1) \cdot T_{in}$. Immediately a question arises: how often do the parts leave the system or, in other words how fast do the parts spread through the system? We have just calculated that the first part needed $t_U + t_3 + t_4$ time units to get from the input to the output. But what about the second, the third and the following parts? We understand from the model that machine M_1 needs t_1 time units after it begins the processing of the first part to become ready for processing the second one. If $T_{in} < t_1$ then the second part should wait some time to be processed by the machine. As parts are arriving with period T_{in}, after some time the machine will get swamped with parts. At this point we may conclude that t_1 should be less than T_{in} in order to allow continuous part flow through the system, but further discussion shows that this is not necessarily true. Even if the condition for regular work of the first machine is satisfied, the same overflow effect will happen with machine M_2 if $T_{in} < t_2$. Since $T_{in} > t_1$, parts arrive at machine M_2 with period T_{in}. Following the same reasoning as for M_1 we realize that t_2 should be less than T_{in}.

This simple example highlights the importance of system analysis when it comes to DES design. How can the max-plus model be used to determine system properties that would reveal conditions that should be satisfied in order to make the system react according to the desired criteria? Let us suppose that each time one part leaves the system, described with Equation (4.16), another part enters the system. As a consequence we have

$$u(k) = y(k-1)$$

Including this into Equaiton (4.16) and having in mind that the model holds for any k, we can write

$$\mathbf{x}(k) = \mathbf{A} \otimes \mathbf{x}(k-1) \oplus \mathbf{B} \otimes \mathbf{y}(k-1)$$
$$\mathbf{y}(k-1) = \mathbf{C} \otimes \mathbf{x}(k-1)$$

Further, we can include $\mathbf{y}(k-1)$ in the equation for $\mathbf{x}(k)$, thus obtaining

$$\mathbf{x}(k) = \mathbf{A} \otimes \mathbf{x}(k-1) \oplus \mathbf{B} \otimes \mathbf{C} \otimes \mathbf{x}(k-1)$$

The manipulation gives

$$\mathbf{x}(k) = \overline{\mathbf{A}} \otimes \mathbf{x}(k-1), \quad \mathbf{x}(0) = \mathbf{x}_0$$
$$\mathbf{y}(k) = \mathbf{C} \otimes \mathbf{x}(k)$$
(4.17)

where

$$\overline{\mathbf{A}} = \mathbf{A} \oplus \mathbf{B} \otimes \mathbf{C}$$

What we get as the result of assumption that $\mathbf{u}(k)=\mathbf{y}(k-1)$ is an autonomous system (4.17). In fact, system (4.17) is a closed-loop form of Equation (4.16) with unity feedback [23]. This feedback can be visualized if one imagines a pallet that moves through the system together with a part. As soon as the part leaves the system the pallet is released and immediately fed to the input. This representation can be further exploited if we delay the trip of the pallet, which in max-plus corresponds to

$$\mathbf{u}(k) = \mathbf{C} \otimes \mathbf{y}(k-1) \qquad (4.18)$$

It is apparent that matrix $\overline{\mathbf{A}}$ can be determined as

$$\overline{\mathbf{A}} = \mathbf{A} \oplus \mathbf{B} \otimes \mathbf{G} \otimes \mathbf{C}$$

Releasing the pallet when the part departs from the system is a very restrictive strategy in the sense of resources utilization. The number of parts that can be simultaneously processed within the system is equal to the number of available pallets. If there is only one pallet at disposition then the entire system will work with only one part. The opposite is the case when tens of pallets are prepared while the system can handle only a few parts at a time, which means that many pallets will remain unused.

An alternative strategy is to release the pallet when some event in the system is started. This strategy does not wait for the part to come to the output in order to allow the next part to be fed into the system. In max-plus form this can be described as

$$\mathbf{u}(k) = \mathbf{K} \otimes \mathbf{x}(k-1) \qquad (4.19)$$

The problem of finding the number of pallets that provide the desired performance of the system is similar to the problem of parts arrival rate determination. One way or the other we have to ascertain some inherent property of the system that would give us an indication of how to feed the parts into the system, *i.e.* how often the parts leave the system.

4.3.2 Periodic Behavior of DEDS in Max-plus

We consider an autonomous system of the form

$$\begin{aligned} \mathbf{x}(k) &= \mathbf{A} \otimes \mathbf{x}(k-1), \quad \mathbf{x}(0) = \mathbf{x}_0 \\ \mathbf{y}(k) &= \mathbf{C} \otimes \mathbf{x}(k) \end{aligned} \qquad (4.20)$$

Usually, the symbol $\otimes$ is omitted, *i.e.* Equation (4.20) is written as

$$\begin{aligned} \mathbf{x}(k) &= \mathbf{A}\mathbf{x}(k-1), \quad \mathbf{x}(0) = \mathbf{x}_0 \\ \mathbf{y}(k) &= \mathbf{C}\mathbf{x}(k) \end{aligned}$$

Periodic activities of the system presume that the difference between two consecutive occasions of event x_i is constant. If we denote this constant difference with symbol λ, then

$$x_i(k+1) - x_i(k) = \lambda$$
$$x_i(k+2) - x_i(k+1) = \lambda$$
$$\ldots$$
$$x_i(k+r) - x_i(k+(r-1)) = \lambda$$
$$\ldots$$

From this set of equations it is easy to show that

$$x_i(k+1) = \lambda + x_i(k)$$
$$x_i(k+2) = 2\lambda + x_i(k)$$
$$x_i(k+3) = 3\lambda + x_i(k)$$
$$\ldots$$

Since all events in the system have cyclic behavior, a general form of the equations above is

$$x_i(k+r) = r\lambda + x_i(k), \quad i = 1, 2, \ldots, n, \quad k \geq k_0 \qquad (4.21)$$

where k_0 is the number of the part that is processed when the system starts periodic activity (after the transient state has finished). Equation (4.21) demonstrates that the inherent property of the system is determined by parameter λ, *i.e.* the time period between departures of two consecutive parts from the system (production cycle) is equal to λ. We may set arrival rate(s) of parts to be as small as possible but the system cannot process the parts faster than what is determined by the production cycle.

At this point, two issues have to be addressed: a) how to calculate λ and b) is λ unique or are there several values that satisfy Equation(4.21)? In other words is it possible for a system to have events whose cycle periods differ? When one considers the former question it appears that it is natural to have more than one cycle period in the system. We saw that two machines, which have been used throughout this chapter as an example, have different cycles (t_1 and t_2). This demonstrates that the manufacturing system designer is the one who actually enforces the operational cycles to the system resources in order to get the final product. As we shall see later, depending on the system structure and desired performance, in some systems this cycle is unique for all resources, while in others there can be more than one.

From (4.21) it seems that λ can be calculated very easily,

$$\lambda = \frac{x_i(k+r) - x_i(k)}{r}, \quad i = 1, 2, \ldots, n, \quad k \geq k_0 \qquad (4.22)$$

The problem with this equation is that in order to get λ, values of $x_i(1)$, $x_i(2)$, $x_i(3)$, ... have to be determined. The second difficulty is that since k_0 is not known in advance we have to execute the max-plus model until cyclic activity is reached, which might be a tedious job. Furthermore, some of the events can start to cycle at $k = k_0$ while others remain in a transition phase (when the system has more than one λ), which means that one should proceed until all events demonstrate periodicity. All these facts show that, although λ can be calculated from Equation (4.21) we should try to find another method.

By definition, the cycle λ represents the difference between two consecutive occasions of an event in the system that has periodic activities. Hence, its value is related to *time*. If we recall the procedure for obtaining an autonomous system Equation (4.17), we remember the assumption that each time a part leaves the system a pallet is relocated to the system input. As a consequence, the pallet holding the part travels along some *circular path* and periodically visits resources, every time with a new part to be processed. The circular path σ has the corresponding weight σ_w and length σ_ℓ. In our case, weight represents the time needed for the pallet to pass the path. If we assume that there are enough pallets, and that the system capacity cannot be influenced by their number, it can be shown that the average time between two successive processings on the resource that belongs to the circular path is equal to the mean weight of the path, $\overline{\sigma}_w$, as defined in Definition 4.1.7. Since the pallets may travel along several paths on their way from the input to the output, if we want to find the production cycle, we have to find the "slowest" path (the path that has the largest mean weight). This value is exactly equal to the maximum cycle mean as defined in Definition 4.1.8.

The previous discussion has shown how to determine the production cycle λ from the adjacency matrix of the graph that represents the system. The weights of circular paths of length $\sigma_\ell = r$ can be found as diagonal elements of matrix $\mathbf{A}^r$, calculated according to Equation (4.12). In order to find the maximum cycle mean of the system, all diagonal entries of matrices $\mathbf{A}, \mathbf{A}^2, \mathbf{A}^3, ..., \mathbf{A}^n$ need to be compared, which yields

$$\lambda = \bigoplus_{i=1}^{n} \left(\frac{trace(\mathbf{A}^i)}{i} \right) \tag{4.23}$$

where division is performed in the standard way and

$$trace(\mathbf{A}) = \bigoplus_{j=1}^{n} a_{jj} \tag{4.24}$$

Relation (4.23) gives the correct value for λ only when the weighted adjacency matrix $\mathbf{A}$ corresponds to a *strongly connected graph*. The obtained value represents the *unique production cycle* of the system. However, one should use Equation (4.23) with caution since this equation assumes that each arc on the circular path can hold a part that is not necessarily a truth as we show in Example 4.3.1.

The strong connectivity of the graph may be tested in several ways. By using powers of adjacency matrix $\mathbf{G}$, obtained by Equation (4.2), we can repeat the multiplication $n-1$ times and then check if

$$g_{ij} + g_{ij}^2 + g_{ij}^3 + \ldots + g_{ij}^{n-1} \neq 0 \;,\; \forall i,j \qquad (4.25)$$

When Equation (4.25) is true, the corresponding graph is strongly connected. The max-plus version of the previous relation has the form

$$(\mathbf{A} \oplus \mathbf{A}^2 \oplus \mathbf{A}^3 \oplus \ldots \oplus \mathbf{A}^{n-1})_{ij} \neq \varepsilon \;,\; \forall i,j \qquad (4.26)$$

where powers of $\mathbf{A}$ are calculated according to Equation (4.12). If string composition is used for the test, then it should be

$$s_{ij} + s_{ij}^2 + s_{ij}^3 + \ldots + s_{ij}^{n-1} \neq 0 \;,\; \forall i,j \qquad (4.27)$$

for the corresponding graph to be strongly connected. In Equation (4.27) 0 is a null string and powers of s_{ij} are obtained from Equation (4.6), while additions are carried out as series string compositions.

By knowing the production cycle λ we can find not only the system throughput, which is defined as $1/\lambda$, but also the utilization of each resource in the system. If we define the *resource cycle*, denoted T_{Mi}, as a time required for a part to be processed by the resource (resource operation) plus the time required for the resource to prepare for work on the part (resource setup(s)), then *resource utilization* is calculated as

$$\eta_i = \frac{T_{\text{Mi}}}{\lambda} \qquad (4.28)$$

A discrete event dynamic system characterized with a graph that is not strongly connected may have more than one cycle mean. This fact is evidence that the system is composed of subsystems that can achieve cyclic activities with different periods. Working out these periods might be difficult, depending on the system structure. However, Equation (4.22) can be a good start. Those who are interested in this subject may wish to consult [20–22].

4.3.3. Buffers in Max-plus Algebra

Although it appears that the determination of the model (4.16) is straightforward once the manufacturing system is designed and its tasks are defined, there are several issues in max-plus DEDS modeling that have to be elaborated further. Let us mention a few of them. For instance, what would happen with the model if there exist bounded buffers between the machines, or how do the initial conditions of machines (idle or work-in-process) influence the model? Moreover, as the elements of matrices in Equation (4.16) can vary with time (in a deterministic or stochastic manner), the question is how does max-plus algebra handle systems that are not time invariant and/or deterministic? It is beyond the scope of this text to elaborate on these topics. Here we just briefly explain a case when a finite-capacity buffer is positioned between two machines. The problem of initial conditions will be addressed in the example that follows at the end of the chapter.

The model derived in the previous example shows that event x_1 does not depend on event x_2 ($a_{12} = \varepsilon$ in matrix $\mathbf{A}$): the processing of parts in M_1 continues no matter what the status of machine M_2 is. It is obvious that when the processing time of M_2 is longer than the processing time of M_1 added to the time needed for the part to get from M_1 to M_2, the second machine in line will be flooded with parts. Hence, the obtained model is valid only if there is a buffer with infinite depth between machines. In order to cover a realistic situation we have to incorporate finite-capacity buffers in the max-plus model of the system.

Let us place a buffer with a finite capacity N between two machines, M_i and M_{i+1}, connected in series (M_i is the predecessor of M_{i+1}). After being processed in M_i a part enters the buffer. When M_{i+1} is idle it takes the part from the buffer. Machine M_i cannot proceed with the processing of parts once the buffer is full. Now, let us assume that the processing time of M_{i+1} is greater than the processing time of M_i. In this situation the buffer becomes occupied sooner or later. If the buffer is full (containing N parts), machine M_i can start processing one more part and then it needs to wait until M_{i+1} becomes ready to free one place in the buffer. We can write:

$$x_i(k) = \max\left(x_i(k-1) + t_i, x_{i+1}(k-(N+1))\right) \quad (4.29)$$

where t_i is the processing time of M_i.

Usually, equations that describe the dynamic behavior of systems by using max-plus operations cannot be directly transferred to the form of Equation (4.16). A more general max-plus form should be used:

$$\begin{aligned}\mathbf{x}(k) &= \mathbf{A}_0 \otimes \mathbf{x}(k) \oplus \mathbf{A}_1 \otimes \mathbf{x}(k-1) \oplus \ldots \\ &\ldots \oplus \mathbf{A}_p \otimes \mathbf{x}(k-p) \oplus \mathbf{B} \otimes \mathbf{u}(k), \quad \mathbf{x}(0) = \mathbf{x}_0\end{aligned} \quad (4.30)$$

$$\mathbf{y}(k) = \mathbf{C} \otimes \mathbf{x}(k).$$

As Equation (4.30) is implicit in $\mathbf{x}(k)$, in order to get form (4.16), the following substitution should be made:

$$\begin{aligned}\mathbf{x}(k) &= \mathbf{A}_0 \otimes \left[\mathbf{A}_0 \otimes \mathbf{x}(k) \oplus \ldots \oplus \mathbf{A}_p \otimes \mathbf{x}(k-p) \oplus \mathbf{B} \otimes \mathbf{u}(k)\right] \\ &\oplus \mathbf{A}_1 \otimes \mathbf{x}(k-1) \oplus \ldots \oplus \mathbf{A}_p \otimes \mathbf{x}(k-p) \oplus \mathbf{B} \otimes u(k) \\ &= \mathbf{A}_0^2 \otimes \mathbf{x}(k) \oplus \left[\mathbf{A}_0 \oplus \mathbf{E}\right] \\ &\otimes \left[\mathbf{A}_1 \otimes \mathbf{x}(k-1) \oplus \ldots \oplus \mathbf{A}_p \otimes \mathbf{x}(k-p) \oplus \mathbf{B} \otimes \mathbf{u}(k)\right]\end{aligned} \quad (4.31)$$

where $\mathbf{E}$ is a max-plus identity matrix and $\mathbf{A}_0^2$ is obtained according to Equation (4.12). The elimination of $\mathbf{x}(k)$ from the right requires n substitutions to be carried out. Finally, we get

$$\mathbf{x}(k) = \mathbf{A}_0^{n+1} \otimes \mathbf{x}(k) \oplus \left[\mathbf{A}_0^n \oplus \mathbf{A}_0^{n-1} \oplus ... \oplus \mathbf{A}_0 \oplus \mathbf{E} \right] \otimes \\ \left[\mathbf{A}_1 \otimes \mathbf{x}(k-1) \oplus ... \oplus \mathbf{A}_m \otimes \mathbf{x}(k-m) \oplus \mathbf{B} \otimes u(k) \right] \tag{4.32}$$

As $\mathbf{A}_0^{n+1} = [\varepsilon]$, the first component of Equation (4.32) can be removed, which yields

$$\mathbf{x}(k) = \mathbf{A}_0^* \otimes \left[\mathbf{A}_1 \otimes \mathbf{x}(k-1) \oplus ... \oplus \mathbf{A}_m \otimes \mathbf{x}(k-m) \oplus \mathbf{B} \otimes u(k) \right] \tag{4.33}$$

where

$$\mathbf{A}_0^* = \left[\mathbf{A}_0^n \oplus \mathbf{A}_0^{n-1} \oplus ... \oplus \mathbf{A}_0 \oplus \mathbf{E} \right] \tag{4.34}$$

In Equation (4.33), which is obtained by substitutions, state space vector $\mathbf{x}(k)$ should be redefined to integrate components $\mathbf{x}(k-2)$, $\mathbf{x}(k-3)$, ..., $\mathbf{x}(k-m)$. Thereafter, Equation (4.33) takes the form of Equation (4.16) and max-plus analysis, described in the previous subsection, can be applied.

In conclusion, we need to stress that we have here presented only a small part of max-plus theory concerned with DEDS. Our intention was to provide the reader with a basic knowledge of the subject and to touch upon the potentials of max-plus in DEDS system analysis and design. Because a max-plus linear model has many similarities with the state space linear model of time-driven systems, it is possible to reproduce procedures and methods used in linear system analysis by replacing standard operations with maximization and addition (of course, symbols need to be replaced carefully as their replacement is not straightforward). In this way, the properties of time-driven systems, such as observability, controllability, transfer function, impulse response, *etc.*, can be transposed to event-driven systems.

Example 4.3.1 (DEDS modeling and analysis by max-plus algebra)

Our goal in this example is to determine a max-plus model of the manufacturing system shown in Figure 4.14. The system has three machines and one AGV. Two types of parts are processed, A and B. Part A is transported by the AGV from an input position to machine M_1. When the processing in M_1 is finished the part moves to the second machine, M_2, to be removed from the system by the AGV and put on the part A output place. Part B is fed directly to the machine M_3. Once the processing in M_3 is finished the AGV takes part B and transports it to the part B output place. The operations that can be identified based on the system description are listed in Tables 4.3 and 4.4, accompanied by operational times and machine setup times.

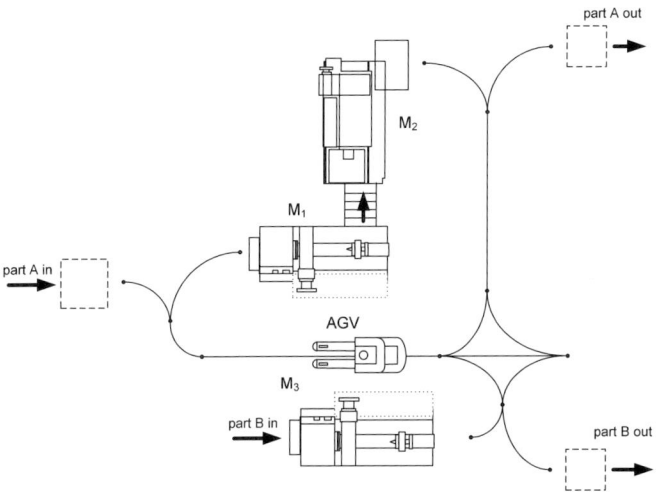

Figure 4.14. Layout of the manufacturing system studied in Example 4.3.1

Table 4.3. Operational times and setup times of machines

Machine	M_1	M_2	M_3
operational time	8	13	15
setup time	2	3	3

Table 4.4. Operational times and set-up times of the AGV

Operation	Transport A to M_1	Transport A from M_2	Transport B from M_3
operational time	5	6	3
setup time	4	6	5

Events of interest (*i.e.* events that should be modeled), shown in Table 4.5, correspond with the beginnings and endings of operations identified in the system. From the system description and Table 4.4 we see that the AGV is a shared resource since it has to perform three different tasks: loading M_1 and unloading M_2 and M_3. As a consequence, two (or even three) events may simultaneously request the AGV. Clearly, the manufacturing system can involve concurrent events and it does not fall in the class of systems that can be described as event graphs, thus it cannot be modeled with max-plus algebra.

Even though the sequence of operations should be strictly defined in order to make the system suitable for max-plus description, in particular circumstances a manufacturing system that employs shared resources does not require beforehand sequencing. This may occur when operational times in the system are structured so that simultaneous requests for a shared resource cannot take place (see Example

3.3.1). This means that because of the natural properties (operational times) of the system, the system works according to a sequence that is its inherent property. However, it is very difficult to establish whether a system has this property, especially if one is dealing with systems that contain many shared resources and part paths. Up to date, time-consuming simulation is the only way to resolve this dilemma.

Table 4.5. Events identified in the system shown in Figure 4.14

Event	Description
x_1	AGV starts to move part A in machine M_1
x_2	transportation of part A to M_1 is finished; release of AGV and start of processing of the part in M1
x_3	processing in M_1 is finished; release of M_1 and start of part A processing in M_2
x_4	AGV starts with M_2 unload; M_2 is released
x_5	transportation of part A to the output place is finished; AGV is released
x_6	start of part B processing in M_3
x_7	AGV starts with M_3 unload; M_3 is released
x_8	transportation of part B to the output place is finished; AGV is released

Although the system in the example is simple, we would not investigate if machine operational times provide sequence(s) without concurrency. We define the sequence of operations since further assumptions regarding parts entering the line would lead to a simultaneous request for shared resource. As the AGV is responsible for three tasks, two sequences are possible: a) loading M_1 – unloading M_2 – unloading M_3, and b) loading M_1 – unloading M_3 – unloading M_2. System layout and operational times suggest it would be reasonable to give priority to sequence b) (we shall leave the investigation of the system with sequence a) to the reader).

Further, we assume that parts A and B are available at any time. This means that immediately after the AGV takes part A from its input place, another part is ready. Also, part B is available for machine M_3 as soon as the processing of the previous part is finished and the machine is ready. As far as system outputs are concerned we assume that as soon as the transportation of a part to its output place is finished, the part leaves the system. These assumptions, related to inputs and outputs, are necessary if we want to model the system according to Equation (4.20) (an autonomous system).

Once we have defined the system and obtained all the necessary data we can create a weighted graph representation of the system, as shown in Figure 4.15.

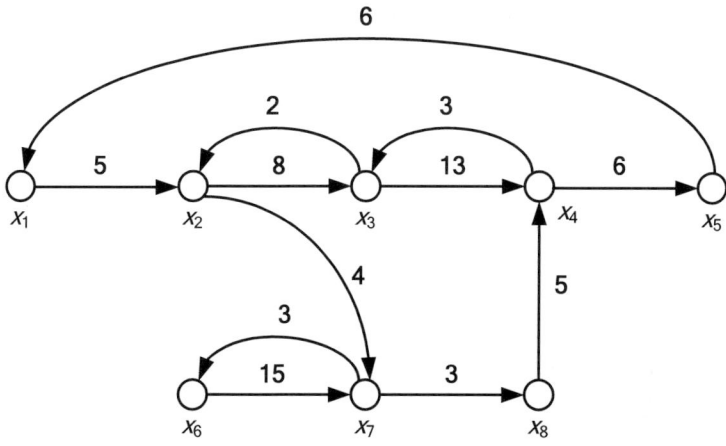

Figure 4.15. A weighted graph of the system studied in Example 4.3.1

The corresponding weighted adjacency matrix has the following form:

$$\mathbf{A} = \begin{bmatrix} \varepsilon & \varepsilon & \varepsilon & \varepsilon & 6 & \varepsilon & \varepsilon & \varepsilon \\ 5 & \varepsilon & 2 & \varepsilon & \varepsilon & \varepsilon & \varepsilon & \varepsilon \\ \varepsilon & 8 & \varepsilon & 3 & \varepsilon & \varepsilon & \varepsilon & \varepsilon \\ \varepsilon & \varepsilon & 13 & \varepsilon & \varepsilon & \varepsilon & \varepsilon & 5 \\ \varepsilon & \varepsilon & \varepsilon & 6 & \varepsilon & \varepsilon & \varepsilon & \varepsilon \\ \varepsilon & \varepsilon & \varepsilon & \varepsilon & \varepsilon & \varepsilon & 3 & \varepsilon \\ \varepsilon & 4 & \varepsilon & \varepsilon & \varepsilon & 15 & \varepsilon & \varepsilon \\ \varepsilon & \varepsilon & \varepsilon & \varepsilon & \varepsilon & \varepsilon & 3 & \varepsilon \end{bmatrix}$$

Although the property of strong connections of the graph in Figure 4.15 may be checked by hand, formal verification can be done by using Equation (4.26). Here we give only the final result:

$$\left(\mathbf{A} \oplus \mathbf{A}^2 \oplus \mathbf{A}^3 \oplus ... \oplus \mathbf{A}^7\right) = \begin{bmatrix} 54 & 49 & 57 & 50 & 44 & 53 & 38 & 49 \\ 47 & 54 & 50 & 49 & 49 & 46 & 43 & 42 \\ 51 & 56 & 54 & 51 & 51 & 48 & 47 & 46 \\ 58 & 59 & 61 & 54 & 48 & 59 & 46 & 53 \\ 48 & 59 & 57 & 54 & 54 & 47 & 50 & 49 \\ 48 & 45 & 45 & 42 & 36 & 54 & 57 & 35 \\ 47 & 58 & 52 & 45 & 51 & 69 & 54 & 44 \\ 48 & 45 & 45 & 42 & 36 & 54 & 57 & 35 \end{bmatrix} \neq \varepsilon, \forall i, j$$

Since all entries of the obtained matrix differ from ε, the weighted graph representing the system is strongly connected, *i.e.* the system has a unique production cycle.

A set of max-plus equations that describe the system dynamics has the following form (see graph in Figure 4.15; we presume that the system starts its activity with all the machines ready and with the AGV set for loading M_1):

$$\begin{aligned}
x_1(k) &= 6x_5(k-1) \\
x_2(k) &= 5x_1(k) \oplus 2x_3(k-1) \\
x_3(k) &= 8x_2(k) \oplus 3x_4(k-1) \\
x_4(k) &= 13x_3(k) \oplus 5x_8(k) \\
x_5(k) &= 6x_4(k) \\
x_6(k) &= 3x_7(k-1) \\
x_7(k) &= 4x_2(k) \oplus 15x_6(k) \\
x_8(k) &= 3x_7(k)
\end{aligned} \qquad (4.35)$$

As we can see, the system description is implicit in $x(k)$, so we should use substitution (4.31) in order to get form (4.20). Matrices $\mathbf{A}_0$ and $\mathbf{A}_1$ can be easily determined from the system equations:

$$\mathbf{A}_0 = \begin{bmatrix} \varepsilon & \varepsilon & \varepsilon & \varepsilon & \varepsilon & \varepsilon & \varepsilon & \varepsilon \\ 5 & \varepsilon & \varepsilon & \varepsilon & \varepsilon & \varepsilon & \varepsilon & \varepsilon \\ \varepsilon & 8 & \varepsilon & \varepsilon & \varepsilon & \varepsilon & \varepsilon & \varepsilon \\ \varepsilon & \varepsilon & 13 & \varepsilon & \varepsilon & \varepsilon & \varepsilon & 5 \\ \varepsilon & \varepsilon & \varepsilon & 6 & \varepsilon & \varepsilon & \varepsilon & \varepsilon \\ \varepsilon & \varepsilon & \varepsilon & \varepsilon & \varepsilon & \varepsilon & \varepsilon & \varepsilon \\ \varepsilon & 4 & \varepsilon & \varepsilon & \varepsilon & 15 & \varepsilon & \varepsilon \\ \varepsilon & \varepsilon & \varepsilon & \varepsilon & \varepsilon & \varepsilon & 3 & \varepsilon \end{bmatrix} \quad \mathbf{A}_1 = \begin{bmatrix} \varepsilon & \varepsilon & \varepsilon & \varepsilon & 6 & \varepsilon & \varepsilon & \varepsilon \\ \varepsilon & \varepsilon & 2 & \varepsilon & \varepsilon & \varepsilon & \varepsilon & \varepsilon \\ \varepsilon & \varepsilon & \varepsilon & 3 & \varepsilon & \varepsilon & \varepsilon & \varepsilon \\ \varepsilon & \varepsilon & \varepsilon & \varepsilon & \varepsilon & \varepsilon & \varepsilon & \varepsilon \\ \varepsilon & \varepsilon & \varepsilon & \varepsilon & \varepsilon & \varepsilon & \varepsilon & \varepsilon \\ \varepsilon & \varepsilon & \varepsilon & \varepsilon & \varepsilon & \varepsilon & 3 & \varepsilon \\ \varepsilon & \varepsilon & \varepsilon & \varepsilon & \varepsilon & \varepsilon & \varepsilon & \varepsilon \\ \varepsilon & \varepsilon & \varepsilon & \varepsilon & \varepsilon & \varepsilon & \varepsilon & \varepsilon \end{bmatrix}$$

By using Equation (4.34) we can find $\mathbf{A}_0^*$ that gives matrix $\mathbf{A}$,

$$A = A_0^* \otimes A_1 = \begin{bmatrix} \varepsilon & \varepsilon & \varepsilon & \varepsilon & 6 & \varepsilon & \varepsilon & \varepsilon \\ \varepsilon & \varepsilon & 2 & \varepsilon & 11 & \varepsilon & \varepsilon & \varepsilon \\ \varepsilon & \varepsilon & 10 & 3 & 19 & \varepsilon & \varepsilon & \varepsilon \\ \varepsilon & \varepsilon & 23 & 16 & 32 & \varepsilon & 26 & \varepsilon \\ \varepsilon & \varepsilon & 29 & 22 & 38 & \varepsilon & 32 & \varepsilon \\ \varepsilon & \varepsilon & \varepsilon & \varepsilon & \varepsilon & \varepsilon & 3 & \varepsilon \\ \varepsilon & \varepsilon & 6 & \varepsilon & 15 & \varepsilon & 18 & \varepsilon \\ \varepsilon & \varepsilon & 9 & \varepsilon & 18 & \varepsilon & 21 & \varepsilon \end{bmatrix}$$

The insertion of given initial conditions, $x(0) = [\varepsilon\ \varepsilon\ \varepsilon\ \varepsilon\ e\ \varepsilon\ e\ \varepsilon]^T$, into Equation (4.20) results in the following evaluation of system states (for easier reading, values of vector $x(k)$ are represented in matrix form):

$$x = \begin{bmatrix} x(1) & x(2) & x(3) & x(4) \\ 6 & 44 & 82 & 120 \\ 11 & 49 & 87 & 125 \\ 19 & 57 & 95 & 133 \\ 32 & 70 & 108 & 146 \\ 38 & 76 & 114 & 152 \\ 3 & 21 & 56 & 94 \\ 18 & 53 & 91 & 129 \\ 21 & 56 & 94 & 132 \end{bmatrix} \ldots$$

From the result we find that, for example, the first four time instants in which event x_5 occurs are 38, 76, 114 and 152.

Once the system states are evaluated, the production cycle λ can be determined according to Equation (4.22). As we have already stated, the problem is that value k_0, for which the system enters periodic behavior, is unknown, thus we start with the first two values of vector x. The difference between $x(2)$ and $x(1)$ gives $\lambda_{21} = [38\ 38\ 38\ 38\ 38\ 18\ 35\ 35]^T$. It may be seen that the components of λ_{21} vary, which clearly indicates that the system is still in transition. Further calculations provide $\lambda_{32} = x(3) - x(2) = [38\ 38\ 38\ 38\ 38\ 35\ 38\ 38]^T$ and $\lambda_{43} = x(4) - x(3) = [38\ 38\ 38\ 38\ 38\ 38\ 38\ 38]^T$. All components of λ_{43} are the same and we conclude that the system starts with cyclic activities at $k = k_0 = 3$ with a unique production cycle $\lambda = 38$ and throughput $1/\lambda = 0.0263$. The reader can check the correctness of the obtained production cycle by using Equation (4.23).

We proceed with the next step in system analysis - the calculation of resources utilization. From Tables 4.3 and 4.4 we find that $T_{M1} = 10$, $T_{M2} = 16$, $T_{M3} = 18$ and $T_{AGV} = 29$, thus

$$\eta_{M1} = \frac{T_{M1}}{\lambda} = \frac{10}{38} = 0.263, \quad \eta_{M2} = \frac{16}{38} = 0.421$$

$$\eta_{M3} = \frac{18}{38} = 0.474, \quad \eta_{AGV} = \frac{29}{38} = 0.763$$

The most exploited resource in the system is the AGV (76.3%), which was expected since it is a shared resource. Still, none of the resources is utilized 100%. Obviously, we must ask: what can be done in order to improve the use of resources and increase the system throughput?

If we look at the system graph representation (Figure 4.15), we will find that path $\sigma_A = (x_1, x_2, x_3, x_4, x_5, x_1)$ has the weight of $\sigma_{Aw} = 38$. When the part moves along this path on the pallet and if there is only one pallet, then the pallet needs 38 time intervals to return to the initial position. That is exactly the value of the production cycle. Now, if we set two pallets on the path, then $38/2 = 19$, i.e. the mean production cycle is reduced by factor 2. So, by inserting 5 pallets since there are 5 arcs on path σ_A (see Definition 4.1.7), we get $38/5 = 7.6$ as the mean production cycle. On the other hand, pallets are physical entities and if our only means of system representation is a graph, it would not be clear whether each arc corresponds to an empty place that can hold a pallet. Moreover, there is another path (besides others) $\sigma_{AGV} = (x_1, x_2, x_7, x_8, x_4, x_5, x_1)$ that has weight $\sigma_{AGVw} = 29$. What influence does the number of pallets have on that path? This brings us back to the question related to the number of pallets, raised in Section 4.3.1. It is obvious that it is impossible to resolve the problem without taking into account the physical limitations of the system hidden in graph representation.

Path σ_A does have 5 arcs but these 5 arcs represent only 3 physical places where the pallets might be positioned: machine M_1, machine M_2 and the AGV. Further, if we put three pallets for part A into the system, two scenarios are possible: a) two pallets are in the machines and the third is on the vehicle on its way to part A output place, and b) two pallets are in the machines and the third is on the vehicle on the way to machine M_1. For scenario a), once the vehicle leaves the part at the output place, it returns to the beginning of the line (according to a predefined sequence) and gets the next part A to be loaded into machine M_1, which leads to scenario b). Analysis of scenario b) reveals that it will end in blocking since machine M_1 already has the pallet so there is no room for the pallet that is transported by the vehicle. Therefore, we can conclude that path σ_A can have at most two pallets.

It is clear that path σ_{AGV} actually represents the route traveled by the vehicle, i.e. one can think of the vehicle as a "pallet". Since there is only one vehicle in the system, the path has only one physical entity where parts can be placed.

Keeping in mind the above discussion, let us write down a max-plus model of the system that encompasses the assumptions that i) one part A is already in machine M_2 waiting to be processed and ii) one part is on the vehicle, ready to be transported to machine M_1. The equation that describes the dynamics of event x_2 (start of processing of the part in M_1) takes this form:

$$x_2(k) = 5x_1(k-1) \oplus 2x_3(k-1)$$

The consequence of the supposition that the vehicle holds a part, can be clearly seen by comparing this equation with Equation (4.35) (where we assumed that the vehicle was empty). The equation states that machine M_1 starts processing the first part, $x_2(1)$, even though the part actually did not enter the system, $x_1(0)$, which is possible since the part was on the vehicle. The same holds for event x_4, i.e. machine M_2 finishes processing the first part, $x_4(1)$, although the part did not enter the machine, $x_3(0)$. This is feasible since according to assumption i) the part was in the machine.

A set of max-plus equations becomes:

$$\begin{aligned} x_1(k) &= 6x_5(k) \\ x_2(k) &= 5x_1(k-1) \oplus 2x_3(k-1) \\ x_3(k) &= 8x_2(k) \oplus 3x_4(k) \\ x_4(k) &= 13x_3(k-1) \oplus 5x_8(k) \\ x_5(k) &= 6x_4(k) \\ x_6(k) &= 3x_7(k) \\ x_7(k) &= 4x_2(k) \oplus 15x_6(k-1) \\ x_8(k) &= 3x_7(k) \end{aligned} \qquad (4.36)$$

A closer look at the graph in Figure 4.15 and Equations (4.35) and (4.36) can affirm a general rule for the holding of parts in the system for max-plus model determination. Namely, if event n_1 is the predecessor of event n_2 and if the task that starts with n_1 and ends with n_2, represented by the arc with weight a, is holding a part, then their dynamics is described as $n_2(k) = a \otimes n_1(k-1)$.

The calculation of matrix **A** from Equation (4.36) obtains:

$$\mathbf{A} = \begin{bmatrix} 29 & \varepsilon & 26 & \varepsilon & \varepsilon & 35 & \varepsilon & \varepsilon \\ 5 & \varepsilon & 2 & \varepsilon & \varepsilon & \varepsilon & \varepsilon & \varepsilon \\ 20 & \varepsilon & 17 & \varepsilon & \varepsilon & 26 & \varepsilon & \varepsilon \\ 17 & \varepsilon & 14 & \varepsilon & \varepsilon & 23 & \varepsilon & \varepsilon \\ 23 & \varepsilon & 20 & \varepsilon & \varepsilon & 29 & \varepsilon & \varepsilon \\ 12 & \varepsilon & 9 & \varepsilon & \varepsilon & 18 & \varepsilon & \varepsilon \\ 9 & \varepsilon & 6 & \varepsilon & \varepsilon & 15 & \varepsilon & \varepsilon \\ 12 & \varepsilon & 9 & \varepsilon & \varepsilon & 18 & \varepsilon & \varepsilon \end{bmatrix}$$

It may be seen that, even though both matrices describe the same system, matrix **A**, which we have just acquired, differs completely from the matrix **A** that corresponds with the system equations (4.35). It may be concluded that the initial conditions, together with the system structure, dictate the form of the system matrix (note that the weighted adjacency matrix remains the same in both cases – it is determined by the system structure).

Let us now see if these changes in matrix **A** influence system behavior. Evaluation of the system states with initial conditions $\mathbf{x}(0) = [e\ \varepsilon\ e\ \varepsilon\ \varepsilon\ e\ \varepsilon\ \varepsilon]^T$, gives:

$$\mathbf{x} = \begin{bmatrix} x(1) & x(2) & x(3) & x(4) \\ 35 & 64 & 93 & 122 \\ 5 & 40 & 69 & 98 \\ 26 & 55 & 84 & 113 \\ 23 & 52 & 81 & 110 \\ 29 & 58 & 87 & 116 \\ 18 & 47 & 76 & 105 \\ 15 & 44 & 73 & 102 \\ 18 & 47 & 76 & 105 \end{bmatrix} \ldots$$

The calculation of the production cycle yields $\lambda_{21} = \mathbf{x}(2) - \mathbf{x}(1) = [29\ 35\ 29\ 29\ 29\ 29\ 29\ 29]^T$ and $\lambda_{32} = \mathbf{x}(3) - \mathbf{x}(2) = [29\ 29\ 29\ 29\ 29\ 29\ 29\ 29]^T$. The system starts with periodic activities for $k_0 = 2$ with the production cycle $\lambda = 29$. This result, when compared to the result obtained from the set of equations (4.35), shows that the transition period of the system has been reduced while the throughput has increased. Although we had set two pallets in path σ_A, the new production cycle was not reduced by a factor of 2 since the other path, σ_{AGV}, became dominant (a new maximum cycle mean).

Utilizations of resources are given below:

$$\eta_{M1} = \frac{T_{M1}}{\lambda} = \frac{10}{29} = 0.345, \quad \eta_{M2} = \frac{16}{29} = 0.552$$

$$\eta_{M3} = \frac{18}{29} = 0.621, \quad \eta_{AGV} = \frac{29}{29} = 1.0$$

Because the production cycle was reduced, utilization of each resource in the system was increased while utilization of AGV attained 100%. By obtaining this result we have reached the physical limitations of the system. Further improvements can be made by including one or more additional vehicle(s).

♦

4.3.4 Deriving Max-plus System Equation from Matrix Model

In this section we draw the connection between the dynamic matrix model of an MS, presented in Section 3.3, and the max-plus system equation. Since the max-plus representation is feasible only for decision-free discrete event systems (event graphs), we consider the dynamic matrix model with no shared resources. For systems with shared resources, a control strategy that provides conflict-free dispatching should be determined prior to transformation of the matrix model to max-plus. As a result, max-plus formulation is a description of the closed-loop system including both the workcell and the controller.

Let us now recall the logical state equation (3.2)

$$\bar{x} = F_v \Delta \bar{v}_c \nabla F_r \Delta \bar{r}_c \nabla F_u \Delta \bar{u}$$

In development of the dynamic matrix model we assumed that parts input and parts output are timeless operations. In order to obtain the max-plus model for the general system, here we define delay matrices T_u and T_y that can be attained in the same way as matrices T_u and T_y, i.e. each entry "1" in F_u and S_y should be replaced with the shift operand representation of the corresponding lifetime. Using these new matrices and by including Equation (3.16) in the logical state equation we obtain

$$\bar{x}(q) = F_v \Delta T_v(q) \bar{x}(q) \nabla F_r \Delta T_r(q) \bar{x}(q) \nabla T_u(q) \bar{u}(q)$$
$$y(q) = T_y(q) x(q)$$

We proceed with the following redefinitions of mathematical operations: logical Δ should be replaced with standard multiplication, standard multiplication with $\otimes$, and logical ∇ with $\oplus$. Then we get

$$x(q) = F_v T_v(q) \otimes x(q) \oplus F_r T_r(q) \otimes x(q) \oplus T_u(q) \otimes u(q)$$
$$y(q) = T_y(q) \otimes x(q)$$

A final form of max-plus is obtained by multiplication of matrices F_x with corresponding matrices T_x, and then by substituting q^{-n} with n, and replacing all occurrences of 0 by ε,

$$x(k) = D_v \otimes x(k) \oplus D_r \otimes x(k) \oplus D_u \otimes u(k)$$
$$y(k) = D_y \otimes x(k)$$
(4.37)

where $x(k)$ gives the time of the kth execution of rules corresponding to the components of the logical state vector, $y(k)$ gives the time of the kth output of finished products.

The max-plus model (4.37) is valid only for systems with no shared resources. As we mentioned, when that system encompasses conflicting rules the dispatching vector should be included into the model. Given that $u_d = S_d \Delta x$ adding conflict resolution vector in Equation (4.37) gives

$$x(k) = D_v \otimes x(k) \oplus D_r \otimes x(k) \oplus D_u \otimes u(k) \oplus D_d \otimes x(k)$$
$$y(k) = D_y \otimes x(k)$$
(4.38)

where D_d is obtained from $F_d T_d(q)$ in the same way as matrices D_v and D_r. A conflict-resolution delay matrix $T_d(q)$ is determined from the dispatching vector release matrix S_d as for the delay matrices T_v and T_r.

It should be noticed that the attained model is implicit in $x(k)$, i.e. it does not account for the available resources or the parts held by operations. As we pointed out earlier, the number of slots in buffers (machines) or number of resources in the resource pool should be incorporated into the model. For Equation (4.38) this can be done in the following way: if resource r, released by the rule x_j, participates in the rule x_i, and if it is able to process N parts simultaneously, then $x_i(k) = d_r \otimes x_j(k-N)$. The same is true for an operation: if operation v, released by the rule x_j, participates in the rule x_i, and if it holds N parts, then $x_i(k) = d_v \otimes x_j(k-N)$.

Example 4.3.2 (Deriving max-plus system equation from the matrix model)

We consider the system shown in Figure 3.2 that is studied in Examples 3.2.1, 3.3.1 and 3.4.1. For a given system matrices and delay matrices we find that

$$D_v = \begin{bmatrix} \varepsilon & \varepsilon & \varepsilon & \varepsilon & \varepsilon & \varepsilon \\ 76 & \varepsilon & \varepsilon & \varepsilon & \varepsilon & \varepsilon \\ \varepsilon & 10 & \varepsilon & \varepsilon & \varepsilon & \varepsilon \\ \varepsilon & \varepsilon & 4 & \varepsilon & \varepsilon & \varepsilon \\ \varepsilon & \varepsilon & \varepsilon & 113 & \varepsilon & \varepsilon \\ \varepsilon & \varepsilon & \varepsilon & \varepsilon & 8 & \varepsilon \end{bmatrix}, D_r = \begin{bmatrix} \varepsilon & 15 & \varepsilon & \varepsilon & \varepsilon & \varepsilon \\ \varepsilon & \varepsilon & 6 & \varepsilon & \varepsilon & 5 \\ \varepsilon & \varepsilon & \varepsilon & 3 & \varepsilon & \varepsilon \\ \varepsilon & \varepsilon & \varepsilon & \varepsilon & 10 & \varepsilon \\ \varepsilon & \varepsilon & 6 & \varepsilon & \varepsilon & 5 \\ \varepsilon & \varepsilon & \varepsilon & \varepsilon & \varepsilon & \varepsilon \end{bmatrix}$$

Let us assume that the dispatching vector release matrix S_d is determined from F_d according to Equation (3.26), as shown below:

$$x_d = \begin{bmatrix} 0 \\ 1 \\ 0 \\ 0 \\ 1 \\ 0 \end{bmatrix}, F_d = \begin{bmatrix} 0 & 0 \\ 1 & 0 \\ 0 & 0 \\ 0 & 0 \\ 0 & 1 \\ 0 & 0 \end{bmatrix} \Rightarrow S_d = \begin{bmatrix} 0 & 0 & 0 & 0 & 1 & 0 \\ 0 & 1 & 0 & 0 & 0 & 0 \end{bmatrix}$$

When there are no delays between components of the conflict-resolution vector, matrix D_d becomes

$$D_d = \begin{bmatrix} \varepsilon & \varepsilon & \varepsilon & \varepsilon & \varepsilon & \varepsilon \\ \varepsilon & \varepsilon & \varepsilon & \varepsilon & e & \varepsilon \\ \varepsilon & \varepsilon & \varepsilon & \varepsilon & \varepsilon & \varepsilon \\ \varepsilon & \varepsilon & \varepsilon & \varepsilon & \varepsilon & \varepsilon \\ \varepsilon & e & \varepsilon & \varepsilon & \varepsilon & \varepsilon \\ \varepsilon & \varepsilon & \varepsilon & \varepsilon & \varepsilon & \varepsilon \end{bmatrix}$$

If we assume that the input and output operations are timeless, then according to the rules, the system input and the system output are calculated by using

$$D_u = \begin{bmatrix} e & \varepsilon & \varepsilon & \varepsilon & \varepsilon & \varepsilon \end{bmatrix}^T$$
$$D_y = \begin{bmatrix} \varepsilon & \varepsilon & \varepsilon & \varepsilon & \varepsilon & e \end{bmatrix}$$

Given that all resources are idle at the beginning, $u(0)=e$ and $\mathbf{u}_{d0} = [e \ \varepsilon]^T$, the initial condition is defined as $\mathbf{x}(0) = [\varepsilon \ \varepsilon \ \varepsilon \ \varepsilon \ \varepsilon \ \varepsilon]^T$, while the system is described with the following set of max-plus equations:

$$x_1(k) = 15x_2(k-1) \oplus u(k)$$
$$x_2(k) = 76x_1(k) \oplus 6x_3(k-1) \oplus x_5(k-1) \oplus 5x_6(k-1)$$
$$x_3(k) = 10x_2(k) \oplus 3x_4(k-1)$$
$$x_4(k) = 4x_3(k) \oplus 10x_5(k-1)$$
$$x_5(k) = x_2(k) \oplus 6x_3(k-1) \oplus 113x_4(k) \oplus 5x_6(k-1)$$
$$x_6(k) = 8x_5(k)$$

4.4 Exercises

1. Find the incidence matrices for the graphs shown in Figure 4.12.
2. Find the critical circuits in the graphs shown in Figure 4.12 by using Definitions 4.1.7 and 4.1.8.
3. Determine the circuits in the graph shown in Figure 4.15 by using string composition. What is the length of the critical circuit?
4. For the given values of cs determine a path (which is not a circuit) with the maximum weight in the graph shown in Figure 4.10 by using max-plus algebra.

c_1	c_2	c_3	c_4	c_5	c_6	c_7	c_8	c_9	c_{10}	c_{11}	c_{12}	c_{13}	c_{14}	c_{15}
10	4	4	6	5	3	6	2	2	8	2	2	4	2	2

5. Determine a max-plus model of the system represented by the graph shown in Figure 4.7. Consider event e_1 as an input u and event e_6 as an output y. Find the maximum allowed arrival rate of parts for given operational and setup times.

t_U	t_{MP1}	t_{M1}	t_T	t_{MP2}	t_{M2}	t_Y
2	12	3	4	17	4	3

References

[1] Biggs NL, Lloyd KE, Wilson RJ. Graph Theory 1736−1936. Clarendon: Oxford University Press, 1976.
[2] Diestel R. Graph theory. Heidelberg New York: Springer, 2000.
[3] Gibbons A. Algorithmic Graph Theory. Cambrifge: Cambridge University Press, 1985.
[4] Godsil CD, Royle G. Algebraic Graph Theory. New York: Springer−Verlag, 2001.
[5] Boffey TB. Graph Theory in Operational Research. London: MacMillan, 1982.
[6] Wysk RA, Yang NS, Joshi S. Detection of Deadlocks in Flexible Manufacturing Cells, IEEE Trans. Rob. Autom. 1991;Vol. 7;No. 6:853−859.
[7] Cormen TH, Leiserson CE, Rivest RL, Stein C. Introduction to Algorithms, 2nd edn. Cambridge: MIT Press, 2001.
[8] Bertsekas DP. An Auction Algorithm for Shortest Paths, SIAM J. on Optimization 1991;Vol. 1: 425−447.
[9] Floyd RW. Algorithm 97: Shortest path, Comm. ACM 1962;Vol. 5;No. 6:345.
[10] Warshall S. A theorem on boolean matrices, Journal of the ACM 1962;Vol. 9;No. 1:11−12.
[11] Dijkstra E. Note on Two Problems in Connection with Graphs, Numerische Mathematik 1959;Vol. 1:269−271.
[12] Wang L. Performance evaluation of switched discrete event systems, IMA preprints 2002:1835.
[13] Baccelli F, Cohen G, Olsder GJ, Quadrat JP. Synchronization and Linearity: An Algebra for Discrete Event Systems. New York: Wiley, 1992.
[14] De Schutter B. Max-Algebraic System Theory for Discrete Event Systems, PhD thesis, Faculty of Applied Sciences, K.U. Leuven, Leuven, Belgium, ISBN 90-5682-016-8, 1996.
[15] Gaubert S, Gunawardena J. A Non-linear Hierarchy for Discrete Event Dynamical Systems, IEE Proc. of the 4th WODES, Cagliary, Italy, 1998.
[16] Gunawardena J. From max-plus algebra to nonexpansive mappings: a nonlinear theory for discrete event systems, Theoretical Computer Science 2003;293:141−167.
[17] Cohen G, Gaubert S, Quadrat JP. Max-plus Algebra and Systems Theory: Where we are and where to go now, Annual Reviews in Control 1999;23:207−219.
[18] de Vries R, De Schutter B, De Moor B. On Max-algebraic Models for Transportation Networks, IEE Proc. of the 4th WODES, Cagliary, Italy, 1998, 457−462.
[19] Gaubert S. Performance Evaluation of (max,+) Automata, IEEE Trans. Aut. Cont. 1995;40;12; 2014−2025.
[20] Terrasson JC, Cohen G, Gaubert S, Mc Gettrick M, Quadrat JP. Numerical computation of spectral elements in max-plus algebra, Proceedings of the IFAC Conference on System Structure and Control (SSC'98), Nantes, France, 1998.

[21] Olsder GJ, Roos C, Van Egmond RJ. An efficient algorithm for critical circuits and finite eigenvectors in the max-plus algebra, Linear Algebra Appl. 1999;295;1−3:231−240.
[22] Dasdan A; Gupta RK. Faster maximum and minimum mean cycle algorithms for system performance analysis, IEEE Trans. CAD of Integr. Circ. Sys. 1998;17;10:889−899.
[23] Cottenceau B, Hardouin L, Boimond JL, Ferrier JL. Model reference control for timed event graphs in dioids, Automatica 2001;37:1451−1458.

5

Manufacturing System Structural Properties in Matrix Form

One fundamental question that needs to be addressed in connection with any FMS dispatching policy is whether or not it is *stable*. Studies of stability for FMS often focus on stability in the sense of *bounded buffers lengths* [1]. In [2], the FBFS policy has been shown to be stable for single-part flowlines with no buffer limits. However, in practice, the buffer lengths are *finite*, and such stability results are inapplicable, since it is not obvious how to keep the buffer lengths below some *fixed finite value*. For finite-buffer multiple re-entrant flowline (MRF) systems [1], which constitute a large class of FMSs, the issue is stability, not in the sense of bounded buffer lengths, but in the sense of absence of deadlock. As we pointed out in previous chapters, a flowline for a given part-class is said to be deadlocked if it holds a part that cannot complete its processing sequence. Many popular dispatching rules can result in deadlock if care is not taken, as has been demonstrated in the examples in Chapter 3. In a finite-buffer system, any dispatching policy for uninterrupted part flow has to essentially take into account the *structure* of the interaction between jobs and resources. Several results based on such a structural approach may be discovered in [3]–[8]. In all of these but [5], Petri-net formalism is used for system modeling.

In this chapter we develop equations to compute structural properties that are essential in stability analysis of the aforementioned MRF class of systems. These equations are based on the matrix model introduced in Chapter 3. First we give the properties that characterize MRF systems, followed by relations that determine *circular waits* among resources (mentioned in string composition Section 4.2). Then we show the correlation between circular waits and certain corresponding structures referred to as *critical siphons*, *critical traps*, and *critical subsystems*. This allows one to obtain computational equations so that NP-hard complexity issues can be avoided. In a separate section we consider and extend matrix formalism to the free-choice multiple re-entrant flowline (FMRF) systems, *i.e.* systems with nondeterministic job routing,

In terms of given constructions, at the end of the chapter we present a minimally restrictive *one-step look-ahead* resource dispatching policy that guarantees the absence of deadlock for MRFs. Deadlock has generally not been a significant problem in traditional manual shop-floor environments because the production operators are often able to recognize deadlock and take corrective measures, such as removing parts from the system or swapping the locations of two or more parts simultaneously. However, as the trend moves toward automation, these seemingly trivial resource-assignment problems have become increasingly important in ensuring a smooth operation of the manufacturing facilities.

We consider the case where the system is *regular,* that is, it cannot contain *key resources* [9] [10] existing in *second-level deadlock* structures [11]. A mathematical test is given to verify that the MRF system is regular. If this is not the case, we can still use matrix formulation, but with a different dispatching policy designed for systems containing second-level deadlock structures.

The chapter is closed with a case study where all the methods and the algorithms presented herein are implemented and realized on the laboratory system.

5.1 Multiple Re-entrant Flowlines – MRF

Notations used in the rest of the chapter are given before formal definition of multiple re-entrant flowlines considered herein. We widen the notions of preset and postset, defined in Definition 4.1.3, to resources in R and jobs in J as follows: for a given logical state vector component x_i we define a preset of x_i, denoted $\bullet x_i$, as a set of resources and jobs that participate in the prerequisite part of rule x_i; a postset of x_i, denoted $x_i \bullet$, is a set of resources and jobs that participate in the consequent part of rule x_i. One can obtain the preset of x_i as $\bullet x_i = sup(\mathbf{r}_{x_i}^d) \cup sup(\mathbf{v}_{x_i}^d)$, where $\mathbf{r}_{x_i}^d$ and $\mathbf{v}_{x_i}^d$ are binary vector equivalents of resources and jobs that contribute in the prerequisite part of rule x_i. Likewise, $x_i \bullet = sup(^d\mathbf{r}_{x_i}) \cup sup(^d\mathbf{v}_{x_i})$. The above definitions of pre and postsets are extended to the set of rules $\Omega = \{x_1, x_2, \ldots, x_k\}$, where $\bullet \Omega = \bullet x_1 \cup \bullet x_2 \cup \ldots \cup \bullet x_k$, and $\Omega \bullet = x_1 \bullet \cup x_2 \bullet \cup \ldots \cup x_k \bullet$.

A preset of resource r_i, denoted $\bullet r_i$, is defined as a set of all rules that release r_i, while a postset $r_i \bullet$ is defined as a set of all rules in which r_i contributes as a prerequisite. Hence, $\bullet r_i = sup(\mathbf{x}_{r_i}^d)$ and $r_i \bullet = sup(^d\mathbf{x}_{r_i})$. Equivalently, $\bullet J_i^k = sup(\mathbf{x}_{J_i^k}^d)$ and $J_i^k \bullet = sup(^d\mathbf{x}_{J_i^k})$. Given a set Ψ that consists of both resources and jobs, we have $\bullet \Psi = sup(\mathbf{x}_\Psi^d)$ and $\Psi \bullet = sup(^d\mathbf{x}_\Psi)$.

For the system studied in Example 3.2.1 one can find that $\bullet x_2 = \{R, MAP\}$ with $\mathbf{r}_{x_2}^d = [0\ 0\ 0\ 1]^T$ and $\mathbf{v}_{x_2}^d = [1\ 0\ 0\ 0\ 0]^T$, while $x_2 \bullet = \{MA, RP1\}$ with $^d\mathbf{r}_{x_2} = [1\ 0\ 0\ 0]^T$ and $^d\mathbf{v}_{x_2} = [0\ 1\ 0\ 0\ 0]^T$. If resource R is considered, then $\bullet R = \{x_3, x_6\}$, $R \bullet = \{x_2, x_5\}$, $\mathbf{x}_R^d = [0\ 0\ 1\ 0\ 0\ 1]^T$,

$^d\mathbf{x}_R = \begin{bmatrix} 0 & 1 & 0 & 0 & 1 & 0 \end{bmatrix}^T$. For $\Psi = \{BP, MB\}$ one has $\bullet \Psi = \{x_3, x_5\}$ and $\Psi \bullet = \{x_4\}$.

The concept of pre and postsets is essential in the analysis of system structural properties and the design of stable dispatching strategies. A formal definition of the MRF class of multiple re-entrant flowlines follows (recall that $P = R \cup J$ is a set of all resource and jobs in the system).

Definition 5.1.1 (multiple re-entrant flowlines – class MRF): MRF is the class of multipart re-entrant flowline systems with the following properties [10]:

i) $\forall p \in P, \bullet p \cap p \bullet \neq \{\varnothing\}$

ii) $\forall x_{in}, x_{in} \bullet \cap R = \{\varnothing\}$ and $\forall x_{out}, \bullet x_{out} \cap R = \{\varnothing\}$

iii) $\forall J_i^k \in J, |R(J_i^k)| = 1$ and $R(J_i^k) \neq R(J_{i+1}^k)$

iv) $\forall J_i^k \in J, |J_i^k \bullet| = 1$

v) $\forall x_i, |\bullet x_i \cap J| \leq 1$

vi) $\exists r \in R, |J(r)| > 1$

In other words, in MRF it is not allowed for an operation or resource to participate in the prerequisite and consequent part of the same rule (i). A rule, denoted x_{in}, that has an input operation in the IF part cannot have resource release in the THEN part, and a rule, denoted x_{out}, that has output operation in the THEN part cannot have resource in the IF part (ii). Statement (iii) in Definition 5.1.1 points out that each operation in the system requires one and only one resource with no two consecutive jobs using the same resource. Furthermore, there are no choice jobs (iv) and no assembly jobs (v). Item (vi) asserts that there are shared resources in the system. Obviously in MRF systems, for any $r \in R$, $J(r) = r \bullet \bullet \cap J = \bullet \bullet r \cap J$ and $R(J_i^k) = \bullet \bullet J_i^k \cap R = J_i^k \bullet \bullet \cap R$.

Subsequent sections of this chapter are dedicated to the analysis of structural properties of MRF systems. Before we proceed, let us repeat the assumptions already stated in Chapter 3:

No pre-emption – once assigned, a resource cannot be removed from a job until it is completed.

Mutual exclusion – a single resource can be used for only one job at a time.

Hold while waiting – a process holds the resources already allocated to it until it has all the resources required to perform a job.

No machine failures.

5.1.1 Circular Waits in MRF Systems

For a class of MRF systems, having in mind the above definition and assumptions, deadlock can occur only if there is a *circular wait relation* (CW) among the resources [12]. Circular wait relations are ubiquitous in re-entrant flowlines and in themselves do not present a problem. However, if a CW develops into *circular blocking*, then one has a deadlock. CWs are key structures in MRF and determination of deadlock avoidance strategies starts with their allocation within the system. In this section we present a digraph matrix procedure to identify all CWs present in an MRF. The following are the formal definitions.

Definition 5.1.1 (wait relation): Given a set of resources R, for any two resources $r_i, r_j \in R$, r_i is said to wait for r_j, denoted $r_i \rightarrow r_j$, if the availability of r_j is an immediate requirement to release r_i, or equivalently, if there exists at least one rule $x_k \in \bullet r_i \cap r_j \bullet$.

The wait relation is similar to the notion of upstream and downstream nodes defined in Definition 4.1.3.

Definition 5.1.2 (circular wait): Circular wait among resources is a set of resources $r_a, r_b, \ldots r_w$, with wait relations among them such that $r_a \rightarrow r_b \rightarrow \ldots \rightarrow r_w$, and $r_w \rightarrow r_a$.

Evidently, a circular wait corresponds to the cyclic path in the graph theory. For MRF systems circular waits are associated with shared resources, which is affirmed in the following lemma.

Lemma 5.1.1 (circular wait contains shared resource): In the MRF system a circular wait C contains at least one shared resource.

Proof:
Let $C=\{r_1, r_2 \ldots, r_q\}$, with $r_1 = r_q$, a circular wait in MRF. Assume there is no shared resource. Then $|J(r_i)|=1$, $\forall i$. Let the timing sequence of jobs performed by resources in C be $s_C =((J(r_1),t_1), (J(r_2),t_2), \ldots, (J(r_q),t_q))$. Then job $J(r_1)$ occurs prior to job $J(r_q)$ which is impossible as $J(r_1) = J(r_q)$ because r_1 is a nonshared resource.
♦

It should be noted that Lemma 5.1.1 is unidirectional, *i.e.* having a shared resource in MRF does not imply the existence of circular wait.

Definition 5.1.3 (simple circular wait): Simple circular wait (sCW), is such that, for some appropriate relabeling, one has $r_1 \rightarrow r_2 \rightarrow \ldots \rightarrow r_q$, with $r_i \neq r_j$ for $i \neq j$.

The above definition states that only one occurrence of a particular resource is allowed in simple circular wait. The importance of sCW will become clear when we introduce key resources and irregular systems.

Manufacturing Systems Structural Properties in Matrix Form   151

According to definition, the release of resource involved in a wait relation is stipulated with the availability of other resources. Since releases and requirements of resources are described with system matrices, as a first step in the determination of simple CWs we use these matrices to identify wait relations among resources, which are given as

$$\mathbf{G}_W = \left(\mathbf{S}_r \Delta \mathbf{F}_r\right)^T \tag{5.1}$$

Operations in Equation (5.1) are carried out in and/or algebra. As both $\mathbf{S}_r$ and $\mathbf{F}_r$ are binary matrices, the wait relation matrix $\mathbf{G}_W$ actually corresponds to the adjacency matrix of a graph that is composed of nodes representing resources connected with arcs that represent wait relations. We call this graph the *wait relation graph*. An element of $\mathbf{G}_W$ is given by $g_{w_{ji}} = \nabla_k s_{ik} \wedge f_{kj}$. As defined, matrix $\mathbf{S}_r$ has element $s_{ik} = 1$ if and only if rule $x_k \in \bullet r_i$. Matrix $\mathbf{F}_r$ has element $f_{kj} = 1$ if and only if rule $x_k \in r_j \bullet$. Thus $g_{w_{ji}} = 1$ if and only if there exists a rule $x_k \in \bullet r_i \cap r_j \bullet$, which is equivalent to $r_i \rightarrow r_j$, i.e. an entry "1" in position $g_{w_{ji}}$ corresponds with an arc from resource r_i to resource r_j.

For the system studied in Example 3.2.1 the wait relation matrix is calculated as

$$\mathbf{G}_W = \left(\mathbf{S}_r \Delta \mathbf{F}_r\right)^T = \left(\begin{bmatrix} 0 & 1 & 0 & 0 & 0 & 0 \\ 0 & 0 & 0 & 0 & 1 & 0 \\ 0 & 0 & 0 & 1 & 0 & 0 \\ 0 & 0 & 1 & 0 & 0 & 1 \end{bmatrix} \begin{bmatrix} 1 & 0 & 0 & 0 \\ 0 & 0 & 0 & 1 \\ 0 & 0 & 1 & 0 \\ 0 & 1 & 0 & 0 \\ 0 & 0 & 0 & 1 \\ 0 & 0 & 0 & 0 \end{bmatrix}\right)^T = \begin{bmatrix} 0 & 0 & 0 & 0 \\ 0 & 0 & 1 & 0 \\ 0 & 0 & 0 & 1 \\ 1 & 1 & 0 & 0 \end{bmatrix} \begin{matrix} \text{MA} \\ \text{MB} \\ \text{B} \\ \text{R} \end{matrix}$$

$$ \text{MA MB B R}$$

From this result we can read four wait relations: MA → R, MB → R, B → MB, and R → B. A corresponding wait relation graph is shown in Figure 5.1.

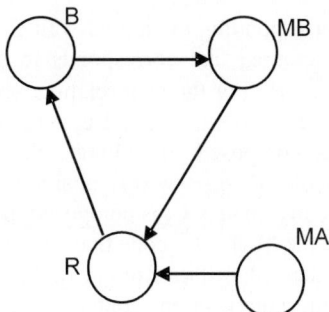

Figure 5.1. A wait relation graph of the system in Example 3.2.1

Having determined the wait relation matrix, the procedure we follow in identification of sCW is the string composition, defined and analyzed in Section 4.2. First, a wait relation matrix **G** is transformed into string matrix **S**. Then, for each power of **S** diagonal elements are identified, representing simple CWs. The question is how far we should go with powers of **S**? We showed in Section 4.2. that when the graph has n nodes, the nth composition gives a cycle that includes all nodes (if one exists). Hence, the string composition should be completed when the power of **S** is equal to the number of resources in the system. However, that might give an incorrect result, since the system could have a so-called *cyclic CW (CCW)*, *i.e.* a CW that is composed of two or more CWs.

This situation is demonstrated in Figure 5.2. A wait relation graph consists of 4 resources, thus one should calculate S^2, S^3 and S^4 in order to get CWs. A string composition reveals two CWs: MA $\rightarrow$ R and MB $\rightarrow$ R $\rightarrow$ B. Nevertheless, there exists a third CW composed of these two: MB $\rightarrow$ R $\rightarrow$ MA $\rightarrow$ R $\rightarrow$ B that remains hidden.

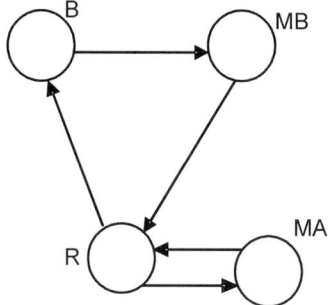

Figure 5.2. A wait relation graph with cyclic circular wait

Cyclic CWs are important because common shared resources among CWs might compose particular structures that must be considered in deadlock-free dispatching strategy design. Thus, the entire set of the system CWs should include the simple CWs plus cyclic CWs composed of unions of nondisjoint simple CWs.

Let us assume that the set of resources $C_i = \{r_a, r_b, \ldots r_m\}$, $C_i \subset R$, is a CW. Then, a binary vector $\mathbf{c}_i$ corresponding to circular wait C_i is defined as $sup(\mathbf{c}_i) = C_i$. In addition, the binary vector $\mathbf{c}_{si}$ that corresponds to shared resources in C_i is determined as $sup(\mathbf{c}_{si}) = C_i \cap R_s$. For the wait relation graph depicted in Figure 5.1 one has $C = \{MB, B, R\}$, $\mathbf{c} = [0\ 1\ 1\ 1]^T$, and $\mathbf{c}_s = [0\ 0\ 0\ 1]^T$. Given an MS, its circular wait matrix **C** is composed of columns that represent circular waits vectors, that is, an entry of "1" on the (i,j) position means that resource i is included in CW j. Equivalently, matrix $\mathbf{C}_s$ is composed of binary vectors $\mathbf{c}_{si}$.

In Figure 5.3, we show the MATLAB® code that calculates all CWs from the sets of simple CWs; it uses a Gurel algorithm from [13]. An input into the algorithm is matrix **C** obtained by string composition, containing simple CWs. As an output from the algorithm, we attain the new matrix **C** containing all CWs in the system, and matrix ζ that provides the set of composed CWs from unions of simple CWs comprised in columns of the input matrix **C**. An entry "1" on ζ_{ij} means that a simple CW i is included in the composed CW j.

Example 5.1.1 (circular waits in MRF)

We consider the manufacturing system depicted in Figure 5.4. Two part types, A and B, are processed in the workcell that consists of four machines and an automated guided vehicle. Part A visits resources in the following order: AGV, M4, M1, AGV, while part B path is: AGV, M2, AGV, M3, M2. Clearly, AGV and M2 are shared resources. The resource set and job set are defined as $R = \{M1, M2, M3, M4, R\}$ and $J = \{AP1, M4P, M1P, AP2, AP3, M2P1, AP4, M3P, M2P2\}$.

```
function [C,Fi]=Gurel(C);
%cyclic CW determination
%input: sCW matrix C
%output: complete matrix C

k=1;
L=size(C,2);
NS=0:L-1;
NT=1:L;
Fi=eye(L);
LT=L;
z=1;

while z
  z=0;
  LTEMP=LT;
  for i=1:L-k
    NUM=0;
    for j=i+1:L-k+1

      for l=NS(j)+1:NT(j)
        if any((C(:,i)&C(:,j)))&~all((C(:,i)&C(:,j))==C(:,i))
          Cp=C(:,i)|C(:,j);
          a=Fi(:,i)|Fi(:,j);
          C(:,LT+1)=Cp;
          Fi=[Fi a];
          LT=LT+1;
          NUM=NUM+1;
        end
      end
    end
    NT(i)=LT;
    NS(i)=LT-NUM;
  end
  if LT>LTEMP
    k=k+1;
    z=1;
    break
  end
end
```

Figure 5.3. MATLAB® code for calculation of circular waits

154 Manufacturing Systems Control Design

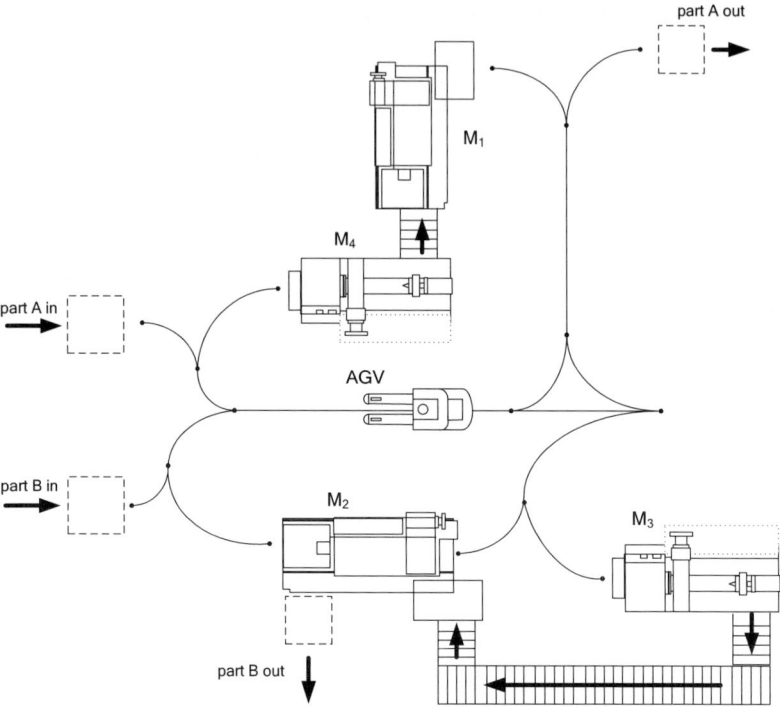

Figure 5.4. A manufacturing system from Example 5.1.1

The workcell can be described with 11 rules (since our purpose is to demonstrate circular waits calculation, herein we do not elaborate on the evaluation of the matrix model). From the system layout and parts paths we determine matrices $\mathbf{S}_r$ and $\mathbf{F}_r$, which have the following form

$$\mathbf{S}_r = \begin{bmatrix} 0 & 0 & 0 & 1 & 0 & 0 & 0 & 0 & 0 & 0 \\ 0 & 0 & 0 & 0 & 0 & 0 & 1 & 0 & 0 & 1 \\ 0 & 0 & 0 & 0 & 0 & 0 & 0 & 0 & 1 & 0 \\ 0 & 0 & 1 & 0 & 0 & 0 & 0 & 0 & 0 & 0 \\ 0 & 1 & 0 & 0 & 1 & 0 & 1 & 0 & 1 & 0 & 0 \end{bmatrix} \begin{matrix} \text{M1} \\ \text{M2} \\ \text{M3} \\ \text{M4} \\ \text{AGV} \end{matrix}, \quad \mathbf{F}_r = \begin{bmatrix} 0 & 0 & 0 & 0 & 1 \\ 0 & 0 & 0 & 1 & 0 \\ 1 & 0 & 0 & 0 & 0 \\ 0 & 0 & 0 & 0 & 1 \\ 0 & 0 & 0 & 0 & 0 \\ 0 & 0 & 0 & 0 & 1 \\ 0 & 1 & 0 & 0 & 0 \\ 0 & 0 & 0 & 0 & 1 \\ 0 & 0 & 1 & 0 & 0 \\ 0 & 1 & 0 & 0 & 0 \\ 0 & 0 & 0 & 0 & 0 \end{bmatrix}$$

According to Equation (5.1), a wait relation matrix is given as

$$\mathbf{G_W} = \begin{bmatrix} 0 & 0 & 0 & 1 & 0 \\ 0 & 0 & 1 & 0 & 1 \\ 0 & 0 & 0 & 0 & 1 \\ 0 & 0 & 0 & 0 & 1 \\ 1 & 1 & 0 & 0 & 0 \end{bmatrix} \begin{matrix} \text{M1} \\ \text{M2} \\ \text{M3} \\ \text{M4} \\ \text{AGV} \end{matrix}$$

with the corresponding wait relation graph shown in Figure 5.5.

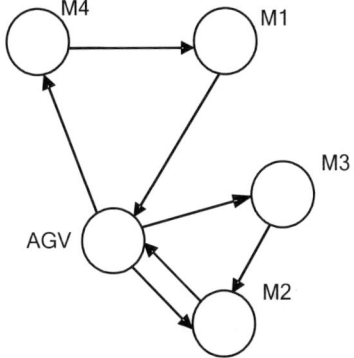

Figure 5.5. A wait relation graph of the workcell shown in Figure 5.4

Three simple CWs can be recognized from the graph: M2 → AGV, M4 → M1 → AGV, M3 → M2 → AGV. As a result of string composition we get the circular wait matrix

$$\mathbf{C} = \begin{matrix} & c_1 & c_2 & c_3 \\ & 0 & 1 & 0 \\ & 1 & 0 & 1 \\ & 0 & 0 & 1 \\ & 0 & 1 & 0 \\ & 1 & 1 & 1 \end{matrix}$$

Execution of the algorithm given in Figure 5.3 reveals all CWs in the system

$$\mathbf{C} = \begin{matrix} & c_1 & c_2 & c_3 & c_4 & c_5 \\ & 0 & 1 & 0 & 1 & 1 \\ & 1 & 0 & 1 & 1 & 1 \\ & 0 & 0 & 1 & 0 & 1 \\ & 0 & 1 & 0 & 1 & 1 \\ & 1 & 1 & 1 & 1 & 1 \end{matrix}, \quad \mathbf{\varsigma} = \begin{matrix} & c_1 & c_2 & c_3 & c_4 & c_5 \\ & 1 & 0 & 0 & 1 & 0 \\ & 0 & 1 & 0 & 1 & 1 \\ & 0 & 0 & 1 & 0 & 1 \end{matrix} \begin{matrix} c_1 \\ c_2 \\ c_3 \end{matrix}$$

From the newly obtained matrix **C** one can find that the system has five CWs, three simple and two cyclic. According to matrix $\boldsymbol{\zeta}$, $C_4=\{M1, M2, M4, AGV\}$ is composed of C_1 and C_2, while $C_5=\{M1, M2, M3, M4, AGV\}$ is composed of C_2 and C_3. Projections of circular waits onto the set of shared resources are $\mathbf{c}_{s1} = \mathbf{c}_{s3} = \mathbf{c}_{s4} = \mathbf{c}_{s5} = [0\ 1\ 0\ 0\ 1]^T$, and $\mathbf{c}_{s2} = [0\ 0\ 0\ 0\ 1]^T$.

♦

5.1.2 Resource Loops in MRF Systems

In our previous discussions it has been pointed out that inappropriate assignment of jobs could lead an MS into irregular states. A first step toward solution of the dispatching problem is determination of circular waits. In this section we analyze these structures in more detail and relate them with job set *J*.

If we consider nonshared resource r_{ns}, then it can be in one of two states, "idle" or "operational". The supervisor's task is to set the resource in one of these two states, *i.e.* from the controller point of view resource r_{ns} is a *binary loop*. The same is true for shared resource r_s, with one difference only; r_s can be in one of three or more states, as it has more than one job to perform. Again, the supervisor selects one of several shared resource states that swap inside the loop. Therefore, resource loop $L(r)$, defined in Definition 3.1.1, is an important MS structure, especially when it belongs to the resource involved in CW.

In order to find a binary vector **p**, which is the projection of the resource loop onto resource and job sets, let us recall the recursive matrix model described in Section 3.2.4, particularly Equations (3.9) and (3.10). Fulfillment of a rules change state of job vector according to $\left[\mathbf{S}_v - \mathbf{F}_v^T \right] \cdot \mathbf{x}(k)$, whereas the state of the resource vector is changed by term $\left[\mathbf{S}_r - \mathbf{F}_r^T \right] \cdot \mathbf{x}(k)$. Since MRF systems are composed of resource loops, any variation in job vector should be balanced by a corresponding change in resource vector [13], that is

$$\left[\mathbf{S}_v^T - \mathbf{F}_v \right] \cdot \mathbf{v} + \left[\mathbf{S}_r^T - \mathbf{F}_r \right] \cdot \mathbf{r} = \begin{bmatrix} \mathbf{W}_v & \mathbf{W}_r \end{bmatrix} \cdot \begin{bmatrix} \mathbf{v} \\ \mathbf{r} \end{bmatrix} = \mathbf{W} \cdot \mathbf{p} = 0 \quad (5.2)$$

or equivalently

$$\mathbf{W}_v \cdot \mathbf{v} = -\mathbf{W}_r \cdot \mathbf{r} \quad (5.3)$$

In order to construct a special left inverse of $\mathbf{W}_v$, required for solving this equation for **v**, we should modify the system matrices in the following way: delete rows of $\mathbf{F}_v$ and $\mathbf{F}_r$ and delete columns of $\mathbf{S}_v$ and $\mathbf{S}_r$ that correspond to the rules with output operations in the consequent part. Let us denote these new matrices as $\hat{\mathbf{F}}_v, \hat{\mathbf{F}}_r, \hat{\mathbf{S}}_v$ and $\hat{\mathbf{S}}_r$. Then

$$\mathbf{v} = -\hat{\mathbf{W}}_v^{-1} \cdot \hat{\mathbf{W}}_r \cdot \mathbf{r} \tag{5.4}$$

where $\hat{\mathbf{W}}_v = \hat{\mathbf{S}}_v^T - \hat{\mathbf{F}}_v$ and $\hat{\mathbf{W}}_r = \hat{\mathbf{S}}_r^T - \hat{\mathbf{F}}_r$.

Deleting rows of $\mathbf{F}_v$ and $\mathbf{S}_v$ makes matrix $\hat{\mathbf{W}}_v$ square. This is allowed, as the deleted rows of $\mathbf{W}_v$ are linear combinations of the remaining rows. One can see from the structure of $\hat{\mathbf{W}}_v$ that its inverse exists: according to our discussion on the special structure of the system matrices (causal ordering of jobs), $\hat{\mathbf{W}}_v$ is a block diagonal matrix, with each diagonal block corresponding to one part path and having a lower triangular form. Binary vectors $\mathbf{p}_i$, representing resource loops, can be obtained from Equation (5.4) for $\mathbf{r} = \mathbf{e}_i$, $i = 1, 2, \ldots, n$, where $\mathbf{e}_i$ is the the ith column of $n \times n$ identity matrix $\mathbf{I}$, and n is the number of resources. Finally, a resource-loops matrix $\mathbf{P}$, with columns formed of resource-loop vectors $\mathbf{p}_i$, is calculated as

$$\mathbf{P} = \begin{bmatrix} -(\hat{\mathbf{S}}_v^T - \hat{\mathbf{F}}_v)^{-1} \cdot (\hat{\mathbf{S}}_r^T - \hat{\mathbf{F}}_r) \\ \mathbf{I} \end{bmatrix} \tag{5.5}$$

To confirm the described procedure we consider the system shown in Figure 3.2. Its modified system matrices and corresponding $\hat{\mathbf{W}}_v$ and $\hat{\mathbf{W}}_r$ are given below

$$\hat{\mathbf{F}}_v = \begin{bmatrix} \text{MAP} & \text{RP1} & \text{BP} & \text{MBP} & \text{RP2} \\ 0 & 0 & 0 & 0 & 0 \\ 1 & 0 & 0 & 0 & 0 \\ 0 & 1 & 0 & 0 & 0 \\ 0 & 0 & 1 & 0 & 0 \\ 0 & 0 & 0 & 1 & 0 \end{bmatrix}, \quad \hat{\mathbf{F}}_r = \begin{bmatrix} \text{MA} & \text{MB} & \text{B} & \text{R} \\ 1 & 0 & 0 & 0 \\ 0 & 0 & 0 & 1 \\ 0 & 0 & 1 & 0 \\ 0 & 1 & 0 & 0 \\ 0 & 0 & 0 & 1 \end{bmatrix} \quad \hat{\mathbf{W}}_v = \begin{bmatrix} 1 & 0 & 0 & 0 & 0 \\ -1 & 1 & 0 & 0 & 0 \\ 0 & -1 & 1 & 0 & 0 \\ 0 & 0 & -1 & 1 & 0 \\ 0 & 0 & 0 & -1 & 1 \end{bmatrix}$$

$$\hat{\mathbf{S}}_v = \begin{bmatrix} 1 & 0 & 0 & 0 & 0 \\ 0 & 1 & 0 & 0 & 0 \\ 0 & 0 & 1 & 0 & 0 \\ 0 & 0 & 0 & 1 & 0 \\ 0 & 0 & 0 & 0 & 1 \end{bmatrix}, \quad \hat{\mathbf{S}}_r = \begin{bmatrix} 0 & 1 & 0 & 0 & 0 \\ 0 & 0 & 0 & 0 & 1 \\ 0 & 0 & 0 & 1 & 0 \\ 0 & 0 & 1 & 0 & 0 \end{bmatrix} \quad \hat{\mathbf{W}}_r = \begin{bmatrix} -1 & 0 & 0 & 0 \\ 1 & 0 & 0 & -1 \\ 0 & 0 & -1 & 1 \\ 0 & -1 & 1 & 0 \\ 0 & 1 & 0 & -1 \end{bmatrix}$$

By applying Equations (5.4) and (5.5) we can calculate the resource-loops matrix

$$\mathbf{v} = \begin{bmatrix} 1 & 0 & 0 & 0 \\ 0 & 0 & 0 & 1 \\ 0 & 0 & 1 & 0 \\ 0 & 1 & 0 & 0 \\ 0 & 0 & 0 & 1 \end{bmatrix} \mathbf{r} \Rightarrow \mathbf{P} = \begin{array}{c} \\ p_1 \\ p_2 \\ p_3 \\ p_4 \end{array} \begin{array}{cccccccc} \text{MAP} & \text{RP1} & \text{BP} & \text{MBP} & \text{RP2} & \text{MA} & \text{MB} & \text{B} & \text{R} \\ \begin{bmatrix} 1 & 0 & 0 & 0 & 0 & 1 & 0 & 0 & 0 \\ 0 & 0 & 0 & 1 & 0 & 0 & 1 & 0 & 0 \\ 0 & 0 & 1 & 0 & 0 & 0 & 0 & 1 & 0 \\ 0 & 1 & 0 & 0 & 1 & 0 & 0 & 0 & 1 \end{bmatrix} \end{array}^T$$

According to Figure 3.2 and the system description, the system jobs J = {MAP, RP1, BP, RP2, MBP} and resources R = {MA, MB, B, R} outline four resource loops; L(MA)={MA, MAP}, L(MB)={MB, MBP}, L(B)={B, BP} and L(R)={R, RP1, RP2}, which is confirmed by the obtained matrix **P**.

5.1.3 Siphons and Traps in MRF Systems

Circular wait is a structural property of the system. As such it is the result of a system layout design. On the other hand circular blocking is a phenomenon caused by unsuitable assignment of tasks performed by resources involved in circular wait. Now we introduce MS structures that connect circular wait and circular blocking.

Definition 5.1.4 (a siphon): A *siphon* is a set $S \subset P$ such that

$$\bullet S \subset S \bullet$$

The notion of siphon is well known in the Petri-net theory, in particular its relation with a deadlock analysis. We shall use a siphon for the same purpose, but in the context of matrix-based MS supervisory design. Having in mind that $P = R \cup J$ the above definition of a siphon emphasizes that a set of resources and/or jobs is a siphon if the set of rules in which they participate in the subsequent part is a subset of rules in which they appear in the prerequisite part.

Generally, a siphon is defined as $\bullet S \subseteq S \bullet$. This definition permits a resource loop to be a part of a siphon. In the analysis of deadlock we are concerned with a siphon in which $\bullet S$ is a strict subset of $S \bullet$ as defined in Definition 5.1.4. From now on this type of siphon is called a *critical siphon*.

A *trap* characterizes the MS structural property that is in some way the inverse of a siphon.

Definition 5.1.5 (a trap): A *trap* is a set $Q \subset P$ such that

$$Q \bullet \subset \bullet Q$$

In other words, a set of resources and/or jobs is a trap if the set of rules in which they participate in the prerequisite part is a subset of rules in which they appear in the consequent part. Definition 5.1.5 stands for a so-called *critical trap* (as in the case of a siphon, a trap is generally defined as $Q \bullet \subseteq \bullet Q$).

The importance of a critical siphon in MS analysis becomes clear when we closely look at its definition through the matrix-based formalism. First, we define a *siphon vector* **s** as $sup(\mathbf{s}) = S$, where S is assumed to be a critical siphon. Then, let us suppose that for given **s** one has $m_S(k) = \mathbf{s}^T \cdot \mathbf{m}(k) = 0$, that is, all components of system vector **m**, that correspond to resources and/or jobs belonging to critical siphon S, attain the value 0 at some instant k (it should be noted that in structural analysis the system vector represents an autonomous system, *i.e.* $\mathbf{m} = [\mathbf{v}_c \ \mathbf{r}_c]^T$). In that case rules that have those components in the prerequisite part ($S\bullet$) cannot be executed. On the other hand, since S is a critical siphon, rules that release resources and start jobs in S ($\bullet S$), according to definition, are a subset of those that have not been activated. Therefore, *once all components of system vector **m** that correspond to resources and/or jobs belonging to critical siphon S, attain the value 0 (we say that S is empty) they will remain 0 indefinitely*. This is an essential property of a critical siphon. Namely, an empty critical siphon remains empty for ever.

The question is in what way the fact that part of the system resources is not available for an indefinite period could influence other resources and jobs in the system? A deadlock situation, demonstrated in Example 3.3.1, together with the above discussion on the critical siphon, suggest that a circular wait is somehow related to the critical siphon. Their connection is additionally confirmed with the graphs shown in Figure 3.8 where the components of the system vector corresponding to resources involved in circular wait attained the value 0 once circular blocking occurred. To maintain the correlation between a CW and a critical siphon straightforwardly we shall extend some already-used notations. These refinements are needed later for the definition and development of MS structures in matrix form.

The number of idle resources in CW, C_i in sample k, is calculated as $m_{C_i}(k) = \mathbf{c}_i^T \cdot \mathbf{r}_c(k)$, *i.e.* by multiplication of a circular wait vector and an idle resource vector. The value of $m_{C_i}(k)$, called the *content of CW*, is changing in accordance with $\mathbf{r}_c(k)$, which is driven by Equation (3.12), *i.e.* by the set of rules. We identify two sets of rules related with each CW C.

Definition 5.1.6 (CW adding rules): For a given CW C a set of *CW adding rules* is defined as $X_C^+ = \bullet C \setminus C \bullet$.

Definition 5.1.7 (CW clearing rules): For a given CW C a set of *CW clearing rules* is defined as $X_C^- = C \bullet \setminus \bullet C$.

The rules that belong to $X_{C_i}^+$ increase, while rules in $X_{C_i}^-$ decrease $m_{C_i}(k)$ each time they are executed. According to the notation introduced at the beginning of Section 5.1, the preset and postset of CW C can be written in vector form as

$$\bullet C = sup(\mathbf{x}_C^{d\,\mathrm{T}}) = sup(\mathbf{c}^{\mathrm{T}} \Delta \mathbf{S}_{\mathrm{r}})$$
$$C\bullet = sup(^d\mathbf{x}_C^{\mathrm{T}}) = sup(\mathbf{c}^{\mathrm{T}} \Delta \mathbf{F}_{\mathrm{r}}^{\mathrm{T}})$$
(5.6)

By using Equation (5.6) one is able to determine a vector representation of CW adding and CW clearing rules,

$$X_C^+ = sup(\mathbf{x}_C^+) = sup(\mathbf{x}_C^d - \mathbf{x}_C^d \wedge {}^d\mathbf{x}_C)$$
$$X_C^- = sup(\mathbf{x}_C^-) = sup({}^d\mathbf{x}_C - {}^d\mathbf{x}_C \wedge \mathbf{x}_C^d)$$
(5.7)

where operation $\mathbf{a} \wedge \mathbf{b}$ represents an element-by-element logical AND operation between vectors $\mathbf{a}$ and $\mathbf{b}$. Note that for two binary vectors $\mathbf{a}$ and $\mathbf{b}$ with support sets A and B one has $A \setminus B \Rightarrow \mathbf{a} \wedge \overline{\mathbf{b}} = \mathbf{a} \wedge (1-\mathbf{b}) = \mathbf{a} - \mathbf{a} \wedge \mathbf{b}$. When there is more than one CW in the system, one has matrices $\mathbf{X}_C^+$ and $\mathbf{X}_C^-$ formed by vectors $\mathbf{x}_C^+$ and $\mathbf{x}_C^-$ as their rows.

Having defined circular wait adding and clearing rules, we continue our investigation of the correlation between a CW and a critical siphon. Let us first check if CW in MRF is a critical siphon. We have to show that $\bullet C \subset C\bullet$. It is known that for every $r_i \in C$, there exists $r_j \in C$, $i \neq j$, such that $\bullet r_i \cap r_j \bullet \neq \emptyset$. So, if $r_i \in C \cap R_{ns}$, then $|\bullet r_i| = 1$, and there exists some $r_j \in C$, $i \neq j$, such that $\bullet r_i \in \{r_j\bullet\}$. If, on the other hand $r_i \in C \cap R_s$ (in Lemma 5.1.1 we proved that each CW in MRF contains a shared resource), then $|\bullet r_i| > 1$. Hence, there exists some rule(s) $x_k \in \bullet r_i$ such that $x_k \notin \{r_j\bullet\}$ for any $r_j \in C$, $i \neq j$. In other words, there are rules that release resources in C and do not have any resource from C in prerequisite part. Therefore, $\bullet C \not\subset C\bullet$, i.e. circular wait is not a critical siphon. The other way to this conclusion follows directly from Equation (5.6). Due to the specific structure of matrices $\mathbf{S}_r$ and $\mathbf{F}_r$, imposed by the MRF system definition, some components of vector $\mathbf{x}_C^d - {}^d\mathbf{x}_C$ are positive, which means that $\bullet C \not\subset C\bullet$.

Evidently, some additional elements are needed in order to create a critical siphon around CW. That is, $Z \cup C = S$, where Z is a set of the system components in the subsequent part of rules $C\bullet$ that at the same time belong to the prerequisite part of rules that release resources in C and do not have any resource from C in the prerequisite part. Again, the specific characteristics of MRF systems help in identification of set Z elements. First, we show that set Z does not comprise any resource. Let rule $x_k \in \bullet r_i$ be such that $x_k \notin \{r_j\bullet\}$ for any $r_j \in C$. If we assume that there exists resource $r_k \in Z$ such that $\bullet r_i \cap r_k\bullet \neq \emptyset$, then according to the CW definition this resource should belong to C that contradicts the assumption that $r_k \in Z$. Therefore, set Z contains only system jobs.

Secondly, let us determine which jobs should be included in a set Z to form siphon S. For each rule $x_k \in \bullet r_i$ such that $x_k \notin \{r_j\bullet\}$ for any $r_j \in C$ we have to find a job J_k such that $x_k \in J_k\bullet$. A set of jobs that satisfies this requirement is defined below.

Definition 5.1.8 (siphon job set): For a given CW C a *siphon job set* is defined as $J_S(C) = J(C) \cap \bullet X_C^+$.

It is worth noting that all the jobs in $J_S(C)$ are performed by the shared resources contained in C. In a matrix form a siphon job set is found as a support of a siphon job vector $\mathbf{v}_{SC}$, obtained by the following relation

$$J_S(C) = sup(\mathbf{v}_{SC}^T) = sup(\mathbf{x}_C^{+T} \Delta \mathbf{F}_v) \tag{5.8}$$

A siphon job vector can be determined directly from the system matrices by including Equations (5.6) and (5.7) in Equation (5.8)

$$\mathbf{v}_{SC} = \mathbf{F}_v^T \Delta \mathbf{S}_r^T \Delta \mathbf{c}_s \wedge \overline{\mathbf{F}_v^T \Delta \mathbf{F}_r \Delta \mathbf{c}} \tag{5.9}$$

Let us take a closer look at the structure of $\mathbf{v}_{sc}$. Matrix element $S_r(i,k) = 1$ if and only if rule $x_k \in \bullet r_i$. $F_v(k,j) = 1$ if and only if $x_k \in v_j \bullet$. Therefore, $\mathbf{F}_v^T \Delta \mathbf{S}_r^T(i,j) = 1$ if and only if there exists some rule $x \in \bullet r_i \cap v_j \bullet$. Postmultiplication by $\mathbf{c}_s$ selects only the shared resources in C. Hence, $\mathbf{F}_v^T \Delta \mathbf{S}_r^T \Delta \mathbf{c}_s$ corresponds to the set $J(C) = \bigcup_{r_i \in C \cap R} J(r_i)$, i.e. the set of all jobs performed by the shared resources in C.

Matrix element $\mathbf{F}_v^T \Delta \mathbf{F}_r (i, j) = 1$ if and only if there exists some rule $x \in r_i \bullet \cap v_j \bullet$. Post-multiplication by $\mathbf{c}$ selects only resources in C. Therefore $\mathbf{F}_v^T \Delta \mathbf{F}_r \Delta \mathbf{c}$ computes jobs that participate in the prerequisite part of rules that also have resources in C as prerequisites. The element-by-element matrix "and" operation between $\mathbf{F}_v^T \Delta \mathbf{S}_r^T \Delta \mathbf{c}_s$ and negated $\mathbf{F}_v^T \Delta \mathbf{F}_r \Delta \mathbf{c}$ then selects jobs of shared resources in C that participate in the prerequisite part of rules that have no resources in C as prerequisites, namely set $J_S(C)$.

Therefore, a critical siphon of CW C is defined as

$$S_C = C \cup J_S(C) \tag{5.10}$$

or in vector form

$$\mathbf{s}_C = \begin{bmatrix} \mathbf{v}_{SC} \\ \mathbf{c} \end{bmatrix} \tag{5.11}$$

As a result, we see that each CW in the MRF system is associated with its critical siphon through the siphon job set. It is important to note from Equation

(5.11) that occupation of all resources in CW C, i.e. $m_C(k) = 0$, does not necessarily mean that the critical siphon is empty since it might happen that $\mathbf{v}_{SC}^T \cdot \mathbf{v}_c(k) \neq 0$. On the contrary, an increase of the work-in-progress, which is the main purpose of most dispatching strategies, requires to keep $m_C(k)$ close to 0 most of the time. A problem arises when jobs performed by the resources in CW are dispatched so that $\mathbf{v}_{SC}^T \cdot \mathbf{v}_c(k) = 0$ when $m_C(k)$ becomes 0. In that case the critical siphon becomes empty. Therefore, execution of the afore-mentioned CW adding and CW clearing rules should be further studied since it changes $m_{S_C}(k)$ not only by changing $m_C(k)$ but also by assigning jobs in $J(C)$. In keeping track of the $m_{S_C}(k)$ it is useful to regard each CW C as a distribution center, with $m_C(k)$ defined as its *kanban content*, and jobs in $J(C)$ as receivers of services provided by the distribution center.

To provide a deeper insight into the structure of jobs associated with CW, in forthcoming definitions a job set $J(C)$ is additionally partitioned into subsets. Each definition is followed by the corresponding relation in a vector form.

Definition 5.1.9 (trap job set): For a given CW C a *trap job set* is defined as $J_Q(C) = J(C) \cap X_C^- \bullet$.

$$J_Q(C) = sup(\mathbf{v}_{QC}^T) = sup(\mathbf{x}_C^{-T} \Delta \mathbf{S}_v^T)$$

$$\mathbf{v}_{QC} = \overline{\mathbf{F}_v^T \Delta \mathbf{F}_r \Delta \mathbf{c}_s} \wedge \overline{\mathbf{F}_v^T \Delta \mathbf{S}_r^T \Delta \mathbf{c}}$$

(5.12)

Hence, a critical trap of CW C is given as

$$Q_C = C \cup J_Q(C) \tag{5.13}$$

or in vector form

$$\mathbf{q}_C = \begin{bmatrix} \mathbf{v}_{QC} \\ \mathbf{c} \end{bmatrix} \tag{5.14}$$

Opposite to the siphon, the main property of a trap is that *once any of the components of system vector $\mathbf{m}$ that correspond to resources and/or jobs belonging to critical trap Q attain a value >0, a trap content will remain >0 indefinitely*. In other words, the trap content cannot be cleared.

Generally in MRF systems a job could belong to both a siphon job set and a trap job set. Their differentiation is made in the next three definitions.

Definition 5.1.10 (siphon-trap job set): For a given CW C a *siphon-trap job set* is defined as $J_{SQ}(C) = J_S(C) \cap J_Q(C)$.

$$J_{SQ}(C) = sup(\mathbf{v}_{SQC}) = sup(\mathbf{v}_{SC} \wedge \mathbf{v}_{QC}) \tag{5.15}$$

Definition 5.1.11 (strictly siphon job set): For a given CW C a *strictly siphon job set* is defined as $J_{0S}(C) = J_S(C) \setminus J_{SQ}(C)$.

$$J_{0S}(C) = sup(\mathbf{v}_{0SC}) = sup\left[\mathbf{v}_{SC} - (\mathbf{v}_{SC} \wedge \mathbf{v}_{SQC})\right] \tag{5.16}$$

Definition 5.1.12 (strictly trap job set): For a given CW C a *strictly trap job set* is defined as $J_{0Q}(C) = J_Q(C) \setminus J_{SQ}(C)$.

$$J_{0Q}(C) = sup(\mathbf{v}_{0QC}) = sup\left[\mathbf{v}_{QC} - (\mathbf{v}_{QC} \wedge \mathbf{v}_{SQC})\right] \tag{5.17}$$

A particularly important job set, as far as a siphon is concerned, is one that comprises all jobs whose assignment does not change $m_{S_C}(k)$.

Definition 5.1.13 (neutral job set): For a given CW C a *neutral job set* is defined as $J_N(C) = J(C) \setminus \left[J_{0S}(C) \cup J_{0Q}(C)\right]$.

$$J_N(C) = sup(\mathbf{v}_{NC}) = sup\left[\mathbf{v}_C - \left[\mathbf{v}_C \wedge (\mathbf{v}_{0QC} + \mathbf{v}_{0SC})\right]\right] \tag{5.18}$$

Definition 5.1.14 (strictly neutral job set): For a given CW C a *strictly neutral job set* is defined as $J_{0N}(C) = J_N(C) \setminus J_{SQ}(C)$.

$$J_{0N}(C) = sup(\mathbf{v}_{0NC}) = sup\left[\mathbf{v}_{NC} - (\mathbf{v}_{NC} \wedge \mathbf{v}_{SQC})\right] \tag{5.19}$$

Having partitioned jobs performed by resources in CW, one is able to determine in which way execution of CW adding and CW clearing rules change their content. However, as we showed, CW is not a siphon, hence, rules that increase or decrease the CW content do not necessarily increase or decrease the content of the associated siphon. Therefore, the other set of rules, we call them *precedent rules* and *posterior rules*, are those that need to be controlled in order to maintain the siphon content on the desired level.

Definition 5.1.15 (precedent rules): For a given CW C and associated siphon S_C a set of *precedent rules* is defined as $X_{S_C}^- = S_C \bullet \setminus \bullet S_C$.

Definition 5.1.16 (posterior rules): For a given CW C and associated siphon S_C a set of *posterior rules* is defined as $X_{S_C}^+ = \bullet S_C \setminus S_C \bullet$.

Execution of any rule that belongs to $X_{S_C}^+$ increases, while execution of $x \in X_{S_C}^-$ decreases the siphon content. Rules that do not change $m_{S_C}(k)$ are fed into a so-called set of *neutral rules* $X_{S_C}^0$. In the next section we make an observation that is significant in devising deadlock-free job-dispatching policies.

5.1.4 Critical Subsystems in MRF Systems

Manipulation with the sets defined previously gives the following relation

$$J(C) = J_{SQ}(C) \cup J_{0N}(C) \cup J_{0S}(C) \cup J_{0Q}(C) \qquad (5.20)$$

It should be noted that sets on the right-hand side of Equation (5.20) are disjoint. By using Definition 5.1.11 the above equation attains the following form

$$J(C) = J_S(C) \cup J_0(C) \qquad (5.21)$$

where $J_0(C) = J_{0N}(C) \cup J_{0Q}(C)$ is a so-called *critical subsystem*, represented in vector form as

$$J_0(C) = sup(\mathbf{v}_{0C}) = sup\left[\mathbf{v}_{0NC} + \mathbf{v}_{0QC}\right] \qquad (5.22)$$

As a critical subsystem and a siphon job set are disjoint sets, Equation (5.21) actually means that the CW content, once distributed, is held by jobs in either $J_S(C)$ or $J_0(C)$. Since

$$C \cup J(C) = \bigcup_{r \in C} L(r) \qquad (5.23)$$

by including Equations (5.10) and (5.21) in Equation (5.23) one obtains

$$S_C \cup J_0(C) = \bigcup_{r \in C} L(r) \qquad (5.24)$$

As we show later, the above equation is essential in the siphon content calculation. Also, it allows us to determine the critical subsystem directly from the critical siphon,

$$\begin{bmatrix} \mathbf{v}_{0C} \\ \mathbf{0}_n \end{bmatrix} = \mathbf{P}_{\Delta}\mathbf{c} \wedge \overline{\mathbf{s}_C} \ . \tag{5.25}$$

Operation $\mathbf{P}_{\Delta}\mathbf{c}$ computes resource loops covering the critical siphon S_C. The element-by-element "and" of this with negated critical siphon vector $\mathbf{s}_C$ translates as subtracting out from set $sup(\mathbf{P}_{\Delta}\mathbf{c})$ elements of the critical siphon S_C, yielding the set $J_0(C)$. Vector $\mathbf{0}_n$ is a null vector with the number of elements equal to the number of resources.

In order to implement efficient real-time control of an MS, we need to arrange the attained vectors in matrices. This can be done easily by positioning vectors in columns of the corresponding matrix. For example, such a critical siphon matrix $\mathbf{S}_C$ is obtained as $\mathbf{S}_C = [\mathbf{s}_{c1}\ \mathbf{s}_{c2}\ \ldots\ \mathbf{s}_{cw}]$, where $\mathbf{s}_{ci}$, $i=1,w$, are vectors corresponding to critical siphons in the system.

Also, it should be noted that the structural properties do not depend on the input and output matrices $\mathbf{F}_u$ and $\mathbf{S}_y$. Furthermore, due to the specific construction of MRF systems, all MS structures defined so far in this chapter can be determined in a different way, by using different relations.

Example 5.1.2 (critical siphons and critical subsystems in MRF)

As an example of critical siphons and critical subsystems calculation we use the system shown in Figure 5.4. Matrices $\mathbf{F}_r$ and $\mathbf{S}_r$ are already given in Example 5.1.1, here we provide $\mathbf{F}_v$ and $\mathbf{S}_v$.

$$\mathbf{F}_v = \begin{bmatrix} 0 & 0 & 0 & 0 & 0 & 0 & 0 & 0 & 0 \\ 1 & 0 & 0 & 0 & 0 & 0 & 0 & 0 & 0 \\ 0 & 1 & 0 & 0 & 0 & 0 & 0 & 0 & 0 \\ 0 & 0 & 1 & 0 & 0 & 0 & 0 & 0 & 0 \\ 0 & 0 & 0 & 1 & 0 & 0 & 0 & 0 & 0 \\ 0 & 0 & 0 & 0 & 0 & 0 & 0 & 0 & 0 \\ 0 & 0 & 0 & 0 & 1 & 0 & 0 & 0 & 0 \\ 0 & 0 & 0 & 0 & 0 & 1 & 0 & 0 & 0 \\ 0 & 0 & 0 & 0 & 0 & 0 & 1 & 0 & 0 \\ 0 & 0 & 0 & 0 & 0 & 0 & 0 & 1 & 0 \\ 0 & 0 & 0 & 0 & 0 & 0 & 0 & 0 & 1 \end{bmatrix}, \mathbf{S}_v = \begin{bmatrix} 1 & 0 & 0 & 0 & 0 & 0 & 0 & 0 & 0 & 0 & 0 \\ 0 & 1 & 0 & 0 & 0 & 0 & 0 & 0 & 0 & 0 & 0 \\ 0 & 0 & 1 & 0 & 0 & 0 & 0 & 0 & 0 & 0 & 0 \\ 0 & 0 & 0 & 1 & 0 & 0 & 0 & 0 & 0 & 0 & 0 \\ 0 & 0 & 0 & 0 & 1 & 0 & 0 & 0 & 0 & 0 & 0 \\ 0 & 0 & 0 & 0 & 0 & 0 & 1 & 0 & 0 & 0 & 0 \\ 0 & 0 & 0 & 0 & 0 & 0 & 0 & 1 & 0 & 0 & 0 \\ 0 & 0 & 0 & 0 & 0 & 0 & 0 & 0 & 1 & 0 & 0 \\ 0 & 0 & 0 & 0 & 0 & 0 & 0 & 0 & 0 & 1 & 0 \end{bmatrix}$$

Let us recall circular waits vectors $\mathbf{c}$ and $\mathbf{c}_s$, calculated in Example 5.1.1,

$$\mathbf{C} = \begin{matrix} & c_1 & c_2 & c_3 & c_4 & c_5 \\ & \begin{bmatrix} 0 & 1 & 0 & 1 & 1 \\ 1 & 0 & 1 & 1 & 1 \\ 0 & 0 & 1 & 0 & 1 \\ 0 & 1 & 0 & 1 & 1 \\ 1 & 1 & 1 & 1 & 1 \end{bmatrix} \end{matrix}, \quad \mathbf{C}_s = \begin{matrix} & c_{s1} & c_{s2} & c_{s3} & c_{s4} & c_{s5} \\ & \begin{bmatrix} 0 & 0 & 0 & 0 & 0 \\ 1 & 0 & 1 & 1 & 1 \\ 0 & 0 & 0 & 0 & 0 \\ 0 & 0 & 0 & 0 & 0 \\ 1 & 1 & 1 & 1 & 1 \end{bmatrix} \end{matrix}$$

At the beginning we determine the siphon job vectors from Equation (5.9). For C_1 we have

$$\mathbf{F}_v^T \Delta \mathbf{S}_r^T \Delta \mathbf{c}_{s1} = \begin{bmatrix} 1 & 0 & 0 & 1 & 1 & 1 & 1 & 0 & 1 \end{bmatrix}^T$$

$$\overline{\mathbf{F}_v^T \Delta \mathbf{F}_r \Delta \mathbf{c}_1} = \begin{bmatrix} 1 & 1 & 0 & 1 & 0 & 0 & 1 & 0 & 1 \end{bmatrix}^T$$

which yields

$$\mathbf{v}_{SC1} = \begin{bmatrix} 1 & 0 & 0 & 1 & 0 & 0 & 1 & 0 & 1 \end{bmatrix}^T$$

i.e. a critical siphon is $S_{C1}=\{AP1, AP2, AP4, M2P2, M2, R\}$.

For other CWs in the system the siphon job vector can be calculated as well,

$$\mathbf{v}_{SC2} = \begin{bmatrix} 0 & 0 & 0 & 1 & 1 & 0 & 1 & 0 & 0 \end{bmatrix}^T$$
$$\mathbf{v}_{SC3} = \begin{bmatrix} 1 & 0 & 0 & 1 & 0 & 0 & 0 & 0 & 1 \end{bmatrix}^T$$
$$\mathbf{v}_{SC4} = \begin{bmatrix} 0 & 0 & 0 & 1 & 0 & 0 & 1 & 0 & 1 \end{bmatrix}^T$$
$$\mathbf{v}_{SC5} = \begin{bmatrix} 0 & 0 & 0 & 1 & 0 & 0 & 0 & 0 & 1 \end{bmatrix}^T$$

As a result, the critical siphon matrix $\mathbf{S}_C$ is given below

$$\mathbf{S}_C = \begin{bmatrix} 1 & 0 & 0 & 1 & 0 & 0 & 1 & 0 & 1 & 0 & 1 & 0 & 0 & 1 \\ 0 & 0 & 0 & 1 & 1 & 0 & 1 & 0 & 0 & 1 & 0 & 0 & 1 & 1 \\ 1 & 0 & 0 & 1 & 0 & 0 & 0 & 0 & 1 & 0 & 1 & 1 & 0 & 1 \\ 0 & 0 & 0 & 1 & 0 & 0 & 1 & 0 & 1 & 1 & 1 & 0 & 1 & 1 \\ 0 & 0 & 0 & 1 & 0 & 0 & 0 & 0 & 1 & 1 & 1 & 1 & 1 & 1 \end{bmatrix}^T$$

Physical interpretation of matrix $\mathbf{S}_C$ can be done if one recalls a set of resources and a set of jobs defined in Example 5.1.1; R = {M1, M2, M3, M4, R} and J = {AP1, M4P, M1P, AP2, AP3, M2P1, AP4, M3P, M2P2}. Let us check what happens if the content of S_{C5}, which is $\mathbf{s}_{C5}^T \cdot \mathbf{m}(k)$, becomes zero. This would mean that all resources, machines and AGV, perform some operations. Since AP2 is the siphon element, AGV is occupied with either AP1 or AP3. Also, M2P2 is the siphon element, therefore, M2 performs M2P1. If we assume that AGV is occupied with AP1 (carrying part A in M4), then resources are in circular blocking since all machines are occupied and cannot be released because AGV tries to push a new part into an already full system. The assumption that AGV is occupied with AP3 (carrying part B in M2), results in the same conclusion. Other critical siphons can be checked in a similar way.

Next, we calculate critical subsystems by using Equation (5.25) (the determination of other job sets we leave to the reader for exercise). First, resource-loop matrix $\mathbf{P}$ has to be determined from Equation (5.5),

$$\mathbf{P} = \begin{bmatrix} 0 & 0 & 1 & 0 & 0 & 0 & 0 & 0 & 0 & 1 & 0 & 0 & 0 & 0 \\ 0 & 0 & 0 & 0 & 0 & 1 & 0 & 0 & 1 & 0 & 1 & 0 & 0 & 0 \\ 0 & 0 & 0 & 0 & 0 & 0 & 0 & 1 & 0 & 0 & 0 & 1 & 0 & 0 \\ 0 & 1 & 0 & 0 & 0 & 0 & 0 & 0 & 0 & 0 & 0 & 0 & 1 & 0 \\ 1 & 0 & 0 & 1 & 1 & 0 & 1 & 0 & 0 & 0 & 0 & 0 & 0 & 1 \end{bmatrix}^T$$

For C_1 one has

$$\begin{bmatrix} \mathbf{v}_{0C1} \\ \mathbf{0}_n \end{bmatrix} = \mathbf{P}\Delta\mathbf{c}_1 \wedge \overline{\mathbf{s}_{C1}} = \begin{bmatrix} 1 & 0 & 0 & 1 & 1 & 1 & 1 & 0 & 1 & 0 & 1 & 0 & 0 & 1 \end{bmatrix}^T$$

$$\wedge \begin{bmatrix} 0 & 1 & 1 & 0 & 1 & 1 & 0 & 1 & 0 & 1 & 0 & 1 & 1 & 0 \end{bmatrix}^T$$

$$= \begin{bmatrix} 0 & 0 & 0 & 0 & 1 & 1 & 0 & 0 & 0 & 0 & 0 & 0 & 0 & 0 \end{bmatrix}^T$$

which gives

$$\mathbf{v}_{0C1} = \begin{bmatrix} 0 & 0 & 0 & 0 & 1 & 1 & 0 & 0 & 0 \end{bmatrix}^T$$

Hence, a critical subsystem of CW C_1 is $J_0(C_1)$={AP3, M2P1}. Now, if we make a union of this result with S_{C1}, then

$$S_{C1} \cup J_0(C_1) = \{\text{AP1, AP2, AP4, M2P2, M2, R, AP3, M2P1}\}$$

which confirms the result specified in relation (5.24) since,

$$\bigcup_{r\in C} L(r) = L(M2) \cup L(R) = \{M2P1, M2P2, M2\} \cup \{AP1, AP2, AP3, AP4, R\}$$

For other CWs, critical subsystems have the following form,

$$\mathbf{v}_{0C2} = \begin{bmatrix} 1 & 1 & 1 & 0 & 0 & 0 & 0 & 0 & 0 \end{bmatrix}^T$$
$$\mathbf{v}_{0C3} = \begin{bmatrix} 0 & 0 & 0 & 0 & 1 & 1 & 1 & 1 & 0 \end{bmatrix}^T$$
$$\mathbf{v}_{0C4} = \begin{bmatrix} 1 & 1 & 1 & 0 & 1 & 1 & 0 & 0 & 0 \end{bmatrix}^T$$
$$\mathbf{v}_{0C5} = \begin{bmatrix} 1 & 1 & 1 & 0 & 1 & 1 & 1 & 1 & 0 \end{bmatrix}^T$$

♦

Given Equation (5.24), the precedent and posterior rules can be redefined as

$$\begin{aligned} X^-_{S_C} &= S_C \bullet \setminus \bullet S_C = \bullet J_0(C) \setminus J_0(C) \bullet \\ X^+_{S_C} &= \bullet S_C \setminus S_C \bullet = J_0(C) \bullet \setminus \bullet J_0(C) \end{aligned} \quad (5.26)$$

Clearly, an increase of a siphon content decreases the $J_0(C)$ content, and *vice versa*, a decrease of a siphon content increases the $J_0(C)$ content. The precedent and posterior rules are calculated from the system matrices as

$$\begin{aligned} X^-_{S_C} &= \sup(\mathbf{x}^-_{S_C}{}^T) = \sup(\mathbf{v}^T_{0C}\Delta\mathbf{S}_v - \mathbf{v}^T_{0C}\Delta\mathbf{S}_v \wedge \mathbf{v}^T_{0C}\Delta\mathbf{F}^T_v) \\ X^+_{S_C} &= \sup(\mathbf{x}^+_{S_C}{}^T) = \sup(\mathbf{v}^T_{0C}\Delta\mathbf{F}^T_v - \mathbf{v}^T_{0C}\Delta\mathbf{S}_v \wedge \mathbf{v}^T_{0C}\Delta\mathbf{F}^T_v) \end{aligned} \quad (5.27)$$

Now, let us formalize our discussion by definition of circular blocking in MRF systems.

Definition 5.1.17 (circular blocking): A CW C is said to be in *circular blocking* if a) $m_C(k) = 0$, and b) for each $r \in C$ if there exists $J(r)$ such that $v_{J(r)}(k) \neq 0$ (the component of the job-completed vector corresponding to $J(r)$ is not 0) then $J(r) \bullet \in C\bullet$.

The next theorem summarizes the results of analysis related to a circular blocking and its relation with an empty siphon. It is one of the main results presented herein.

Theorem 5.1.1 (circular blocking and empty siphon): Given a system of class MRF, a circular wait C is in a circular blocking if and only if the critical siphon S_C is empty.

Proof:
  Necessity:
  Let $C=\{r_1, r_2 ..., r_q\}$, with $r_1 = r_q$, be a circular wait in circular blocking, i.e. $m_C(k) = 0$, and for each $r \in C$, $\forall J(r)$ if $v_{J(r)}(k) \neq 0$ then $J(r)\bullet \in C\bullet$. Now suppose that critical siphon S_C is not empty. Then there exists $J(r) \in J_S(C)$ such that $v_{J(r)}(k) \neq 0$. By construction of S_C, $J_S(C)\bullet \not\subset C\bullet$, i.e. $J(r)\bullet \not\subset C\bullet$ and therefore C is not in circular blocking, which is a contradiction.
  Sufficiency:
  Let S_C be empty. Since S_C is a siphon, it will remain empty and therefore $m_C(k) = 0$ for any k. Obviously for any $J(r)$ with $v_{J(r)}(k) \neq 0$, it holds that $J(r) \not\in J_S(C)$. Therefore $J(r) \in J_0(C)$ and hence C is in a circular blocking.

♦

This result shows the way out of the quandary noted in MS analysis, where it was realized that an empty content of CW was not necessarily a circular blocking: as we already pointed out, in addition to checking that the CW content is empty, it is necessary to check that the content of certain special jobs is also empty.

5.1.5 Key Resources and Irregular Systems in MRF

There is a specific structural condition in MRF systems that requires extreme care in deadlock-avoidance dispatching. This condition is related to the so-called *second-level deadlock* [11]. A basis for the existence of SLD is the presence of *critical resources*, also known as bottlenecks [10] and *key resources* [9]. It should be noted that bottleneck resources are referred to as the structural bottleneck resources, not the well-known timed bottleneck resources.

Since later in this chapter we introduce a dispatching policy based on the one-step-ahead prediction, it is important to note that in irregular systems a situation may arise, which, though not a circular blocking in an immediate sense is unavoidably going to end up as one within the next few sampling intervals. Even in this situation the results presented so far hold, though a one-step-ahead deadlock-avoidance policy cannot be implemented. Therefore, before a particular dispatching policy is applied, one has to check if a given MRF system is irregular. Key resources can be identified by analyzing interconnections of CWs and their siphons, which is demonstrated in the text that follows where we use the system matrices.

To confirm the existence of key resources in the system, we must determine the presence of *cyclic circular wait* (CCW) loops. These structures specify a particular sharing among circular waits, and are a requisite for the existence of key resources. Specific structures are defined next in terms of precedent and posterior rules. In order to identify whether the system has CCW loops, let C_i and C_j be two circular waits with

$$X^+_{S_{Ci}} \cap X^-_{S_{Cj}} \neq \emptyset \text{ and } X^-_{S_{Ci}} \cap X^+_{S_{Cj}} \neq \emptyset \qquad (5.28)$$

170 Manufacturing Systems Control Design

If this is the case, then there exists CCW=$\{C_i, C_j\}$. The matrix test to find CCW among all CWs in the system is

$$\mathbf{C}_{CW} = \left(\mathbf{X}_{S_C}^{-\mathrm{T}} \Delta \mathbf{X}_{S_C}^{+}\right)^{\mathrm{T}} \wedge \left(\mathbf{X}_{S_C}^{-\mathrm{T}} \Delta \mathbf{X}_{S_C}^{+}\right) \quad (5.29)$$

where $\mathbf{X}_{S_C}^{+}$ and $\mathbf{X}_{S_C}^{-}$ are matrices formed of vectors $\mathbf{x}_{S_C}^{+}$ and $\mathbf{x}_{S_C}^{-}$, respectively.

When $\mathbf{C}_{CW} = [\mathbf{0}]$ the system is regular, otherwise an element $\mathbf{C}_{CW}(i,j)=1$ indicates that C_i and C_j form a CCW. Obviously, $\mathbf{C}_{CW}$ is a symmetric matrix. The rules that interconnect such CCWs are needed to determine key resources. We can use matrix $\mathbf{C}_{CW}$ and the precedent and posterior matrices $\mathbf{X}_{S_C}^{-}$ and $\mathbf{X}_{S_C}^{+}$ to identify such rules,

$$\begin{aligned}\mathbf{X}_{CCW}^{-} &= \left(\mathbf{X}_{S_C}^{+} \Delta \mathbf{C}_{CW}\right) \wedge \mathbf{X}_{S_C}^{-} \\ \mathbf{X}_{CCW}^{+} &= \left(\mathbf{X}_{S_C}^{-} \Delta \mathbf{C}_{CW}\right) \wedge \mathbf{X}_{S_C}^{+}\end{aligned} \quad (5.30)$$

We call them *cyclic precedent* and *cyclic posterior* rules, respectively. The set of key resources is determined as follows: let $\{C_i, C_j\}$ be a CCW such that $C_i \cap C_j = \{r_{CCW}\}$. If $X_{S_{C_i}}^{+} \cap X_{S_{C_j}}^{-} \subset r_{CCW} \bullet$ and $X_{S_{C_i}}^{-} \cap X_{S_{C_j}}^{+} \subset \bullet r_{CCW}$, then $\{C_i, C_j\}$ is said to be a *critical CCW* and if r_{CCW} is a single resource (not a resource pool), then it is called a *key resource* (structural bottleneck resource [10]). We can proceed to identify the critical resources using the following straightforward matrix formula

$$\mathbf{R}_{CCW} = \left(\mathbf{F}_r^{\mathrm{T}} \Delta \mathbf{X}_{CCW}^{+}\right) \wedge \left(\mathbf{F}_r^{\mathrm{T}} \Delta \mathbf{X}_{CCW}^{-}\right) \quad (5.31)$$

where matrix $\mathbf{R}_{CCW}$ provides, for each CW, the corresponding vector of key resources shared with other CWs in one or more CCW. If this matrix is zero, there are no key resources in the system.

5.2 Free Choice Multiple Re-entrant Flowlines – FMRF

In this section we extend multiple re-entrant flowlines structural analysis on the systems with jobs that do not have predetermined resources assigned. That is, several resources might be capable and available to perform a specific job (or operation from the set of operations needed to build a product). We call these systems *free-choice multiple re-entrant flowlines* (FMRF). As in MRF systems, dispatching policies should provide conflict- and deadlock-free activities of the system. However, systems without predeterministic routing paths are much more challenging than MRF systems and little work has been done, specifically in the study of blocking phenomena. With the exception of [14] – [16] few other deadlock-avoidance approaches for FMRF systems had been suggested.

In addition to the assumptions made at the beginning of Chapter 3, a general class of FMRF systems has the following nonrestrictive capabilities:

- Some jobs have the option of being machined in a resource from a set of resources (routing of jobs), and each resource might be used to machine different jobs (*i.e.* shared resources),
- Job/part routings are *NOT* deterministic (statement iv) in Definition 5.1.1).

For each job that can be performed by more than one resource, there exists *a material handling buffer (routing resources)* that routes parts. Its role in the FMRF systems is very important, and it is explained in the next example.

A system that satisfies the FMRF assumptions is shown in Figure 5.6. The system consists of 5 machining centers that are capable of performing tasks required to make a final product, and 12 conveyers where semiproducts are placed and then carried from machines to material handling buffers or *vice versa*. A job sequence is defined as $J = \{J_1, J_2, J_3\}$. Assignments of resources are given in Table 5.1.

Table 5.1. Resources assignmets in the system shown in Figure 5.6

	M_1	M_2	M_3	M_4	M_5
J_1	◆		◆	◆	
J_2		◆		◆	
J_3			◆		◆

As we can see, machine M_1 is assigned to job J_1, while machine M_3 is capable of performing two jobs J_1 and J_3, hence, this machine is a shared resource. What differentiates this workcell from the systems discussed so far is the fact that a particular job can be carried out by more than one resource. For example, three resources, M_1, M_3 and M_4 are able to perform job J_1. Therefore, there are many part routes that complete the required job sequence. We mention just a few of them; $M_1 \rightarrow M_2 \rightarrow M_3$, $M_1 \rightarrow M_2 \rightarrow M_5$, $M_3 \rightarrow M_2 \rightarrow M_5$, $M_4 \rightarrow M_4 \rightarrow M_5$, and so on. It is apparent that the description of all possible routes in the form of IF-THEN rules would cause rules explosion. For example, the beginning of part processing can be described with three rules:

IF B_1 holds part **AND** M_1 is ready **THEN** rule 1 is TRUE,
IF rule 1 is TRUE **THEN** start job J_1 in M_1 **AND** release B_1

IF B_1 holds part **AND** M_3 is ready **THEN** rule 2 is TRUE,
IF rule 2 is TRUE **THEN** start job J_1 in M_3 **AND** release B_1

IF B_1 holds part **AND** M_4 is ready **THEN** rule 3 is TRUE,
IF rule 3 is TRUE **THEN** start job J_1 in M_4 **AND** release B_1

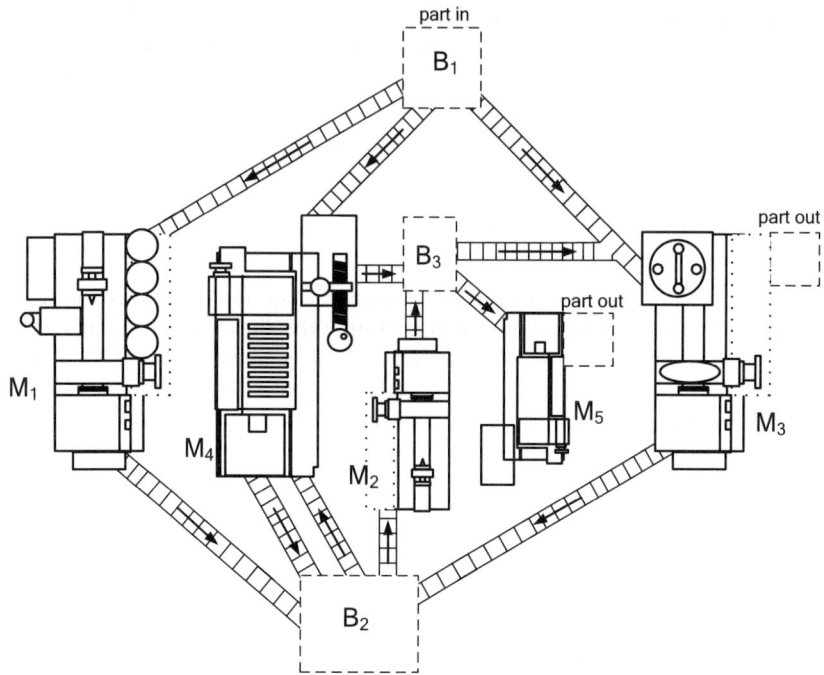

Figure 5.6. An example of a free-choice multiple re-entrant flowline

In order to prevent rule explosion material-handling buffers are included in FMRF systems as some kind of crossroads where decisions regarding part routes are made. Each job that can be performed by more than one resource has a corresponding buffer. In our example B_1, B_2 and B_3 execute the following tasks; B_1 receives row parts upon their entrance into the system and directs them to the first available machine that is able to complete job J_1 (M_1, M_3, M_4), B_2 holds parts upon completion of job J_1 and routes them to the first vacant machine for job J_2 (M_2, M_4) to be finished, and B_3 receives parts to be sent in one of the machines performing J_3 (M_3, M_5). Processed parts then leave the system.

Nondeterministic part routing has a serious impact on IF-THEN rules. Specifically, each job that can be completed by more than one resource can be started by more than one rule. For example, for J_2 there exist two such rules,

IF B_2 *holds part* **AND** M_2 *is ready* **THEN** *rule 1 is TRUE*,
IF *rule 1 is TRUE* **THEN** *start job J_2 in M_2* **AND** *release B_2*

IF B_2 *holds part* **AND** M_4 *is ready* **THEN** *rule 2 is TRUE*,
IF *rule 2 is TRUE* **THEN** *start job J_2 in M_4* **AND** *release B_2*

What is important to note is that these two rules (as well as the three rules stated previously) are in conflict, although a shared resource does not participate in their prerequisite parts. A conflict is caused by free-choice, *i.e.* when both machines, M_2

and M_4, are ready, a part that is held by B_2 "can choose" in which machine to be processed. That is why Equation (3.24) cannot be used for determination of all conflicting rules. Generally, if we denote a *set of material handling buffers* as B, then conflicting rules can be obtained as $B \bullet \cup X_d$, where X_d is a set of rules determined by Equation (3.24). Given that buffer B_2 holds more than one part, the above rules are not in conflict.

5.2.1 Structural Properties of FMRF

To be able to analyze properly FMRF systems, we needed to identify not only the resources that compose each CW, but also the rules that link them. This will give us specific information needed to locate critical siphons and critical subsystems required for the construction of the deadlock policy for FMRF systems. For instance, and related to connectivity between resources and rules, if we define (by duality of $\mathbf{G}_W$)

$$\mathbf{G}_{WX} = \left(\mathbf{F}_r \Delta \mathbf{S}_r \right)^T \tag{5.32}$$

we will get a digraph of rules. Given $\mathbf{G}_{WX}$ one can identify loops among rules by using string algebra. However, by running independently the algorithm for $\mathbf{G}_W$ and $\mathbf{G}_{WX}$ from the resulting rules CWs and resources CWs we might not be able to identify which set of rules CWs correspond to which set of resources CWs. This is why we need a general digraph wait relation matrix

$$\mathbf{G}_W = \begin{bmatrix} \mathbf{0} & \mathbf{S}_r \\ \mathbf{F}_r & \mathbf{0} \end{bmatrix} \tag{5.33}$$

which couples rules and resources. Then, if we use this digraph matrix with a string algebra algorithm to find CWs, we will get both results by obtaining circular waits of resources, denoted $\mathbf{C}_r$, and circular waits of rules, denoted $\mathbf{C}_x$, by obtaining the coupled matrix

$$\mathbf{C} = \begin{bmatrix} \mathbf{C}_r \\ \mathbf{C}_x \end{bmatrix} \tag{5.34}$$

Each ith column from $\mathbf{C}$ contains resources from the ith CW (vector $\mathbf{c}_{ri}$), which accordingly corresponds to the ith CW of rules (vector $\mathbf{c}_{xi}$ – although vector $\mathbf{c}_{xi}$ is a rule vector, we are not changing notation to $\mathbf{x}$ since CWs are denoted with the letter $\mathbf{c}$ throughout the text). The dimensions of $\mathbf{C}$ are $(n+m) \times c$, where c is the total number of CWs, n is the number of resources, and m is the number of rules. Execution of the algorithm given in Figure 5.3 calculates the final matrix $\mathbf{C}$ with the corresponding matrix ζ, thus revealing all the CWs in the system.

It should be noted that for an FMRF system having a simple circular wait C_r, which contains at least one resource $b \in B$, only one rule from $\bullet J_b$ and one rule from $J_b \bullet$ participates in its corresponding C_x, where J_b is a buffer job.

For a given matrix **C**, we can find CW C_r (resources CW) adding and clearing rules that have a slightly different form from those defined in Definitions 5.1.6. and 5.1.7,

$$X_C^+ = \bullet C_r \setminus C_x$$
$$X_C^- = C_r \bullet \setminus C_x \tag{5.35}$$

The preset and postset of CW C_r are determined by the following equations

$$\bullet C_r = sup(\mathbf{x}_{C_r}^{d\ T}) = sup(\mathbf{c}_r^T \Delta \mathbf{S}_r)$$
$$C_r \bullet = sup(^d \mathbf{x}_{C_r}^T) = sup(\mathbf{c}_r^T \Delta \mathbf{F}_r^T) \tag{5.36}$$

Now we can write a vector representation of CW adding and clearing rules,

$$X_C^+ = sup(\mathbf{x}_C^+) = sup(\mathbf{x}_C^d - \mathbf{x}_C^d \wedge \mathbf{c}_x)$$
$$X_C^- = sup(\mathbf{x}_C^-) = sup(^d \mathbf{x}_C - {}^d \mathbf{x}_C \wedge \mathbf{c}_x) \tag{5.37}$$

In MRF analysis we used only these two categories for computation of CW jobs and other structures. For FMRF systems adding and clearing rules are additionally partitioned in *neutral rules*, $X_C^N = X_C^+ \cap X_C^-$, *strictly adding rules*, $X_C^{0+} = X_C^+ \setminus X_C^N$, and *strictly clearing rules* $X_C^{0-} = X_C^- \setminus X_C^N$. In vector form they can be calculated as;

$$X_C^N = sup(\mathbf{x}_C^N) = sup(\mathbf{x}_C^+ \wedge \mathbf{x}_C^-)$$
$$X_C^{0+} = sup(\mathbf{x}_C^{0+}) = sup(\mathbf{x}_C^+ - \mathbf{x}_C^N) \tag{5.38}$$
$$X_C^{0-} = sup(\mathbf{x}_C^{0-}) = sup(\mathbf{x}_C^- - \mathbf{x}_C^N)$$

As in the standard MRF, jobs performed by resources in CW play an essential role in supervision of an FMRF system. Since the properties of all structures related to the CW (siphons, traps, critical subsystems, *etc.*) were described in previous subsections, here we skip explanations and give only final results.

A set of jobs performed by resources in CW C is defined as

$$J(C) = sup(\mathbf{v}_C^T) = sup(\mathbf{x}_{C_r}^{d\ T} \Delta \mathbf{S}_v^T) = sup(\mathbf{x}_{C_r}^{d\ T} \Delta \mathbf{F}_v) \tag{5.39}$$

Having defined $J(C)$ one is able to determine a siphon job set as

$$J_S(C) = sup(\mathbf{v}_{SC}^T) = sup\left[\mathbf{v}_C^T \wedge \overline{\mathbf{c}_x^T \Delta \mathbf{F}_v} \right] \qquad (5.40)$$

The key approach in siphon job set determination in FMRF systems is the same as in the case of MRF. That is, one needs to calculate all the jobs satisfying the existence of postset rules as adding rules of CW. However, the problem in FMRF systems is that not all jobs contain unique postset rules, due to the incorporation of the material-handling buffer set B into the system. Now, we can make two remarks: first, all rules from C_x, corresponding to resource CW C_r, are not adding rules. Secondly, all clearing rules from CW C_r have postset jobs from set $J(C)$. Therefore, by eliminating all preset jobs from C_x, and considering only those intersecting set $J(C)$, preset jobs from the adding rules set will be selected.

A trap job set in FMRF is defined as $J_Q(C) = \left(J(C) \cap X_C^{0-} \bullet \right) \backslash \left(C_x \bullet \cap \bullet C_x \right)$.
Comparing this equation with the one in Definition 5.1.9, one can notice similarity. Specifically, the trap job set, $J_Q(C)$, in FMRF contains the same elements as in MRF systems, excluding jobs J_{bi} for the case of routing resources b_i are included in C. In matrix form a trap job set is

$$J_Q(C) = sup(\mathbf{v}_{QC}^T)$$
$$= sup\left[\left(\mathbf{x}_C^{0-} \right)^T \Delta \mathbf{S}_v^T - \left(\left(\mathbf{x}_C^{0-} \right)^T \Delta \mathbf{S}_v^T \wedge \mathbf{c}_x^T \Delta \mathbf{F}_v \wedge \mathbf{c}_x^T \Delta \mathbf{S}_v^T \right) \right] \qquad (5.41)$$

The rest of the job sets, siphon-trap job set, strictly siphon job set, *etc.*, are defined equally for MRF and FMRF systems. Hence, Equations (5.15)–(5.19) can be used for their determination. Furthermore, calculation of CCWs and the regularity test remain the same as for MRF systems.

A matrix relation for a critical subsystem follows from Equations (5.21) and (5.40):

$$J_0(C) = sup(\mathbf{v}_{0C}^T) = sup(\mathbf{c}_x^T \Delta \mathbf{F}_v) \qquad (5.42)$$

Example 5.2.1 (critical siphons and critical subsystems in FMRF)

We consider an MS system described with the following matrices:

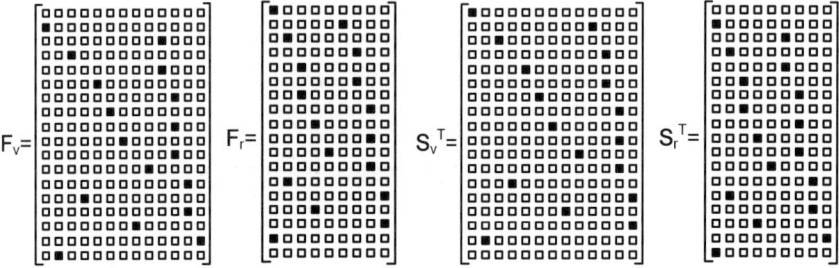

Table 5.2. Resources assignments for the system in Example 5.2.1

	M_1	M_2	M_3	M_4	M_5
J_1	♦				
J_2		♦	♦		
J_3			♦	♦	♦
J_4		♦		♦	
J_5	♦				

Resources assignments for a job sequence $J = \{J_1, J_2, J_3, J_4, J_5\}$ are given in Table 5.2. The system has 5 machines and 4 material handling buffers. A set of resources and a set of jobs is defined as $R = \{M1, M2, M3, M4, M5, B1, B2, B3, B4\}$ and $J = \{M1J1, M1J5, M2J2, M2J4, M3J2, M3J3, M4J3, M4J4, M5J3, B1P, B2P, B3P, B4P\}$. The circular wait matrix (5.34) is given as

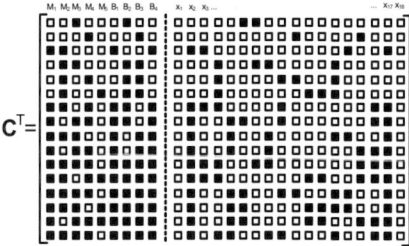

From the rows in matrix $\mathbf{C}^T$, we can observe the resources and rules that compose sixteen simple circular waits in considered FMRF system. For example, from the

first row, we can see that CW C_{r1} is composed of resources M3 and B2, with corresponding rules CW C_{x1} comprising rules x_6 and x_7.

The cyclic circular waits are given by matrix ζ,

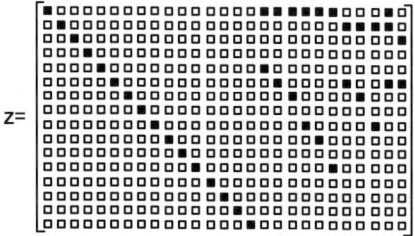

A total of twenty seven CWs is identified, consisting of sixteen simple CWs, and eleven cyclic CWs. For example, the 17th column from ζ stands for a CCW composed of the first and fifth CW from matrix **C**. It is composed of resources M3, B2, M2, M4, and B3 (resource B2 is common to both CWs).

Next, we calculate siphon job sets and critical subsystems by using Equations (5.40) and (5.42), respectively,

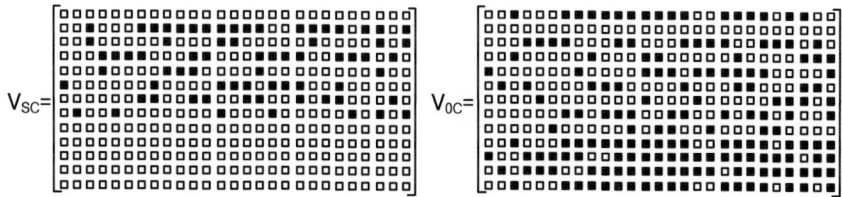

Matrices $\mathbf{V}_{SC}$ and $\mathbf{V}_{0C}$ are composed of rows corresponding with siphon job vectors and critical subsystem vectors. At the end of the example we give results attained by the regularity test presented in Section 5.1.5.

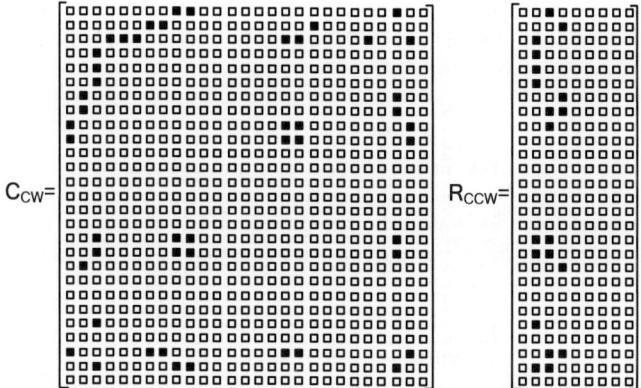

Since $\mathbf{C}_{CW}$ is not a zero matrix, we can conclude that our system is an irregular system with key resources M_2, M_3, and M_4 (see matrix $\mathbf{R}_{CCW}$).

♦

5.3 Matrix Controller Design in MRF Systems

According to Theorem 5.1.1, resources in CW get into circular blocking when the corresponding critical siphon becomes empty. Since in MRF systems circular blocking is equivalent to a deadlock, our main concern in deriving a deadlock free dispatching policy is to keep all critical siphons in the system full. However, as we mentioned in Section 5.1.3, an increase of the work-in-progress requires that system resources are busy most of the time. Balancing between these two marginal conditions, $m_{S_C}(k) > 0$ and $m_C(k) \Rightarrow 0$, is what makes a particular dispatching policy efficient. In this section we use the results of MRF structural analysis in order to devise a maximally permissive one-step look-ahead dispatching strategy that avoids deadlock in regular MRF systems. We also show how circular waits in irregular systems can be kept away from circular blocking. At the end of the section we describe a scheduling strategy for FMRF systems that is based on matrix formalism and so-called *time windows*.

5.3.1 Deadlock Avoidance in MRF Systems

At the beginning of the determination of a dynamic deadlock-free dispatching policy, let us remember that

$$\begin{bmatrix} \mathbf{W}_v & \mathbf{W}_r \end{bmatrix} \cdot \begin{bmatrix} \mathbf{v} \\ \mathbf{r} \end{bmatrix} = \mathbf{W} \cdot \mathbf{p} = 0$$

From this equation, explained in detail in Section 5.1.2, the following relation can be attained,

$$\mathbf{p}^T \cdot \mathbf{m}(k) = m_p(k) = const. \tag{5.43}$$

i.e. content of resource loop is constant. Implementation of this result on Equation (5.23) gives

$$\sum_{r \in C} \mathbf{p}_r^T \cdot \mathbf{m}(k) = const. \tag{5.44}$$

where $\mathbf{p}_r$ are resource loop vectors that correspond to resources in CW C. Since Equation (5.44) holds for any k and if we assume that all resources in C are idle for $k = 0$, then

$$\sum_{r \in C} \mathbf{p}_r^T \cdot \mathbf{m}(0) = \mathbf{c}^T \cdot \mathbf{r}_c(0) = m_C(0) \tag{5.45}$$

which finally yields,

$$\sum_{r \in C} \mathbf{p}_r^T \cdot \mathbf{m}(k) = m_C(0) \tag{5.46}$$

This result is important since it states that the content of resource loops that belong to the resources involved in CW is equal to the number of resources involved in the circular wait, which is a design parameter and it is known in advance.

Further combination of the above equation and Equation (5.24) has an even more significant outcome,

$$\mathbf{s}_C^T \cdot \mathbf{m}(k) + \mathbf{v}_{0C}^T \cdot \mathbf{v}_c(k) = \sum_{r \in C} \mathbf{p}_r^T \cdot \mathbf{m}(k) = m_C(0) \tag{5.47}$$

or in a different form,

$$m_{S_C}(k) = m_C(0) - \mathbf{v}_{0C}^T \cdot \mathbf{v}_c(k) \tag{5.48}$$

Hence, as long as

$$m_C(0) > v_{J_{0C}}(k) \tag{5.49}$$

where $v_{J_{0C}}(k) = \mathbf{v}_{0C}^T \cdot \mathbf{v}_c(k)$, a critical siphon will not be empty, i.e. $m_{S_C}(k) > 0$.

According to Equation (5.22) the content of critical subsystem, $v_{J_{0C}}(k)$, is increased by 1 each time a job that belongs to $J_{0N}(C)$ or $J_{0Q}(C)$ is dispatched. On the other hand, execution of $J_i \in J_{0N}(C)$ does not influence $m_{S_C}(k)$. In summary, the effect that jobs dispatching has on critical siphon content is such that; i) $m_{S_C}(k)$ is decreased by 1 for $J_i \in J_{0Q}(C)$, ii) $m_{S_C}(k)$ is increased by 1 for $J_i \in J_{0S}(C)$, and iii) $m_{S_C}(k)$ remains unchanged for $J_i \in J_{0N}(C)$.

The other point that should be noted is that for a given part path in the MRF system with sequential shared resources, jobs that belong to trap and neutral job sets are visited by parts before jobs from the siphon job set. As a consequence, when trap or neutral jobs are dispatched parts are pushed into the system, while execution of siphon jobs pulls parts out from the system (Figure 5.7).

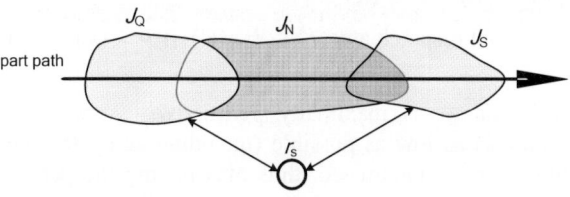

Figure 5.7. Job sets in MRF systems

The above-mentioned property of MRF systems is a basis for multipart scheduling rules that replicate two well-known strategies – FBFS and LBFS. Although both of them could lead a system into a deadlock, the next two theorems ensure stability.

Theorem 5.3.1 (stable LBFS dispatching): Given a regular MRF system, deadlock will not occur if and only if a dispatching policy is used such that:

i) whenever a multitude of jobs $\{J_i^k\}$ are activated simultaneously (conflict), they are dispatched according to the following: for every CW C such that $\{J_i^k\} \cap J(C) \neq \emptyset$ priority is given to jobs $J_i^k \in J_{0S}(C) \cup J_{0N}(C)$, and

ii) does not dispatch any $J_i^k \in J_{0Q}(C)$ if $v_{J_{0C}}(k) = m_C(0) - 1$.

The first part of the theorem handles conflict situations in the way that jobs that pull parts out from the system are preferred. This corresponds to the LBFS strategy. The second part of the theorem ensures a deadlock-free behavior of the system by disallowing execution of jobs that decrease $m_{S_C}(k)$ when the critical subsystem content is on the lower limit. A practical implementation of this part of the theorem requires slight modification due to the existence of the hidden parts (see Section 3.4). According to the theorem, the supervisor has to track the contents of all critical subsystems, but information obtained from sensors in sampling interval k (vector $\mathbf{v}_c(k)$) could come too late for appropriate actions. This is why, as described in Chapter 3, instead of $\mathbf{v}_c(k)$, vector $\mathbf{v}_c^s(k)$, obtained from Equation (3.21), should be used in the determination of $v_{J_{0C}}(k)$.

The deadlock-free dispatching policy stated next defines the generalized kanban strategy.

Theorem 5.3.2 (stable FBFS dispatching): Given a regular MRF system, deadlock will not occur if and only if a dispatching policy is used such that:

i) whenever a multitude of jobs $\{J_i^k\}$ are activated simultaneously (conflict), they are dispatched according to the following: for every CW C such that $\{J_i^k\} \cap J(C) \neq \emptyset$ priority is given to jobs $J_i^k \in J_{0Q}(C) \cup J_{0N}(C)$, and

ii) does not dispatch any $J_i^k \in J_{0Q}(C)$ if $v_{J_{0C}}(k) = m_C(0) - 1$.

This control strategy is maximally permissive. Moreover, by keeping the kanban content $m_C(k)$ as low as possible (including zero), the work-in-process in the critical subsystem is maximized, thus maximizing the per cent utilization of resources. The difference between the standard FBFS and the one introduced in

Theorem 5.3.2 is that exploitation of the standard policy forces parts into the system all the time, while the stable FBFS drives parts forward as long as a particular part of the system is full, and then starts to pull parts out from the system.

The supervisor, which dispatches jobs according to strategies given in the above theorems, can be realized in the form of the control vector, whose components are incorporated in the logical state vector equation through matrix $\mathbf{F}_d$, introduced in Section 3.4 and calculated from the conflict-rules vector according to Equation (3.25). It can be shown that $\bullet \mathbf{v}_{0QC} \subset \mathbf{x}_d$ and $\bullet \mathbf{v}_{0SC} \subset \mathbf{x}_d$. However, depending on the system structure, $\bullet \mathbf{v}_{0NC} \subset \mathbf{x}_d$ is not necessarily true, thus, $\bullet \mathbf{v}_{0NC}$ should be added to the conflict-rules vector. Regarding the control vector, it follows from the theorems that for the online deadlock-avoidance implementation only particular parts of the system, namely jobs, are important. Hence, the control vector (3.22) can be determined from

$$\mathbf{u}_d(k) = h\left(\mathbf{m}(k), \mathbf{v}_c^s(k)\right)$$

It is evident that both strategies, stable LBFS and stable FBFS, give the same result when no conflict occurs in the system. In that case only a situation described with the second rule, common to both theorems, could happen, which is presented in Figure 3.12 (Example 3.4.1), where the results obtained with the dispatching that is equivalent to the stable LBFS policy, are shown. Specifically, the critical subsystem from this example is $J_0(C)=\{RP1, BP, MBP\}$, with $m_C(0) = 4$. The dispatching proposed in the example takes actions exactly according to Theorem 5.3.2.

Results attained with stable FBFS and stable LBFS policies are depicted in Figures 5.8 and 5.9, respectively. As may be seen, the throughput of the system remained unchanged but resource utilization is improved in the case of FBFS dispatching. When LBFS is used the buffer never reaches its full capacity (2 parts). The system is stable in both cases.

5.3.2 Deadlock Avoidance in Irregular Systems

The dispatching strategies given in Theorems 5.3.1 and 5.3.2 can be implemented in irregular systems as well, with an additional verification that is stated in the next theorem.

Theorem 5.3.3 (stable dispatching in irregular system): Given an irregular MRF system, with C_1 and C_2 forming a CCW with a key resource, then a deadlock will not occur if and only if the last idle resource in CCW is not a key resource.

Evidently, one-step look-ahead control strategies, exemplified in the previous section, cannot cope with the condition illustrated in the theorem. There are two reasons for this. First, the supervisor that implements dispatching according to Theorem 5.3.3 should track the number of available resources in CCW, and

182   Manufacturing Systems Control Design

secondly, if two resources in CCW are idle then in the worst case $v_{J_{0CCW}}(k) = m_{CCW}(0) - 2$, hence, activation of any $J(CCW)$ is allowed.

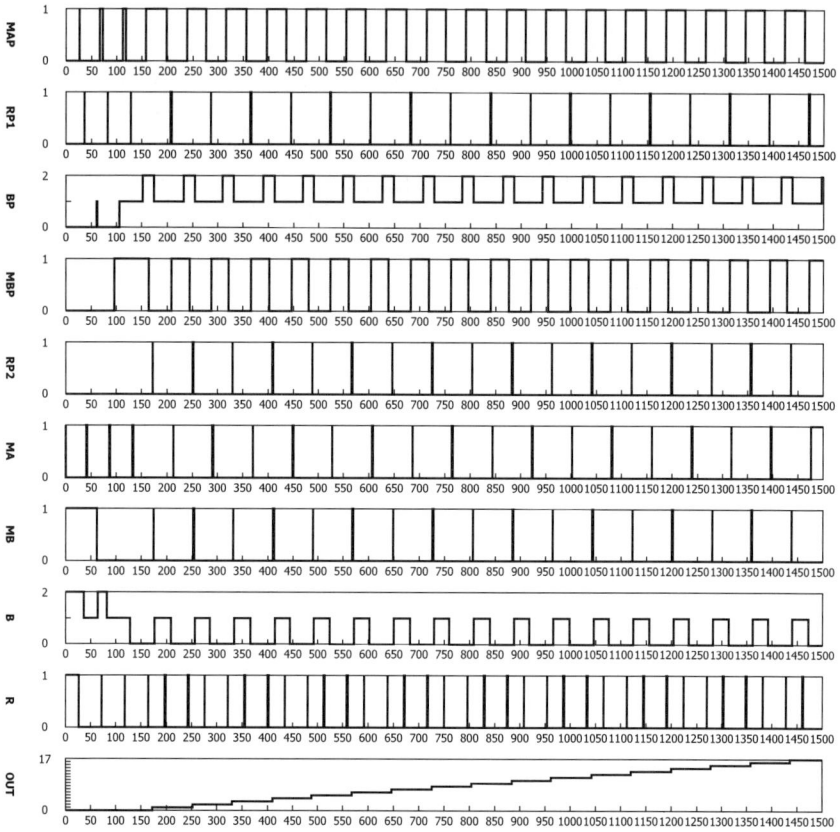

Figure 5.8. Response of the system from Example 3.4.1 with stable FBFS dispatching

We illustrate the occurrence of the second-level deadlock in the following example. Let us consider a CCW that is composed of two CWs, $C_1 = \{ r_a, r_k, r_c, \ldots \}$ and $C_2 = \{ r_d, r_k, r_b, \ldots \}$, as shown in Figure 5.10. Further, assume that resources in the CCW are related by the following equations, $r_k \bullet \cap \bullet r_a = x_1$, $r_k \bullet \cap \bullet r_d = x_2$, $r_b \bullet \cap \bullet r_k = x_3$, $r_c \bullet \cap \bullet r_k = x_4$. Now, let us suppose that the two remaining idle resources are r_c and r_k, and the prerequisites for rules x_1, x_2 and $x_m \in r_c \bullet$ are met. This is a situation in which not only conditions related to the content of critical subsystems of all three CWs should be checked, but also the condition regarding the number of idle resource should be taken into consideration.

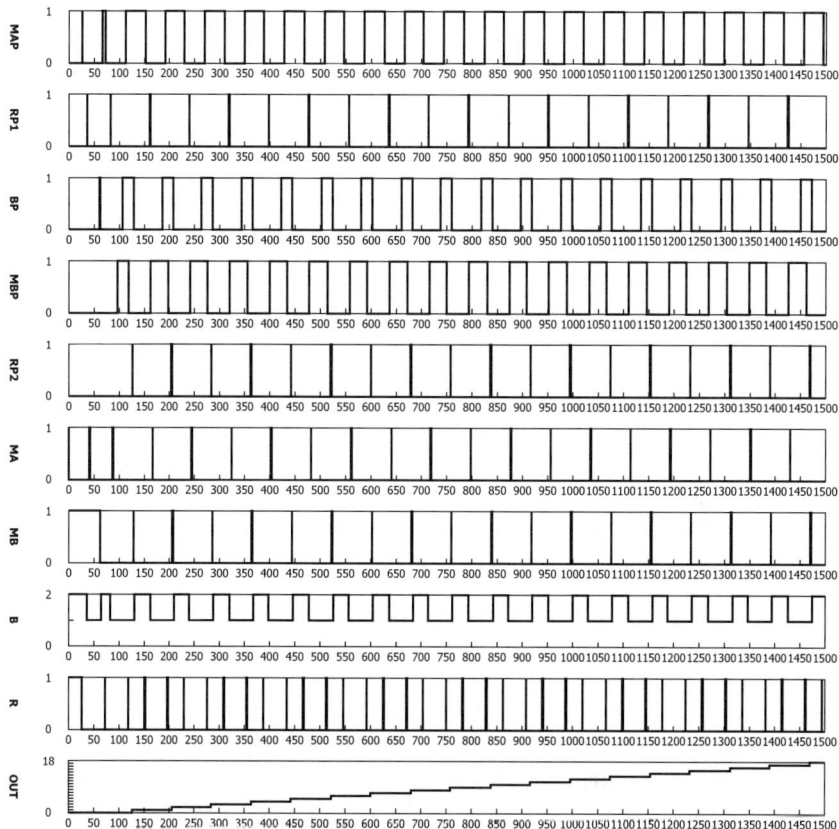

Figure 5.9. Response of the system from Example 3.4.1 with stable LBFS dispatching

Three decisions could be made; i) execution of rule x_1 – in this case the key resource becomes occupied but resource r_a is released. Although two resources in CCW, r_a and r_c, remain idle, the critical siphon S_{C2} is empty, thus, x_1 should not be executed due to the condition $v_{J_{0C2}}(k) < m_{C_2}(0)$, ii) execution of rule x_2 – in this case the key resource becomes occupied and resource r_d is released. Now, C_1 has idle resource r_c and C_2 has available resource r_d. Since both of them belong to CCW, it has idle resources too. Therefore, conditions $v_{J_{0Ci}}(k) < m_{C_i}(0)$ are not violated. Furthermore, the key resource is not the last available resource in CCW, iii) execution of rule x_m – in this case the key resource remains the last idle resource in CCW while preconditions for both rules, x_1 and x_2, are still satisfied. However, execution of any of them would make both critical siphons, S_{C1} and S_{C2} empty. This is known as the second-level deadlock, that is, the consequences of a decision regarding job activation (rule x_m) are becoming evident two sampling intervals later, when it is too late to correct an already-completed action.

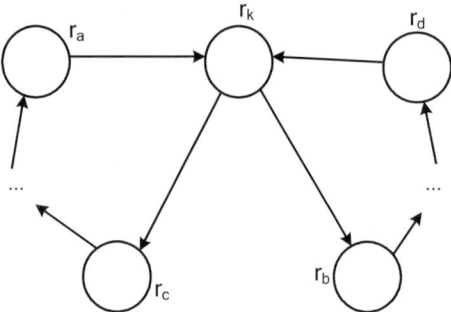

Figure 5.10. Two CWs in a CCW

Then, an obvious solution, suggested in Theorem 5.3.3, is to deny having the key resource as the last available resource for any critical CW (and any CW). We should always give priority to usage of the key resource (in our demonstration that corresponds to execution of rule x_2) over all other resources available in the critical CW (in our case execution of rule x_m), while at the same time the conditions stated in Theorems 5.3.1 and 5.3.2 should be checked.

5.3.3 Deadlock Avoidance in FMRF Systems

The dispatching strategies based on the content of critical subsystems and the number of available resources in CCW could be used in FMRF as well. In Section 5.2.1 it has been shown how to determine the structural properties of a free-choice MRF system that are required for successful deadlock avoidance. However, the existence of alternative part paths, we call them *resource sequences*, that execute the same job sequence, opens space for implementation of more sophisticated routing and dispatching strategies. The supervisory control techniques proposed so far in the book suggested solutions that were *static*. That is, once proposed a resource sequence (RS) remained unchanged until the part has left the system. A method described in the text that follows offers a possibility for *dynamic* change of the resource sequence [17], *i.e.* depending on the *priority*, the resource sequence can be changed before the part reaches the output of the system.

Based on the string composition the proposed method finds the candidate resource sequence by matrix composition and then *time windows* are used for checking if the determined resource sequence is feasible. The viability of a particular RS is evaluated by time-windows insertion that is followed by a *time-windows overlap* (conflict) test. In the case of an overlap, the algorithm iteratively reinserts time windows until there are no overlaps or the overlap is present on the first resource visited by the part, which means that the candidate RS is not feasible. The procedure is repeated for all candidate RSs. As a result, the set of executable sequences is formed and the final task of the algorithm is to choose the optimal one in terms of the time required for the part to get from the input to the output of the system.

Introduction of dynamic scheduling into the FMRF systems has many advantages: increase of system throughput, reduction of operational costs, consistent execution of predetermined tasks, *etc*. However, this requires superior control strategies that are able to solve problems such as a conflict and a deadlock. Usually, routing and scheduling algorithms should be executed online in a very short time, which is challenging since the problem is NP-hard.

Various methods for dynamic routing and scheduling, especially in autonomous guided vehicle (AGV) systems, are currently in use [18] – [22], and still researchers are working on new methods in their quest for faster and computationally efficient algorithms. Generally, two approaches can be recognized in the literature: static routing and dynamic routing. While the first approach is concerned only with the spatial dimension of the routing problem (determination of resource sequences in the space domain), the second approach perceives routing as a time-space problem (determination of part paths feasible in space and time). In some cases the time-space approach could be seen as static routing. Since the goal of routing is to find the optimal sequence, many algorithms are based on Dijkstra's shortest-path algorithm [23].

Since the number of processed parts changes with time, by elongation of the time windows the proposed method assures that the shortest RS becomes feasible, thus providing collision-free and deadlock-free paths for all processed parts.

Usually, when it comes to the mathematical analysis and control algorithm design, a manufacturing shop-floor layout is represented by a graph. In the approach described herein, the start and stop of a particular operation are represented by nodes n_i, while part processing is represented by a weighted arc, a_i. Since we are concerned with dynamic scheduling based on time windows, it is natural to choose the processing time as a variable that represents the arc (resource) weight. A nominal weight (minimal processing time) of resource r_j for part p_i is denoted $\hat{w}_{ij}$.

A graph adjacency matrix is typically defined with respect to nodes. Herein we define an *arc adjacency matrix* since time windows are associated with arcs. Having a directed graph, $G = (N, A)$, an *arc adjacency matrix* **G** is defined as a matrix with the number of rows and columns equal to the number of arcs in G, with element g_{ij} equal to 1 if arc a_i is upstream of arc a_j, otherwise it is 0.

In FMRF systems a required job sequence J can be executed by several resource sequences. Such resource sequences, for example, {M_1, M_2, M_3, M_2, M_1} and {M_1, M_2, M_3, M_4, M_1} complete the required job sequence J = {J_1, J_2, J_3, J_4, J_5} in Example 5.2.1. A set of active resource sequences is defined as $\Pi_a = \{\pi_i : \pi_i \in \Pi\}$, where Π is a set of all possible resource sequences that execute the required job sequence.

An RS π_i is defined in the following way: $\pi_i = \left(o_i, d_i, {}^i\hat{\sigma}, P_{i0}, p_i\right)$, where o_i is the first resource (an origin arc) and d_i is the last resource (a destination arc) visited by the part on the sequence π_i, ${}^i\hat{\sigma}$ is the shortest RS (in the sense of processing time) between the first and the last resource, P_{i0} is the initial priority of the sequence (a sequence with the highest priority has the lowest value of P_{i0}), and p_i is a part processed by the sequence π_i. On its route from the origin to the destination, a part visits a set of resources represented by arcs, $\sigma = \{r_j : r_j \in R\}$. The *weight of the path* σ is equal to the release time of the last resource of the sequence, i.e. $W(\sigma) = {}^{out}t_{d_\sigma}$. A set of all RSs that connect origin arc and destination arc of sequence π_i is $\Sigma_i = \{{}^i\sigma_1, {}^i\sigma_2 ... {}^i\sigma_q\}$.

Given that in the case of dynamic routing the path ${}^i\hat{\sigma}$, as well as the RS priority, can be changed during mission execution, a mission is defined in the following way:

$$\pi_i(t) = \left(o_i, d_i, {}^i\hat{\sigma}(t), P_i(t), p_i\right) \tag{5.50}$$

The mission priority $P_i(t)$ is calculated according to the relation:

$$P_i(t) = \begin{cases} \min\left[\dfrac{t_{di} - t}{W({}^i\hat{\sigma}) - t}, P_{i0}\right] & \text{for } W({}^i\hat{\sigma}) \neq \infty \\ -\infty & \text{for } W({}^i\hat{\sigma}) = \infty \end{cases} \tag{5.51}$$

where t_{di} is due time of mission π_i.

Determined in this way, the priority of the RS with the part that is far from its destination is higher than the priority of the RS that has the part already close to its goal. In addition, the RS whose due time is close to expiration has a higher priority than the RS that has enough time to meet its due time. Initial RS' priorities, assigned by the dispatching controller, are recalculated each time the request for a new part processing arrives or current sequences become unviable. In this way the influence of livelock is reduced as a sequence with low initial priorities would not wait in a queue indefinitely. Care should be taken since more than one RS might have priority $-\infty$. In that case, the priorities of the sequences could be arranged according to FIFO.

We assume that a part can reside only in resources (arcs). A part occupies a particular resource for some time (we suppose that only one part at a time is allowed to be processed by the resource). This time is called a *time window*, defined as

$$w_{ij} = {}^{out}t_{ij} - {}^{in}t_{ij}, \quad w_{ij} \geq \hat{w}_{ij} \tag{5.52}$$

where w_{ij} is a time window of part p_i in resource r_j, ${}^{out}t_{ij}$ is the release time of resource r_j from part p_i, and ${}^{in}t_{ij}$ is an entry time of p_i in resource r_j. Time windows, as well as release times and entry times of resource r_j, can be represented in the form of *time vectors*:

$$\mathbf{w}_j = \begin{bmatrix} w_{ij} \end{bmatrix}, \quad {}^{in}\mathbf{t}_j = \begin{bmatrix} {}^{in}t_{ij} \end{bmatrix}, \quad {}^{out}\mathbf{t}_j = \begin{bmatrix} {}^{out}t_{ij} \end{bmatrix} \tag{5.53}$$

where the 1st component corresponds with the highest priority RS, the nth component with the lowest priority RS and $n = |\Pi_a|$, i.e. the dimension of all three vectors is equal to the number of active RSs. Dimension n varies with time, since the number of active resource sequences is changing dynamically. Also, it should be noted that a part may visit a particular resource two or more times, hence, more than one component of a time vector would correspond to the same sequence, i.e. $n \neq |\Pi_a|$. In that case index ivj corresponds to the vth time window of sequence π_i on resource r_j. The components of vector $\mathbf{w}_j$, that correspond to active sequences that do not use resource j, are set to zero, while the components of vectors ${}^{in}\mathbf{t}_j$ and ${}^{out}\mathbf{t}_j$ that correspond to those sequences are set to ∞.

From time vectors defined as in Equation (5.52) we know which RS visit which resources but we are not able to tell, directly, in which order. For the purpose of the time-window insertion, which is elaborated in more detail later in the text, we have to position components of time vectors in chronological order. Vector $\mathbf{x} = [x_i]$ can be converted into *sorted vector* $\langle \mathbf{x} \rangle = [x_i]$, where $\langle x \rangle_i = x_i \leq \langle x \rangle_{i+1} = x_{i+1}$.

The concept of time windows is shown in Figure 5.11. In the example, sequences π_1 and π_2 have the highest and the lowest priorities, respectively. Time vectors of a given resource a are

$$\mathbf{w}_a = \begin{bmatrix} w_{1a} & w_{7a} & w_{3a} & 0 & 0 & 0 & w_{2a} \end{bmatrix}^T$$

$${}^{in}\mathbf{t}_a = \begin{bmatrix} {}^{in}t_{1a} & {}^{in}t_{7a} & {}^{in}t_{3a} & \infty & \infty & \infty & {}^{in}t_{2a} \end{bmatrix}^T$$

$${}^{out}\mathbf{t}_a = \begin{bmatrix} {}^{out}t_{1a} & {}^{out}t_{7a} & {}^{out}t_{3a} & \infty & \infty & \infty & {}^{out}t_{2a} \end{bmatrix}^T$$

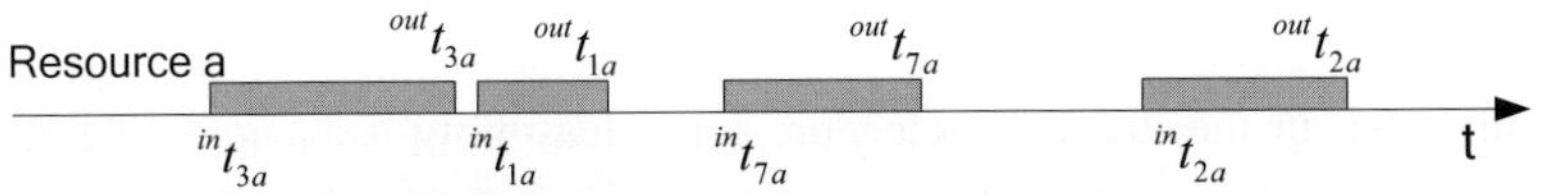

Figure 5.11. The concept of time windows

It should be noted that, although 7 sequences are active, only four of them are using resource a. Sorted time vectors for resource a are written as

$$\langle \mathbf{w}_a \rangle = \begin{bmatrix} w_{3a} & w_{1a} & w_{7a} & w_{2a} & 0 & 0 & 0 \end{bmatrix}^T$$

$$\langle {}^{in}\mathbf{t}_a \rangle = \begin{bmatrix} {}^{in}t_{3a} & {}^{in}t_{1a} & {}^{in}t_{7a} & {}^{in}t_{2a} & \infty & \infty & \infty \end{bmatrix}^T$$

$$\langle {}^{out}\mathbf{t}_a \rangle = \begin{bmatrix} {}^{out}t_{3a} & {}^{out}t_{1a} & {}^{out}t_{7a} & {}^{out}t_{2a} & \infty & \infty & \infty \end{bmatrix}^T$$

When a new part arrives into the system at moment t_m, a supervisor assigns an idle resource, o_m, as the origin of a new resource sequence π_m, which has initial priority P_{m0}. Then, the shortest path for sequence π_m is determined by calculation of powers of vector Σ_m – the row of string matrix $\mathbf{S}$ that corresponds with the origin resource o_m,

$$\Sigma_m^p = \tilde{\Sigma}_m^{p-1} \bullet \mathbf{S} \tag{5.54}$$

The string matrix $\mathbf{S}$ is formed as described in Chapter 4. Having vector Σ_m^p, the weight of each sequence σ_i^p, represented by a string in Σ_m^p, has to be determined and then vector $\tilde{\Sigma}_m^p$ is formed in the following way: if there exists a pth order sequence that connects o_m and d_m, then

$$W(^m\hat{\sigma}^p) = \min_i \left[W(^m\sigma_i^p) \right] \tag{5.55}$$

Furthermore, if $W(^m\hat{\sigma}^p) < W(^m\hat{\sigma})$ then the string that stood for $^m\hat{\sigma}$ is replaced by the string representing $^m\hat{\sigma}^p$. When $W(\sigma_i^p) \geq W(^m\hat{\sigma})$, sequence σ_i^p in vector $\tilde{\Sigma}_m^p$ is replaced by a null string, otherwise the sequence remains the component of the vector. Initially, when a new resource sequence π_m is requested, $^m\hat{\sigma} = \{\emptyset\}$ and $W(^m\hat{\sigma}) = \infty$. Since the weight of the sequence is equal to the release time of the sequence's destination resource, in the following text we describe in detail how the feasibility of sequences and their release times are determined.

A.   *Initialization of time vectors* (Step 1)

The first step in the iterative procedure for a feasibility test and a release time determination of sequences in Σ_m^p, is an initialization of time vectors. Let us choose a candidate sequence $\sigma_i^p \in sup(\Sigma_m^p)$. For each resource $r_j \in \sigma_i^p$ its time vectors are initialized as

$$\mathbf{w}_j = \begin{bmatrix} w_{1j} & w_{2j} & \cdots & \hat{w}_{mj} & 0 & 0 \end{bmatrix}^T$$

$$^{in}\mathbf{t}_j = \begin{bmatrix} ^{in}t_{1j} & ^{in}t_{2j} & \cdots & ^{in}t_{mj} & \infty & \infty \end{bmatrix}^T$$

$$^{out}\mathbf{t}_j = \begin{bmatrix} ^{out}t_{1j} & ^{out}t_{2j} & \cdots & ^{out}t_{mj} & \infty & \infty \end{bmatrix}^T$$

During the process of initialization, components of $^{out}\mathbf{t}_j$ whose values are less than t_m, are set to ∞ (as well as their counterparts in $^{in}\mathbf{t}_j$), since they correspond to sequences that occupied resource r_j prior to the moment a new sequence was requested, hence they do not influence the time-windows settings. When $^{in}t_{ij} \leq t_m$ and $^{out}t_{ij} > t_m$ a part p_i occupies resource r_j at the moment of request t_m and these components of time vectors remain unchanged. Components of vector $\mathbf{w}_j$, which belong to the sequences with lower priorities than sequence π_m, are set to 0. At the same time all components of vectors $^{in}\mathbf{t}_j$ and $^{out}\mathbf{t}_j$ that correspond to these sequences are set to ∞. In this way the time windows of RSs with lower priorities are excluded from consideration, which means that resource is freed for a new mission. Components that belong to RS π_m, $^{in}t_{mj}$ and $^{out}t_{mj}$, are unknown values that have to be determined by dynamic routing.

It is assumed that the part p_m, which is processed by the new RS, occupies o_m at the moment of entrance. Therefore the entry time of o_m is set to be equal to the part-arrival time t_m. A release time of resource o_m depends on the average processing time $\tilde{w}_{mo_m}$. Accordingly, for the origin resource we set

$$^{in}\mathbf{t}_{o_m} = \begin{bmatrix} ^{in}t_{1o_m} & ^{in}t_{2o_m} & \ldots & t_m & \infty & \infty \end{bmatrix}^T$$

$$^{out}\mathbf{t}_{o_m} = \begin{bmatrix} ^{out}t_{1o_m} & ^{out}t_{2o_m} & \ldots & t_m + \tilde{w}_{mo_m} & \infty & \infty \end{bmatrix}^T$$

B. Insertion of time windows (Step 2)

Having time windows of all resources that belong to the candidate sequence initialized, starting from the second resource of the sequence, we are looking on each resource $r_j \in sup(\sigma_i^p)$ for the first available time window that fulfils two requirements: a) it is wide enough to accommodate part p_m for a predetermined period, and b) its entry time $^{in}t_{mj}$ is set after the release time of the upstream resource $^{out}t_{mi}$, $i \to j$.

When

$$\left[\langle ^{in}t_j \rangle_1 - t_m \right] > \hat{w}_{mj} + \varepsilon_{mj} \text{ and } \left[\langle ^{in}t_j \rangle_1 - (\hat{w}_{mj} + \varepsilon_{mj}) \right] > {}^{out}t_{mi} \text{ for } i \to j \qquad (5.56)$$

then

$$^{in}t_{mj} = {}^{out}t_{mi}$$
$$^{out}t_{mj} = {}^{in}t_{mj} + \hat{w}_{mj}$$
(5.57)

otherwise the index of the first available time window is determined by

$$p = \arg\min_{\ell}\left\{\left\langle {}^{in}t_j\right\rangle_{\ell}:\right.$$
$$\left[\left\langle {}^{in}t_j\right\rangle_{\ell+1} - \max\left(\left\langle {}^{out}t_j\right\rangle_{\ell}, {}^{out}t_{mi}\right)\right] > \hat{w}_{mj} + 2\varepsilon_{mj}, \ i \to j, \ \ell = 1, \ n-1\right\}$$
(5.58)

where n is the number of time-vector components that are $\neq \infty$ and ε_{mj} is a *safety processing time* of part p_m in resource r_j. The safety time depends on the processing time uncertainty. Its value is usually 1–5% of $\hat{w}_{mj}$.

Once p is determined, the entry and release times of part p_m on resource r_j are calculated as

$$^{in}t_{mj} = \max\left(\left\langle {}^{out}t_j\right\rangle_{p-1} + \varepsilon_{mj}, \ {}^{out}t_{mi}\right)$$
$$^{out}t_{mj} = {}^{in}t_{mj} + \hat{w}_{mj}$$
(5.59)

It may happen that the time-windows distribution on resource r_j is so dense that $^{in}t_{mj}$ cannot be determined, *i.e.* none of the relations in Equations (5.57) and (5.59) give an answer as to where to insert a time window for a new mission. In that case a new time window is set after the last time window on resource r_j, i.e.

$$^{in}t_{mj} = \left\langle {}^{out}t_j\right\rangle_n + \varepsilon_{mj}$$
$$^{out}t_{mj} = {}^{in}t_{mj} + \hat{w}_{mj}$$
(5.60)

An example of time-windows insertion is shown in Figure 5.12. At the moment t_m a new sequence is requested. The first resource to process a part is resource c, *i.e.* resource c becomes an origin resource of sequence π_m. Let one of the candidate sequences, obtained by the string composition, be $\sigma_1^2 = \{c, b, a\}$.

First, the initial values of the sorted time vectors are determined according to the initialization procedure (Step 1):

Manufacturing Systems Structural Properties in Matrix Form   191

$$\langle \mathbf{w_a} \rangle = [w_{7a} \quad w_{2a} \quad 0 \quad 0 \quad 0 \quad 0 \quad 0]^T$$

$$\langle {}^{in}\mathbf{t_a} \rangle = [{}^{in}t_{7a} \quad {}^{in}t_{2a} \quad \infty \quad \infty \quad \infty \quad \infty \quad \infty]^T$$

$$\langle {}^{out}\mathbf{t_a} \rangle = [{}^{out}t_{7a} \quad {}^{out}t_{2a} \quad \infty \quad \infty \quad \infty \quad \infty \quad \infty]^T$$

$$\langle \mathbf{w_b} \rangle = [w_{2b} \quad w_{4b} \quad 0 \quad 0 \quad 0 \quad 0 \quad 0]^T$$

$$\langle {}^{in}\mathbf{t_b} \rangle = [{}^{in}t_{2b} \quad {}^{in}t_{4b} \quad \infty \quad \infty \quad \infty \quad \infty \quad \infty]^T$$

$$\langle {}^{out}\mathbf{t_b} \rangle = [{}^{out}t_{2b} \quad {}^{out}t_{4b} \quad \infty \quad \infty \quad \infty \quad \infty \quad \infty]^T$$

$$\langle \mathbf{w_c} \rangle = [w_{7c} \quad 0 \quad 0 \quad 0 \quad 0 \quad 0 \quad 0]^T$$

$$\langle {}^{in}\mathbf{t_c} \rangle = [{}^{in}t_{7c} \quad \infty \quad \infty \quad \infty \quad \infty \quad \infty \quad \infty]^T$$

$$\langle {}^{out}\mathbf{t_c} \rangle = [{}^{out}t_{7c} \quad \infty \quad \infty \quad \infty \quad \infty \quad \infty \quad \infty]^T$$

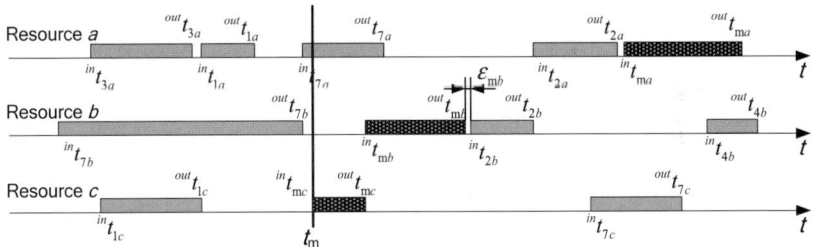

Figure 5.12. Time-windows insertion

We can see that the components of vectors $\langle {}^{in}\mathbf{t_a} \rangle$ and $\langle {}^{out}\mathbf{t_a} \rangle$ that correspond to RSs π_1 and π_3 are set to ∞ since the part processed by those two sequences have occupied resource a prior to the request for π_m. In the same way, missions π_7 and π_1 are removed from the time vectors of resources b and c, respectively. The components of RS π_7, that occupied resource a at the moment of request, remain unchanged.

Having initialized vectors, we can start with time-windows insertion. First we set ${}^{in}t_{mc} = t_m$ and ${}^{out}t_{mc} = t_m + \tilde{w}_{mc}$ for the origin resource. The next resource of the sequence is resource b. According to Equation (5.56) we check if

$$\left[\left\langle {}^{in}t_b\right\rangle_1 - t_m\right] = \left[{}^{in}t_{2b} - t_m\right] > \hat{w}_{mb} + \varepsilon_{mb} \text{ and}$$

$$\left[\left\langle {}^{in}t_b\right\rangle_1 - (\hat{w}_{mb} + \varepsilon_{mj})\right] = \left[{}^{in}t_{2b} - (\hat{w}_{mb} + \varepsilon_{mj})\right] > {}^{out}t_{mc}$$

From Figure 5.12 we can see that both conditions are satisfied which yields ${}^{in}t_{mb} = {}^{out}t_{mc}$ and ${}^{out}t_{mb} = {}^{in}t_{mb} + \hat{w}_{mb}$. We proceed to the next resource of the sequence, resource a;

$$\left[\left\langle {}^{in}t_a\right\rangle_1 - t_m\right] = \left[{}^{in}t_{7a} - t_m\right] < 0$$

hence, one of the conditions in Equation (5.56) is not satisfied so we have to find the first free time window by using Equation (5.58). The number of components of the sorted time vector $\left\langle {}^{in}t_a\right\rangle$ that are $\neq \infty$ is 2, i.e. $n = 2$. For $\ell = 1$ we obtain

$$\left[\left\langle {}^{in}t_a\right\rangle_2 - \max\left\{\left\langle {}^{out}t_a\right\rangle_1, {}^{out}t_{mb}\right\}\right] =$$
$$\left[{}^{in}t_{2a} - \max\left\{{}^{out}t_{7a}, {}^{out}t_{mb}\right\}\right] =$$
$$\left[{}^{in}t_{2a} - {}^{out}t_{mb}\right] < \hat{w}_{ma} + 2\varepsilon_{ma}$$

thus a new time window cannot be placed before ${}^{in}t_{2a}$. Since w_{2a} is the last time window on resource a, w_{ma} is set after it, which gives

$${}^{in}t_{ma} = \left\langle {}^{out}t_a\right\rangle_2 + \varepsilon_{ma} = {}^{out}t_{2a} + \varepsilon_{ma}, \quad {}^{out}t_{ma} = {}^{in}t_{ma} + \hat{w}_{ma}$$

By this action all time windows of sequence $\sigma_1^2 = \{c, b, a\}$ have been inserted with no overlaps.

C. *Time-windows elongation and overlaps* (Step 3)

As assumed earlier, a part can reside only in a resource, therefore, immediately upon leaving one resource it enters the next one, i.e. the following equation should be fulfilled for all resources of the sequence:

$${}^{in}t_{mj} = {}^{out}t_{mi}, \quad i \to j \tag{5.61}$$

It may be seen from the previous example (Figure 5.12) that the inserted time windows do not satisfy Equation (5.61). Although ${}^{in}t_{mb} = {}^{out}t_{mc}$, this is not the case for resources b and a, ${}^{in}t_{ma} \neq {}^{out}t_{mb}$. In order to check if sequence, σ_i^p, is feasible, first we have to expand the inserted time windows to meet requirement (5.61). The time-window elongation on resource r_j yields:

$$w_{mj} = \hat{w}_{mj} + {}^{in}t_{mi} - {}^{out}t_{mj}, \quad j => i \tag{5.62}$$

A time window can be widened in two ways, by changing the duration of the processing time of a particular resource or by holding a part in a resource before processing and/or after processing. As a consequence, a resource release time is changed,

$$\overset{\rightharpoonup}{{}^{out}t_{mj}} = w_{mj} + {}^{in}t_{mj} \tag{5.63}$$

thus changing the time vectors of resource r_j

$$\mathbf{w}_j = \begin{bmatrix} w_{1j} & w_{2j} & w_{3j} & \ldots & w_{mj} & 0 & 0 \end{bmatrix}^T$$

$${}^{in}\mathbf{t}_j = \begin{bmatrix} {}^{in}t_{1j} & {}^{in}t_{2j} & {}^{in}t_{3j} & \ldots & {}^{in}t_{mj} & \infty & \infty \end{bmatrix}^T$$

$${}^{out}\overset{\rightharpoonup}{\mathbf{t}_j} = \begin{bmatrix} {}^{out}t_{1j} & {}^{out}t_{2j} & {}^{out}t_{3j} & \ldots & {}^{out}\overset{\rightharpoonup}{t_{mj}} & \infty & \infty \end{bmatrix}^T$$

The time-window elongation can cause an *overlap*, which is equivalent to a *conflict*; a situation when two (or more) parts request a resource over the same time period. The situation when an overlap takes place after applying time-window elongation is shown in Figure 5.13. Since ${}^{in}t_{ma} \neq {}^{out}t_{mb}$, a newly inserted time window on resource b, which belongs to sequence π_m, has been widened. This action caused an overlap with the time window of mission π_2, which indicates that if the processing of part p_m on resource b is prolonged in order to be finished just at the moment when resource a is ready to receive the part, then it will collide with part p_2.

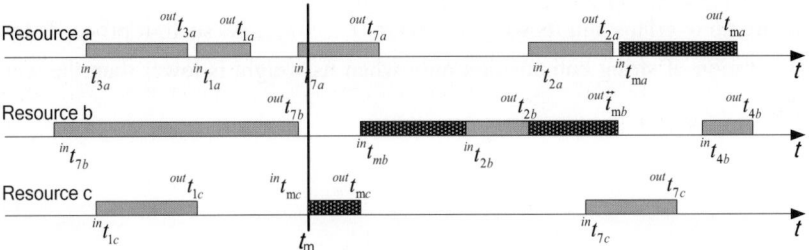

Figure 5.13. Time-windows overlap

Having in mind this situation, once all time windows that belong to a particular sequence are extended according to Equation (5.62), new time vectors should be checked for overlaps, starting from the origin resource of the sequence. If

$$\left\{\left\langle {}^{in}t_j\right\rangle_\ell : \left[\left\langle {}^{in}t_j\right\rangle_{\ell+1} - \left\langle {}^{out}\vec{t}_j\right\rangle_\ell\right] < 0, \ell = 1, n-1\right\} = \emptyset \tag{5.64}$$

then there are no overlaps on resource r_j.

When Equation (5.64) is not satisfied, the first resource with an overlap should be detected and the time windows should be reinserted, starting from the resource with an overlap all the way to the last resource of the sequence. A new time window is inserted on the resource with an overlap by using Equation (5.59), only this time index p is calculated according to the following relation

$$p = \arg\min_\ell \left\{\left\langle {}^{in}t_j\right\rangle_\ell : \right.$$

$$\left[\left\langle {}^{in}t_j\right\rangle_{\ell+1} - \max\left(\left\langle {}^{out}t_j\right\rangle_\ell, {}^{out}t_{mi}\right)\right] > \hat{w}_{mj} + 2\varepsilon_{mj}, \, i \to j, \, \ell = q, \, n-1\right\} \tag{5.65}$$

where q corresponds with the last time window involved in the overlap, i.e.

$$q = \arg\max_\ell \left\{\left\langle {}^{in}t_j\right\rangle_\ell : \left[\left\langle {}^{in}t_j\right\rangle_{\ell+1} - \left\langle {}^{out}\vec{t}_j\right\rangle_\ell\right] < 0, \ell = 1, n-1\right\} \tag{5.66}$$

Since a new time window cannot be placed upstream of the time window q, in Equation (5.65) only those time windows that follow after q are checked. When time windows are reinserted on all resources they should be checked for overlaps and the procedure repeats until a) there are no overlaps or b) overlap occurs on the origin resource. In case a) the sequence σ_i^p is feasible and its weight is equal to the destination resource-release time. In case b) the sequence σ_i^p is not feasible and its weight is set to ∞, hence, the sequence is removed from $\tilde{\Sigma}_m^p$.

The described procedure gives a final form of time windows for the sequence $\sigma_1^2 = \{c, b, a\}$, as shown in Figure 5.14. It can be seen that this RS is feasible since there are no overlaps and its weight is $W(\sigma_1^2) = {}^{out}t_{ma}$. As such, it proceeds to the next iteration of string composition only when its weight is lower than the weight of ${}^m\hat{\sigma}$.

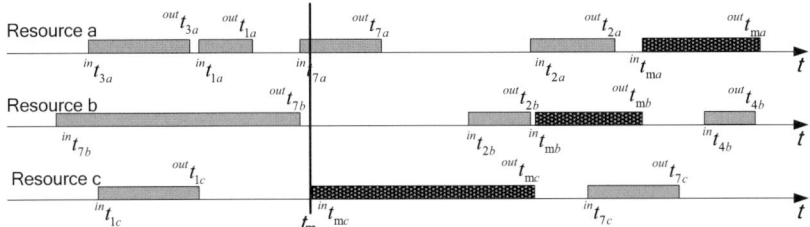

Figure 5.14. Reinserted time windows with no overlaps

Example 5.3.1 (deadlock avoidance in FMRF – multi-AGV routing)

As an example of the dynamic deadlock avoidance we consider routing and scheduling in a multi-AGV system shown in Figure 5.15. Implementation of the time-windows approach for a dynamic multi-AGV routing problem is done in a way that arcs are considered as resources that are used by the vehicles, which are seen as parts passing through the system. The layout depicted in Figure 5.15 comprises 3 vehicles that have to execute sequences (pass particular arcs) in order to move from one point to the other. The highest-priority sequence π_1 is executed by vehicle 5 (V5) that carries pallets from an unloading station to a packing station, vehicle 2 (V2) executes the medium-priority sequence π_2 and vehicle 3 (V3) is assigned to the lowest-priority sequence π_3.

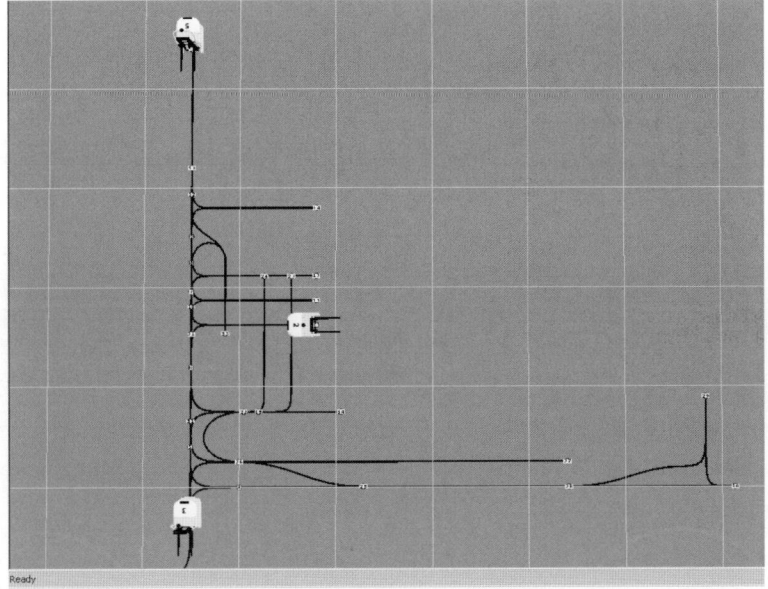

Figure 5.15. A multi-AGV system layout

The shortest paths for all three sequences for a of single-vehicle system are shown in Figures 5.16, 5.17. and 5.18. It can be seen that sequence π_1 and sequence π_3 have the same shortest path, only arcs are visited in reverse order. Figure 5.19 shows the final result of a dynamic deadlock-avoidance algorithm. It can be seen that only the highest-priority sequence is routed through its shortest path, $^1\hat{\sigma}^{10} = \{35, 13, 12, 11, 10, 9, 8, 7, 2, 1, 0\}$.

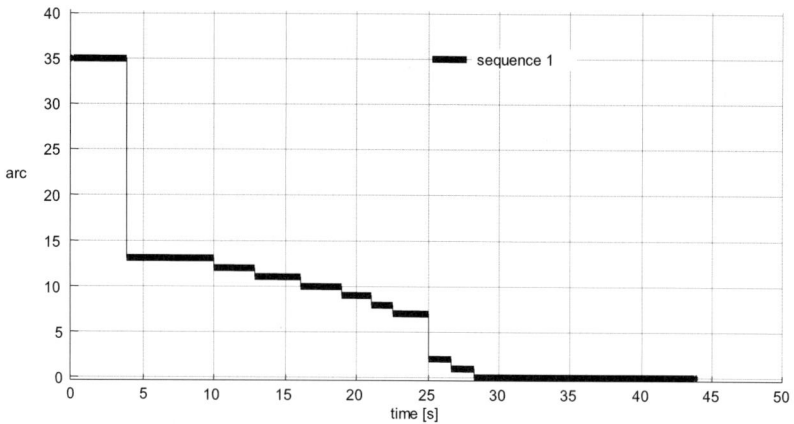

Figure 5.16. The shortest path for sequence π_1

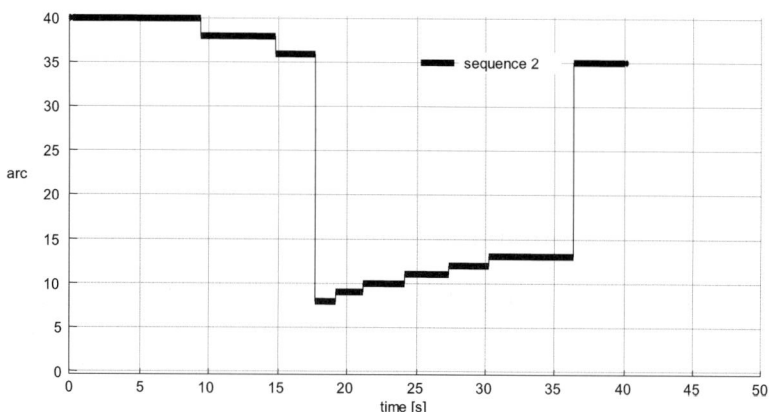

Figure 5.17. The shortest path for sequence π_2.

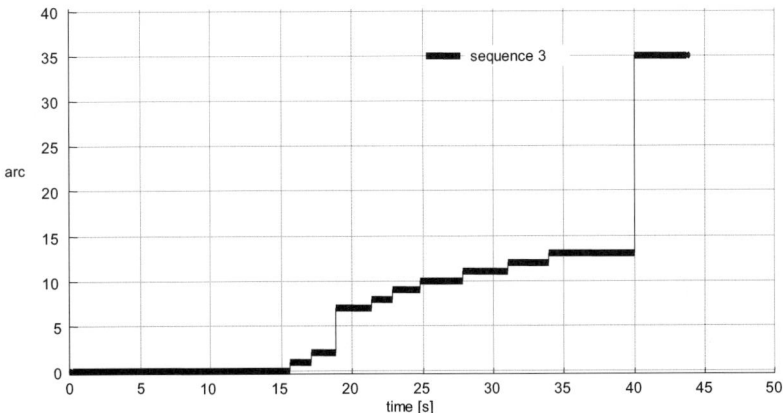

Figure 5.18. The shortest path for sequence π_3

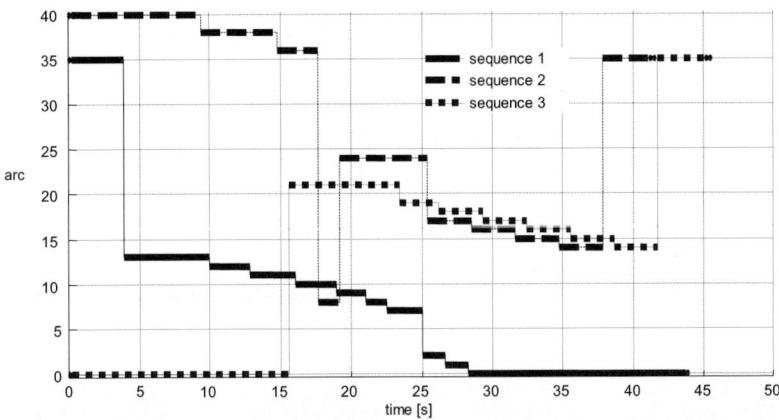

Figure 5.19. The final result of routing algorithm – all three sequences with no deadlocks

Other sequences are detoured to avoid head-on and deadlocks. Sequence π_2 is changed so that vehicle V2 takes arc 24 (instead of arc 9) after arc 8, and the final result is $^2\hat{\sigma}^9 = \{40, 38, 36, 8, 24, 17, 16, 15, 14, 35\}$. As we already mentioned sequence π_3 corresponds to reverse sequence π_1. In order to avoid vehicles head-on, original sequence π_3 is changed to $^3\hat{\sigma}^8 = \{0, 21, 19, 18, 17, 16, 15, 14, 35\}$.

The presented method for dynamic deadlock avoidance is the core of the multi-AGV industrial environment supervisor [24]. The supervisor enables real-time control and simulation of shop-floor layouts that may contain a number of manufacturing cells and a number of AGVs that commutate between dynamically

198 Manufacturing Systems Control Design

determined starting and end nodes. Figure 5.20 shows simulation screenshots of the system presented in Figure 5.15. Vehicles execute the sequences depicted in Figure 5.19.

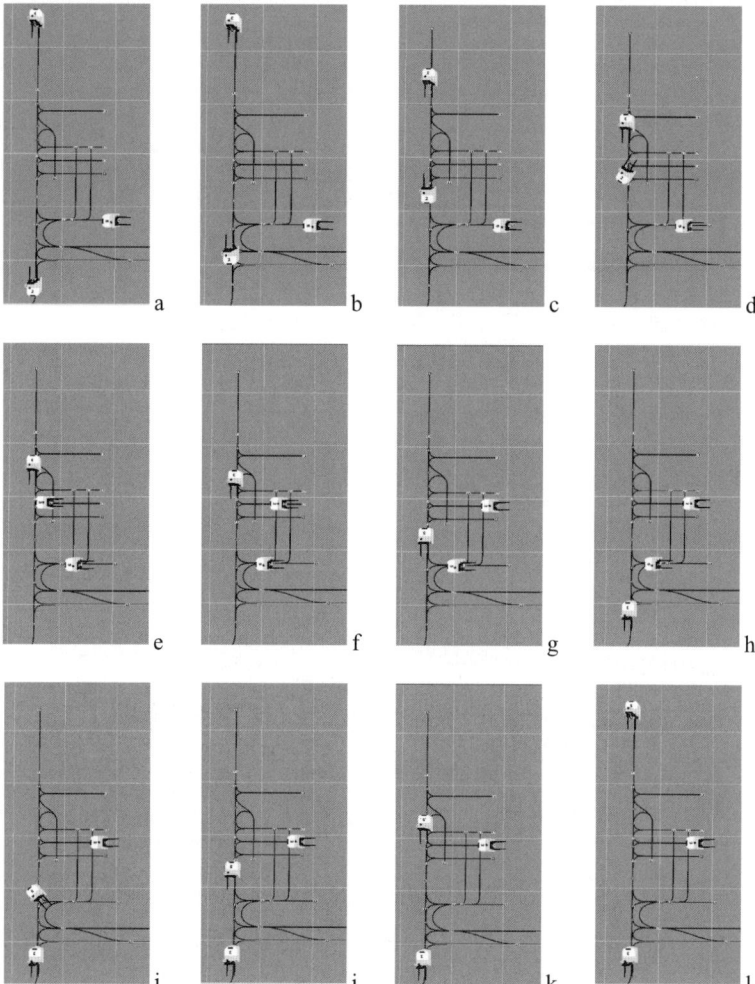

Figure 5.20. Simulation screenshots of the system presented in Figure 5.15

The dynamic deadlock avoidance described herein may be seen as a variation of a well-known label-setting algorithm. The difference lies in the fact that standard label-setting algorithms proceed in the next iteration only with a dominant (optimal) label while in our case all feasible sequences (labels) are carried to the next step. In this way, a sequence that seemed to be the best choice in one iteration, could be replaced by another sequence during the steps that follow, because of time-windows overlaps that may happen on its successors. Although this variation increases the worst-case computational complexity, implementation of a multi-AGV industrial environment supervisor showed that for real shop-floor layouts the computational time has the same order as standard label-setting algorithms (most of the alternative sequences are eliminated at early stages of the calculation).

♦

5.4. A Case Study: Deadlock Avoidance in PLC-controlled FMS

In this section we demonstrate the MS supervisor design based on the matrix controller and realized on the industrial PLC Simatic S7-216. The laboratory setup is shown in Figure 5.21.

The setup contains two educational robots, Rhino XR-3 and Rhino XR-4 (XR3, XR4), three belt conveyers (T1, T2, T3), one x-y transporter (XY), one carousel (CR) and one gravitational buffer (GS). Two part types, A and B, are handled by the system. Processed part types visit several resources on their way through the system. A part A enters the system when it is put on the conveyer T1 (Figure 5.22). When the part gets to the end of the conveyer, XR3 transfers it to the XY. Upon the arrival at the opposite side of the transporter, the part is picked by XR4 and placed on the conveyer T2, which carries the part to the output.

Figure 5.21. The setup of the laboratory MS (a two-robot material-handling cell)

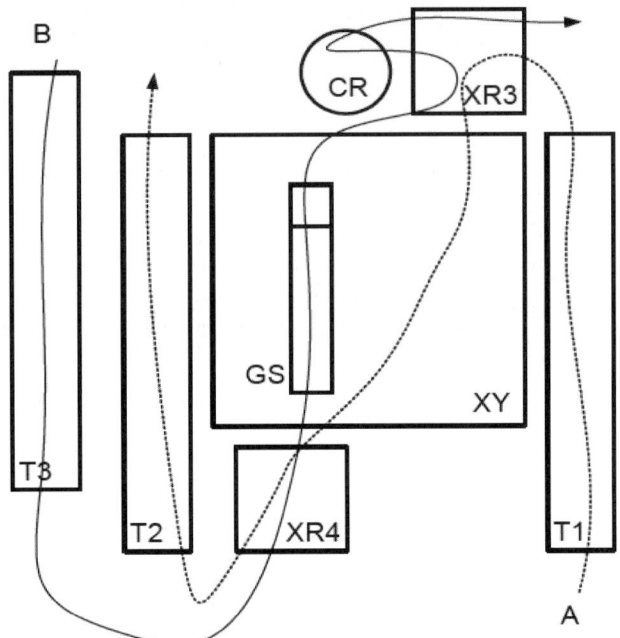

Figure 5.22. A top-view layout of the laboratory FMS with designated parts paths

Likewise, a part *B* enters the system (Figure 5.22) when it is put on the conveyer T3, which brings the part to its end point. Once the part is at the right position, it is lifted by XR4 to the GS. When the part reaches the bottom of the buffer it is removed by XR3 and placed on the CR. The CR rotates the part, which is finally removed from the system by XR3. In our experiments, the capacity of the buffer GS is 1, while the capacity of the carousel is 3.

We start the supervisor design with the matrix-model determination. From the system layout and description, we may distinguish 11 operations, five on the part *A* and six on the part *B*. These operations are carried out by eight resources. Robots XR3 and XR4 are shared resources - XR3 has to perform three tasks; XR31 – moving the part *A* from T1 to XY, XR32 – moving the part *B* from GS to CR, XR33 – moving the part *B* from CR to the system output, while XR4 has two operations; XR41 – moving the part *A* from XY to T3, XR42 – moving the part *B* from T3 to GS. A set of jobs and set of resources are defined as $J = \{T1P, XR31, XYP, XR41, T2P, T3P, XR42, GSP, XR32, CRP, XR33\}$ and $R = \{XR3, XR4, T1, XY, T2, T3, GS, CR\}$.

By identifying the relations among operations and resources, and the sequence of operations, we can define the system matrices that describe the FMS behavior.

$F_v=$ [matrix], $F_r=$ [matrix], $F_u=$ [matrix]

$S_v=$ [matrix], $S_r=$ [matrix], $S_y=$ [matrix]

First we find CWs by using Equation (5.1) and Gurel's algorithm shown in Figure 5.3,

$$G_W = \begin{bmatrix} 0 & 0 & 0 & 1 & 0 & 0 & 0 & 1 \\ 0 & 0 & 0 & 0 & 1 & 0 & 1 & 0 \\ 1 & 0 & 0 & 0 & 0 & 0 & 0 & 0 \\ 0 & 1 & 0 & 0 & 0 & 0 & 0 & 0 \\ 0 & 0 & 0 & 0 & 0 & 0 & 0 & 0 \\ 0 & 1 & 0 & 0 & 0 & 0 & 0 & 0 \\ 1 & 0 & 0 & 0 & 0 & 0 & 0 & 0 \\ 1 & 0 & 0 & 0 & 0 & 0 & 0 & 0 \end{bmatrix}, \quad C = \begin{bmatrix} 1 & 1 & 1 \\ 0 & 1 & 1 \\ 0 & 0 & 0 \\ 0 & 1 & 1 \\ 0 & 0 & 0 \\ 0 & 0 & 0 \\ 0 & 1 & 1 \\ 1 & 0 & 1 \end{bmatrix}$$

There exist three CWs, two simple, $C_1 = \{XR3, CR\}$ and $C_2 = \{XR3, XR4, XY, GS\}$ and one that is a union of these two, $C_3 = \{XR3, XR4, XY, GS, CR\}$. The corresponding critical siphons and critical subsystems are given in a matrix form;

$$S_C = \begin{bmatrix} 0 & 1 & 0 & 0 & 0 & 0 & 0 & 0 & 0 & 1 & 1 & 0 & 0 & 0 & 0 & 0 & 1 \\ 0 & 0 & 0 & 1 & 0 & 0 & 0 & 0 & 1 & 0 & 1 & 1 & 1 & 0 & 1 & 0 & 0 & 1 & 0 \\ 0 & 0 & 0 & 1 & 0 & 0 & 0 & 0 & 0 & 1 & 1 & 1 & 0 & 1 & 0 & 0 & 1 & 1 \end{bmatrix}^T$$

$$J_0 = \begin{bmatrix} 0 & 0 & 0 & 0 & 0 & 0 & 0 & 1 & 1 & 0 \\ 0 & 1 & 1 & 0 & 0 & 0 & 1 & 1 & 0 & 0 \\ 0 & 1 & 1 & 0 & 0 & 0 & 1 & 1 & 1 & 0 \end{bmatrix}^T$$

From matrix F_r we can determine the conflicting-rules vector x_d and the dispatching matrix F_d,

$$\mathbf{x}_d = \begin{bmatrix} 0 & 1 & 0 & 1 & 0 & 0 & 0 & 1 & 0 & 1 & 0 & 1 & 0 \end{bmatrix}^T$$

$$\mathbf{F}_d = \begin{bmatrix} 0 & 1 & 0 & 0 & 0 & 0 & 0 & 0 & 0 & 0 & 0 & 0 & 0 \\ 0 & 0 & 0 & 0 & 0 & 0 & 0 & 0 & 0 & 1 & 0 & 0 & 0 \\ 0 & 0 & 0 & 0 & 0 & 0 & 0 & 0 & 0 & 0 & 0 & 1 & 0 \\ 0 & 0 & 0 & 1 & 0 & 0 & 0 & 0 & 0 & 0 & 0 & 0 & 0 \\ 0 & 0 & 0 & 0 & 0 & 0 & 1 & 0 & 0 & 0 & 0 & 0 & 0 \end{bmatrix}^T$$

A control vector has five components that participate in the prerequisite parts of rules x_2, x_4, x_8, x_{10} and x_{12}. A control function h is represented by a set of rules such that the controlled system is free of conflicts and deadlocks. Since the critical subsystem that corresponds to C_3 is the union of $J_0(C_1)$ and $J_0(C_2)$, only the content of those two subsystems is checked, $v_{J_{0C1}}(k) < m_{C_1}(0) = 4$ and $v_{J_{0C2}}(k) < m_{C_2}(0) = 4$. In the case of parallel conflicts of shared resources, a priority is given to jobs on part path A.

As described previously, in order to get the model that describes the system dynamics, we have to determine the duration of each operation performed on the parts. Measurements of the system-resources performances yield the durations of operations and resource-release times that are expressed in Table 5.3 as the number of sampling intervals required for the particular operation (in our case, the sampling interval is $T_d = 0.5$ s).

Table 5.3. Operation times and resource-release times (# of sampling intervals)

resource	T1	T2	T3	XY	GS	CR
operation	T1P	T2P	T3P	XYP	GSP	CRP
duration	70	42	32	16	2	24
release	2	2	2	22	2	2
resource	XR3				XR4	
operation	XR31	XR32	XR33		XR41	XR42
duration	24	18	26		30	36
release	10	10	10		10	10

The next step in the system-controller design is virtual modeling and simulation of the system with FlexMan, which is described in detail in Chapter 7. The results of dynamic simulation are given in Figure 5.23. As one can see, conflicts are successfully handled and the system is deadlock free.

When the required system behavior is confirmed by FlexMan, the PLC code of the tested matrix controller can be generated and downloaded into the PLC. The other possibility is execution of the control algorithm on a PC and communication with the PLC through the OPC server [25]. One way or the other, the main benefit

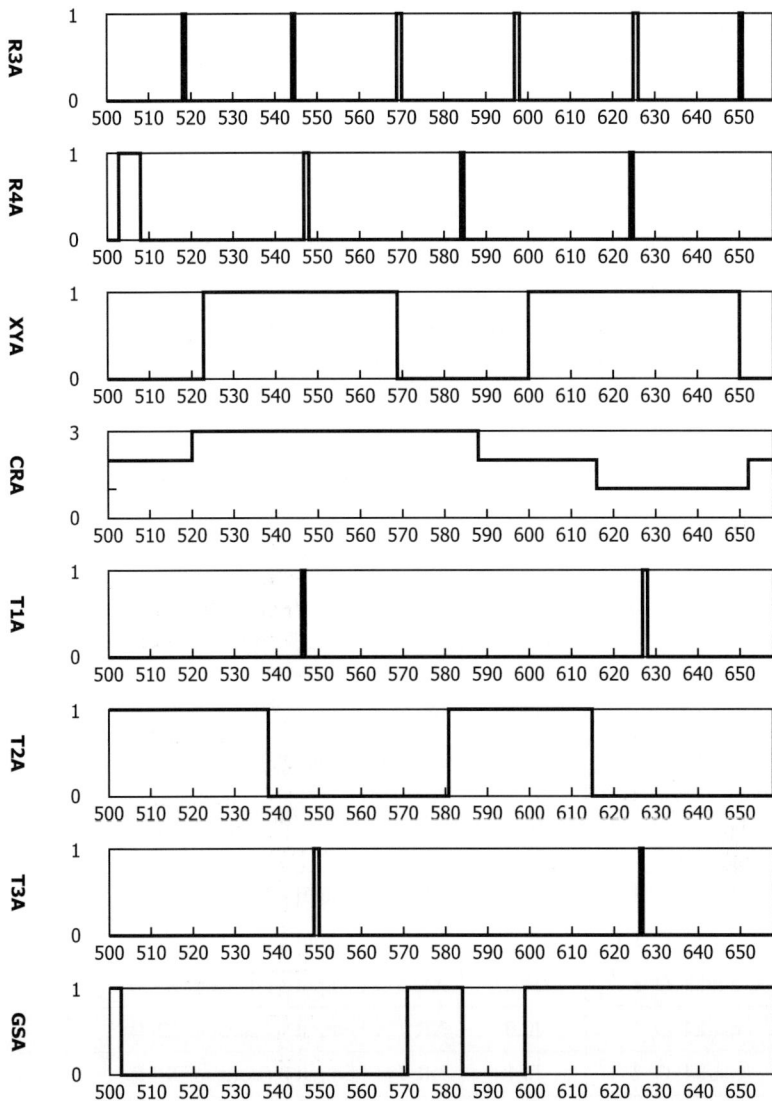

Figure 5.23. Utilization of resources attained by simulation of the system shown in Figure 5.21

of the matrix controller is its straightforward realization on standard industrial programmable logic controllers (PLCs) or on specialized software, such as FlexMan or Petri.NET, a software tool that is described in the next Chapter.

However, due to various types of physical interaction of a PLC and the controlled system, which can result in the form of digital and/or analog signals, serial communication links (RS232, RS485, ...) and local area networks (PROFIBUS, MODBUS...), implementation of the controller requires not only transformation of matrix model into the PLC code, but also *data acquisition and interpretation*. Signals acquired from the PLC interface should be converted into vectors **u**, **y**, **v**$_c$ and **r**$_c$.

The simplest case is the situation when each component of the system vector corresponds to one physical (digital) input. This allows direct mapping of the PLC's memory and the system vector. On the other hand, when several physical inputs take part in the creation of the system vector component, then a special interface function is necessary to map those inputs with the system vector. The same holds for the PLC's outputs. The components of vectors **v**$_s$ and **r**$_s$ should be mapped with digital outputs or the PLC's communication modules. Again, there is no general method for this step in the supervisor design. Relation one-to-one is the easiest case and it permits direct connection of the job-start vector and the resource-release vector with the physical outputs of the PLC.

In our case, information regarding the system status is acquired from the sensors (Figure 5.24) connected to the PLC's digital inputs shown in Table 5.4.

Table 5.4. Digital inputs of the PLC supervisor

Name	Address	Description
IR1	I2.0	IR sensor – part at the beginning of T1
IR2	I2.1	IR sensor – part at the end of T1
IR3	I2.2	IR sensor – part in GS
IR4	I2.3	IR sensor – part at the end of T2
IR5	I2.4	IR sensor – part at the end of T3
IR6	I2.5	IR sensor – part at the beginning of T3
IR7	I0.2	IR sensor – part on XY
mark3_O_1	I1.0	XR 3 ctrl output 1 (status of XR3)
mark3_O_2	I1.1	XR 3 ctrl output 2 (status of XR3)
mark3_O_3	I1.2	XR 3 ctrl output 3 (status of XR3)
mark4_O_1	I0.0	XR 4 ctrl output 1 (status of XR4)
mark4_O_2	I0.1	XR 4 ctrl output 2 (status of XR4)
SW_1	I2.6	switch 1 – part on CR
SW_2	I0.5	switch 2 – XY at XR4
SW_3	I0.6	switch 3 – XY at XR3

In order to form vectors **u**, **y**, **v**$_c$ and **r**$_c$, these inputs are combined and interpreted in an interface function. A part of this function, written in a S7-216 function block diagram, is shown in Figure 5.25 (we assume that the reader has a basic knowledge of PLC programming, for more information see [26]). It can be seen, for example, that **v**$_c$ components XR3_1_c, XR3_2_c and XR3_3_c, which correspond to completion of operations (trajectories) XR31, XR32 and XR33 are comprised of signals from three digital inputs, I1.0, I1.1 and I1.2. The other example is component XYP_c, calculated in Network 12, which is obtained as the logical AND of signals from switch 2 and infrared sensor 7.

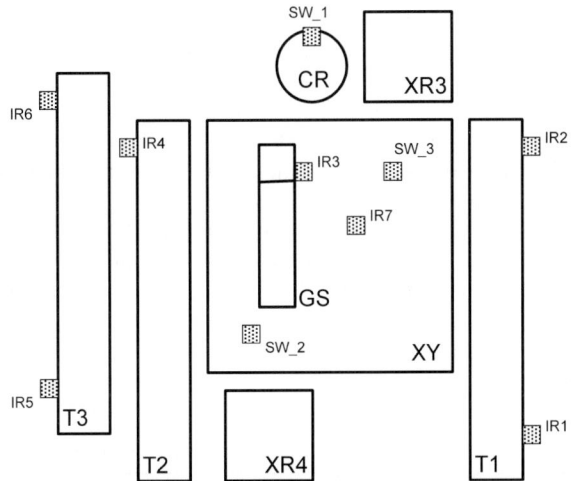

Figure 5.24. Sensor positions in the laboratory FMS

A similar function exists for interpretation of the matrix controller requests for tasks, which are determined as components of vectors $\mathbf{v}_s$ and $\mathbf{r}_s$. The PLC outputs are described in Table 5.5 and part of the interpretation function is shown in Figure 5.26.

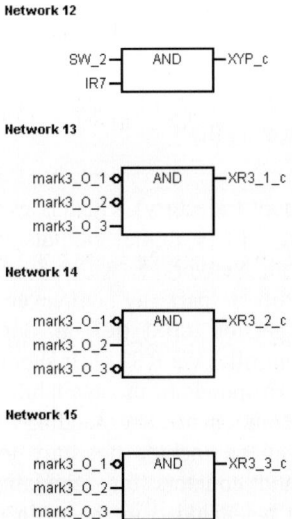

Figure 5.25. A part of the digital inputs interface function realized in FBD

Table 5.5. Digital outputs of the PLC supervisor

Name	Address	Description
mark4_I_1	Q0.0	XR 4 ctrl input 1 – trajectory coding
mark4_I_2	Q0.1	XR 4 ctrl input 2 – trajectory coding
mark3_I_1	Q1.0	XR 3 ctrl input 1 – trajectory coding
mark3_I_2	Q1.1	XR 3 ctrl input 2 – trajectory coding
mark3_I_3	Q1.2	XR 3 ctrl input 3 – trajectory coding

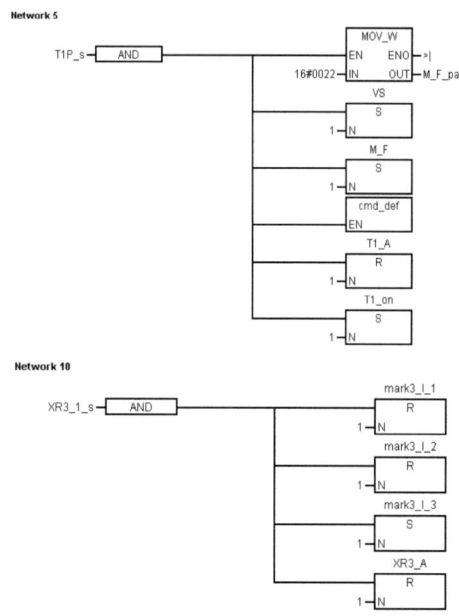

Figure 5.26. A part of the digital outputs interface function realized in FBD

The start of operation (robot trajectory) XR31 is executed by Network 10 that sets (S) and resets (R) the PLC's digital outputs connected with the robot controller. On the other hand, Network 5, which is responsible for the start of operation T1P (transport of a part by conveyer T1), executes a serial communication protocol (function cmd_def in our example) that sends a command to the conveyer controller via RS232. It should be noted that Network 5 resets the component that corresponds to the availability of conveyer T1 (T1_A), while Network 10 resets the component corresponding to robot XR3 (XR3_A).

Sometimes interface functions include not only logical operations but also timers for signals delays and counters for calculation of critical subsystems contents. These utilities can be included in the main control algorithm as well, however, in that case each rule of the matrix controller cannot be directly transferred in the ladder diagram (LD) or the statement list (STL) network. Realization of the matrix controller in STL on PLC S7-216 is given in Figures 5.27.

At each sampling interval Network 1 sets all the components of the control vector to 1. Then in Networks 2 and 3 the controller checks if the critical subsystems are full and resets the corresponding components of the control vector. The following two Networks, 4 and 5, resolve conflicts so that priority is given to the part A path and to the jobs according to Theorem 5.3.2. Once the control vector is determined, PLC executes matrix-controller rules, which is done in Networks 6

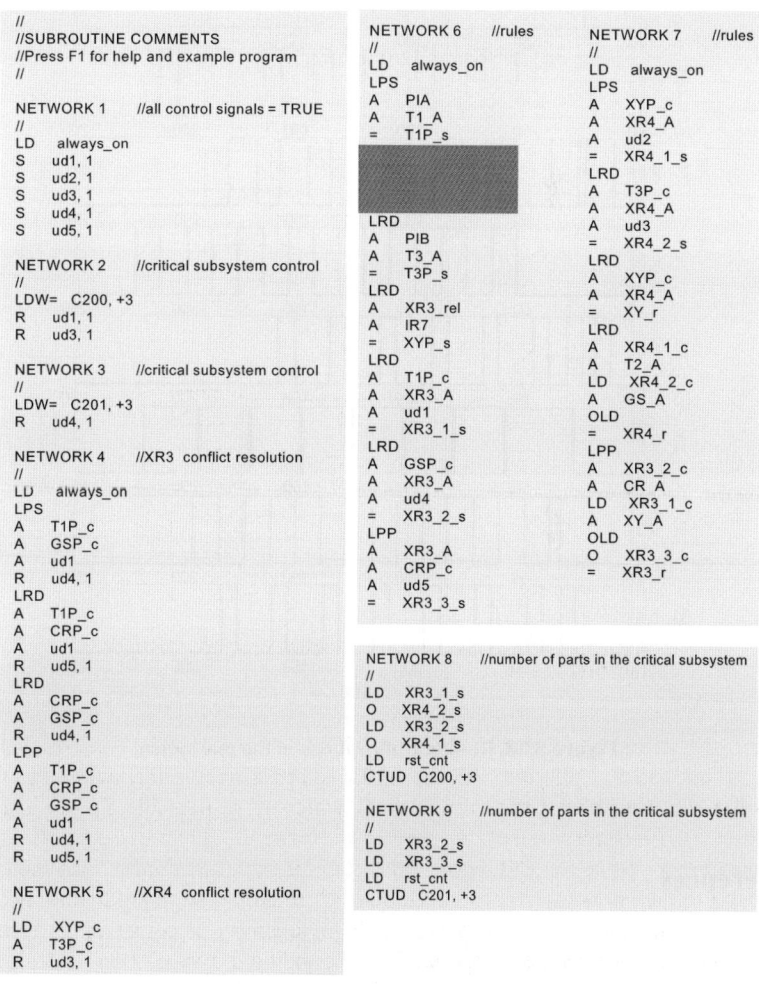

Figure 5.27. The matrix controller realized in STL

and 7. The rules can be read directly from the STL. For example, part of the code, which is marked in Figure 5.27, corresponds to rule x_5 that states that if operation XR41 is completed (XR4_1_c) and resource T2 is available (T2_A) then operation T2P should be started (T2P_s). Finally, the contents of critical subsystems are determined in counters C200 and C201.

A graphical presentation of resources utilizations in a real system is shown in Figure 5.28. The robots' trajectories and idle periods are all shown in one graph, while "1" on other graphs stands for a busy resource.

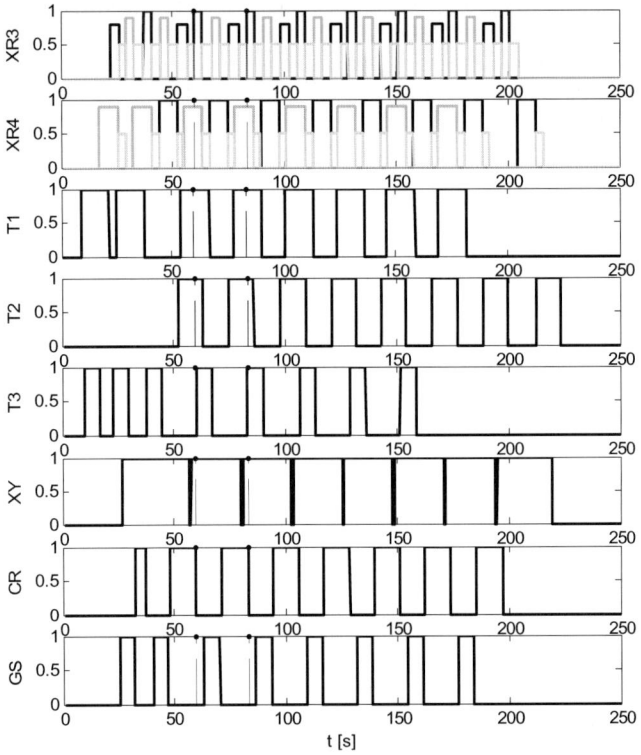

Figure 5.28. Resource utilizations in the real system

References

[1] Kumar PR, Meyn SP. Stability of queuing networks and scheduling polices, IEEE Trans. Aut. Contr. 1995;40:251–260.
[2] Lu SH, Kumar PR. Distributed scheduling based on due dated and buffer priorities, IEEE Trans. Aut.Contr.1991;36:1406–1416.
[3] Gray WS, Mesko JP. Observability functions for linear and nonlinear systems, Systems Control Letters 1999;38:99–113.

[4] Byrnes CI, Martin CF. An integral-invariance principle for non-linear systems, IEEE Trans. Aut. Contr. 1995;40:983–994.
[5] Kato T. Short Introduction to Perturbation Theory for Linear Operators. Berlin: Springer-Verlag, 1982.
[6] Milnor J. Morse Theory. New Jersey: Princeton University Press, 1963.
[7] Moore BC. Principal component analysis in linear systems: Controlability, observability and model reduction, IEEE Trans. Aut. Contr. 1981; AC-26:17–32.
[8] Scherpen JMA. Balancing for nonlinear systems, Proceedings of European Control Conference 1993;4:1838–1843.
[9] Xing KY, Xing KL, Li JM, and Hu BS. Deadlock Avoidance Controller for a class of Manufacturing Systems, Proceedings of the 1996 IEEE International Conference on Robotics and Automation 1996:220–224.
[10] Gurel A, Bogdan S, Lewis FL. Matrix Approach to Deadlock-Free Dispatching in Multi-Class Finite Buffer Flowlines, IEEE Trans. Aut. Cont. 2000;45;11:2086–2090.
[11] Fanti MP, Maione B, Mascolo S, Turchiano B. Event-Based Feedback Control for Deadlock Avoidance in Flexible Production Systems, IEEE Trans. Rob. Autom. 1997;13;3:.
[12] Jeng MD, DiCesare F. Synthesis Using Resource Control Nets for Modeling Shared-Resource Systems, IEEE Trans. Rob. Autom. 1995; RA-11:317–327.
[13] Lewis FL, Gurel A, Bogdan S, Docanalp A, Pastravanu OC. Analysis of Deadlock and Circular Waits using a Matrix Model for Flexible Manufacturing Systems, Automatica 1998;34:9:1083–1100.
[14] Xing KY, Hu BS, Chen HX. Deadlock Avoidance Policy for Petri-Net Modeling of Flexible Manufacturing Systems with Shared Resources, IEEE Trans. Aut. Contr. 1996;41:2:289–295.
[15] Lawley M. Deadlock Avoidance in Manufacturing Systems with Flexible Routing and Mixed Capacity, IEEE International Conference on Systems, Man, and Cybernetics 1998;1:594–599.
[16] Lawley M. Flexible Routing and Deadlock Avoidance in Automated Manufacturing Systems, Proceedings of the 1998 IEEE International Conference on Robotics and Automation 1998:591–596.
[17] Smolic-Rocak N, Bogdan S, Kovacic Z, Petrinec K. Multi AGV control system, Report on research and development assisted in 2004/2005 by SITEK S.p.a., 2005.
[18] Broadbent AJ, Besant CB, Premi SK, Walker SP. Free ranging AGV Systems: Promises, Problems and Pathways, Proc. of the 2nd Int'l Conf. on Automated Materials Handling 1985;221–237.
[19] Daniels SC, Real-time Conflict Resolution in Automated Guided Vehicle Scheduling, Ph.D. thesis 1988, Dept. of Industrial Eng., Penn. State University, USA.
[20] Desaulniers G, Langevin A, Riopel D. Dispatching and conflict-free routing of automated guided vehicles: an exact approach, Int'l J. of Flex. Manuf. Sys. 2003;15:309–331.
[21] Möhring RH, Köhler E, Gawrilow E, Stenzel B, Conflict-free Real-time AGV Routing, Proc. of the of 3rd Int'l C. on Applied Infrastructure Res. 2004;661–675.
[22] Taghaboni-Dutta F, Tanchoco JMA, Comparison of Dynamic Routing Techniques for Automated Guided Vehicle Systems, Int'l J. Product. Res. 1995;33:2653–2669.
[23] Qiu L, Hsu WJ, Huang SY, Wang H, Scheduling and routing algorithms for AGVs: a survey, Int'l J. Produc. Res. 2002;40:745–760.
[24] Petrinec K, Kovacic Z, Marozin A. Simulator of Multi-AGV Robotic Industrial Environments, CD-ROM Proceedings of ICIT03 2003.
[25] OPC Foundation at http://www.opcfoundation.org
[26] Siemens AG, Simatic S7-200 Electronic manuals, 2000.

6
Petri Nets

In 1962 Carl Adam Petri from TU Darmstadt developed one of the most popular DES modeling tools – *Petri nets (PN)* [1]. They provide a mathematical framework for DES analysis, DES supervisory design and DES performance evaluation (static and dynamic). More general than automata (any automaton can be represented as a Petri net, while the opposite is not always true), Petri nets allow description of very complex DES. However, in the case of large DES, PN models tend to become immense and complicated for analysis. The main benefit of PNs is their graphical nature that allows visualization of the modeled system. Namely, a *Petri-net graph* directly embodies many structural properties of the system, which is not the case when an automaton is used for DES modeling. As such, Petri nets are used in a wide variety of applications, from communications to fault-tolerant systems. We shall see later in the text that in the case of an MS modeled by PN, system resources and part paths can both be recognized straight from a corresponding PN graph. However, PN are very difficult to design for specific FMS of reasonable complexity, and to modify if objectives, products, or resources change. A major problem is that PN properties such as reachability must be verified for each given system by using simulation. Moreover, to accommodate manufacturing design algorithms in the PN framework, it is necessary to introduce colored PN, hierarchical PN, generalized PN, multiple types of places, or other esoteric notions that quickly go beyond the experience of the manufacturing engineer and invalidate most PN analysis techniques.

We start this chapter with basic definitions and properties of PNs, followed by a description of MS modeling by Petri nets [2], [3]. Introduction of control places in an uncontrolled PN model of the system is presented next, together with a linear PN controller based on *p-invariants*. In Section 6.3 we describe the relation between PN- and matrix-based modeling of MSs. At the end of this section a PN simulation tool used throughout the chapter is presented (the tool is available for download).

6.1 Basic Definitions

A short preamble to PNs and some of their properties have already been given in Chapter 1. In this section we give formal definitions of terms used in the remainder of the book.

A PN is represented by a *directed bipartite multigraph* containing two types of nodes, *places* (drawn as circles) and *transitions* (drawn as bars or rectangles) connected with *directed arcs*. Arcs, labeled with their *weights*, can join only certain types of nodes. In some applications an arc with weight w is replaced with w parallel arcs with weight 1 (Figure 6.1). We say that the PN is *ordinary* if all its arcs have weight equal to 1. Usually, for convenience, arcs with weight 1 are not labeled.

A particular property that differentiates a PN from an ordinary graph is a *marking m*, which assigns a non-negative integer to each PN place. A marking $m(p_i) = l$ is characterized by l black dots (*tokens*) inside a circle representing place p_i. We say that p_i is *marked* with l tokens. A *marking vector* $\mathbf{m} = [m(p_1)\ m(p_2)\ \ldots\ m(p_n)]^T$ represents a PN state, which means that a state space of PN with n places is described with all n-dimensional marking vectors.

The other property associated with place p_i is its *capacity* $K(p_i)$, which refers to the number of tokens that can be held by the place. Apparently, for practical applications $K(p_i)$ should be bounded by an upper limit. When $K(p_i) < \infty$ for each place in a PN, we say that the PN has a *finite capacity*, as opposed to an *infinite-capacity* PN in which at least one place has $K(p_i) = \infty$. The PN is said to be *safe* if $\forall p_i, K(p_i) = 1$.

The PN graph shown in Figure 6.1 has two transitions and 5 places with marking $m(p_1) = 1$, $m(p_2) = 1$, $m(p_3) = 0$, $m(p_4) = 2$, $m(p_5) = 0$. There are 6 arcs connecting these places with transitions. Their weights are $w(p_1,t_1)=1$, $w(p_2,t_1)=1$, $w(t_1,p_3)=3$, $w(p_3,t_2)=1$, $w(p_4,t_2)=2$ and $w(t_2,p_5)=1$. Place p is called an *input (output)* place of transition t if $w(p,t)\neq 0$ ($w(t,p)\neq 0$). The same holds for the input (output) transition t of place p, i.e. $w(t,p)\neq 0$ ($w(p,t)\neq 0$). A place (transition) that has no input transitions (places) is called *a source*, and a place (transition) without output transitions (places) is called *a sink*. In the PN shown in Figure 6.1 source places are p_1, p_2 and p_4, while the sink place is p_5. We say that a place p_i and transition t_j are involved in a *self-loop* if $w(p_i,t_j)\neq 0$ and $w(t_j,p_i)\neq 0$. A PN with no self-loops is called *pure*.

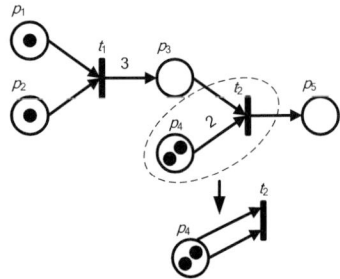

Figure 6.1. An example of a PN graph

Interpretation of places and transitions depends on the application, but in general, places represent *conditions* required for the occurrence of a particular event, for example, resource availability, parts or data readiness, *etc.* Transitions, on the other hand, represent the occurrence of *an event*, such as the start of a task/operation, release of a resource, step in a computation algorithm, *etc.* In this concept the existence of a token in a place is understood as a fulfillment of the condition represented by the place. Since a place can hold more than 1 token, their presence in the place can be taken as the number of processed parts, the number of customers in a queue, the number of available resources, etc.

A mechanism that changes a PN state (marking) is described with two simple *firing rules* given in the following definitions.

Definition 6.1.1 (enabled transition): We say that transition t is *enabled* if each input place p of t is marked with at least $w(p,t)$ tokens.

Definition 6.1.2 (firing of transition): An enabled transition t will *fire* if the event that it represents occurs. In that case i) $w(p,t)$ tokens are removed from each input place p of t, and ii) $w(t,p)$ tokens are added in each output place p of t.

Here we should make an important remark regarding the last definition. In the text that follows we assume that *as soon as a transition is enabled it fires*, meaning that all conditions for the occurrence of an event represented by a particular transition are modeled and included in a PN graph.

Firing of transitions in PN graph is shown in Figure 6.2. Initially, (a) transition t_1 is enabled since $w(p_1,t_1)=m(p_1)=1$ and $w(p_2,t_1)=m(p_2)=1$. On the other hand, $w(p_3,t_2)>m(p_3)=0$ and $w(p_4,t_2)=m(p_4)=2$, hence, transition t_2 is not enabled. When t_1 fires (b) one token is removed from each input place, p_1 and p_2, and 3 tokens are added to output place p_3 as $w(t_1,p_3)=3$. Now $w(p_1,t_1)>m(p_1)=0$ and $w(p_2,t_1)>m(p_2)=0$, thus t_1 is no longer enabled, while t_2 becomes enabled since $w(p_3,t_2)<m(p_3)=3$ and $w(p_4,t_2)=m(p_4)=2$. Firing of t_2 (c) removes one token from p_3 and two tokens from p_4, and adds one token to p_5.

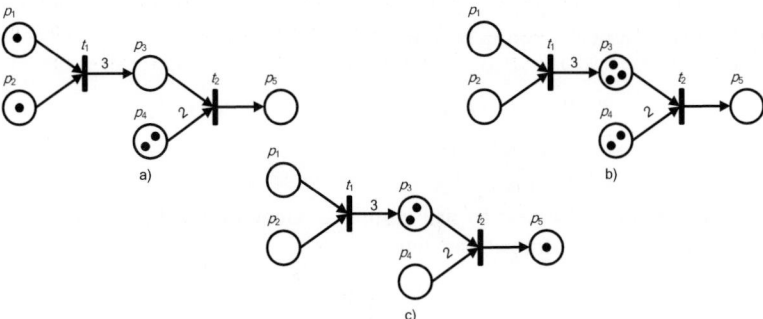

Figure 6.2. Firing of transitions in a PN graph

In order to be able to analyze the evolution of tokens in PN as well as PN structural properties, we have to define a mathematical framework behind the PN graph. A formal definition of a PN is given next.

Definition 6.1.3 (Petri net): A Petri net is a 6-tuple, $PN = \{P, T, \mathbf{I}, \mathbf{O}, \mathbf{M}, \mathbf{m_0}\}$, where,

$P = \{p_1, p_2, p_3, \ldots, p_m\}$ – a finite set of places,
$T = \{t_1, t_2, t_3, \ldots, t_m\}$ – a finite set of transitions,
$\mathbf{I}: P \times T \rightarrow \{0,1\}$ – an *input incidence matrix* – relates places to transitions,
$\mathbf{O}: T \times P \rightarrow \{0,1\}$ – an *output incidence matrix* – relates transitions to places,
$\mathbf{M}: \mathbf{I}, \mathbf{O} \rightarrow \{1, 2, 3, \ldots \}$ – is a weight function,
$\mathbf{m_0}$ – initial value of the marking vector $\mathbf{m}: P \rightarrow \aleph$.

According to the definition, for the PN graph given in Figure 6.2, one has

$$P = \{p_1, p_2, p_3, p_4, p_5\}, \; T = \{t_1, t_2\}, \; \mathbf{m_0} = \begin{bmatrix} 1 & 1 & 0 & 2 & 0 \end{bmatrix}^T$$

$$\mathbf{I} = \begin{bmatrix} 1 & 1 & 0 & 0 & 0 \\ 0 & 0 & 1 & 1 & 0 \end{bmatrix}, \; \mathbf{O} = \begin{bmatrix} 0 & 0 & 1 & 0 & 0 \\ 0 & 0 & 0 & 0 & 1 \end{bmatrix}$$

$$\mathbf{M} = \begin{bmatrix} 1 & 1 & 0 & 0 & 0 & 0 & 3 & 0 & 0 \\ 0 & 0 & 1 & 2 & 0 & 0 & 0 & 0 & 1 \end{bmatrix}$$

Now, let us see if we can write a PN driving mechanism, described by Definition 6.1.2, in the form of algebraic equations. As already explained, firing of t_1 in the PN shown in Figure 6.2, changes the marking of places p_1, p_2 and p_3, while firing of t_2 changes p_3, p_4 and p_5. For place p_3 we can write

$$m_k(p_3) = m_{k-1}(p_3) + w(t_1, p_3) \cdot t_{1,k} - w(p_3, t_2) \cdot t_{2,k}$$

where k is a firing step. When t_j fires in step k, then $t_{jk}=1$, otherwise $t_{jk}=0$. For $k=1$ (firing of t_1) the above equation becomes

$$m_1(p_3) = m_0(p_3) + w(t_1, p_3) \cdot t_{1,1} - w(p_3, t_2) \cdot t_{2,1}$$
$$= 0 + 3 \cdot 1 - 0 \cdot 0 = 3$$

which corresponds with case b) in Figure 6.2. Generally, a PN place could have several input and output transitions, thus,

$$m_k(p_i) = m_{k-1}(p_i) + \sum_{t_j \in T} w(t_j, p_i) \cdot t_{j,k} - \sum_{t_j \in T} w(p_i, t_j) \cdot t_{j,k} \qquad (6.1)$$

or in vector form

$$\mathbf{m}_k = \mathbf{m}_{k-1} + \mathbf{W}^T \mathbf{t} \qquad (6.2)$$

where $\mathbf{W}$ is an *incidence matrix* with $w_{ij} = w(t_j,p_i) - w(p_i,t_j)$, and $\mathbf{t}$ is a *transition vector*. It should be noted that $\mathbf{W} = \mathbf{O} - \mathbf{I}$ for an ordinary PN. A transition vector $\mathbf{t}$ is composed of non-negative integers that correspond with the number of times a particular transition has been fired between markings $\mathbf{m}_k$ and $\mathbf{m}_{k-1}$.

Relation (6.2) is called a *PN state equation* or *PN marking transition equation*. Its similarity with recursive matrix model (3.12) is apparent. We shall discuss this issue in more detail in section 6.3. By using a PN state equation we can mathematically formalize the firing of transitions in the PN graph shown in Figure 6.2,

$$\mathbf{m}_1 = \mathbf{m}_0 + \mathbf{W}^T \mathbf{t}_0 = \begin{bmatrix} 1 \\ 1 \\ 0 \\ 2 \\ 0 \end{bmatrix} + \begin{bmatrix} -1 & 0 \\ -1 & 0 \\ 3 & -1 \\ 0 & -2 \\ 0 & 1 \end{bmatrix} \begin{bmatrix} 1 \\ 0 \end{bmatrix} = \begin{bmatrix} 0 \\ 0 \\ 3 \\ 2 \\ 0 \end{bmatrix}$$

$$\mathbf{m}_2 = \mathbf{m}_1 + \mathbf{W}^T \mathbf{t}_1 = \begin{bmatrix} 0 \\ 0 \\ 3 \\ 2 \\ 0 \end{bmatrix} + \begin{bmatrix} -1 & 0 \\ -1 & 0 \\ 3 & -1 \\ 0 & -2 \\ 0 & 1 \end{bmatrix} \begin{bmatrix} 0 \\ 1 \end{bmatrix} = \begin{bmatrix} 0 \\ 0 \\ 2 \\ 0 \\ 1 \end{bmatrix}$$

We say that the marking (state) $\mathbf{m}_2$ is *reached* from $\mathbf{m}_0$ by *firing sequence* $\sigma = t_1 t_2$, denoted $\mathbf{m}_0 [\sigma > \mathbf{m}_2$. The concept of reachability in PN is very important and we return to this issue later on. It should be noted that the firing sequence σ is only the sequence that can be fired in the PN depicted in Figure 6.2. If we change the initial marking of that PN as shown in Figure 6.3, then both transitions, t_1 and t_2, are enabled and the question is which one fires first? A PN cannot give an answer to that question, that is, definitions of the PN and the PN graph *do not specify firing sequences*, and thorough analysis of the PN requires examination of all possible sequences.

216 Manufacturing Systems Control Design

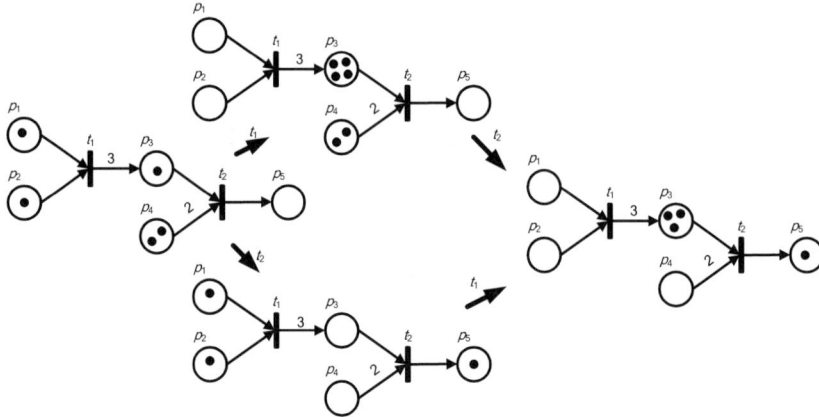

Figure 6.3. Firing of different sequences in a PN graph

In our case, if t_1 fires prior to t_2 one has a sequence $\sigma_1 = t_1 t_2$ with $\mathbf{m}_1$=[0 0 4 2 0]^T and $\mathbf{m}_2$=[0 0 3 0 1]^T. On the other hand, if t_2 fires first, then the sequence is $\sigma_2 = t_2 t_1$ with $\mathbf{m}_1$=[1 1 0 0 1]^T and $\mathbf{m}_2$=[0 0 3 0 1]^T. Although in both cases the initial and final markings are the same, the movement of the marking vector in state space depends on the firing sequence. This example shows that a supervisory mechanism should be added in a PN model in order to obtain the required behavior of the controlled system. Let us examine some properties of a PN before proceeding in that direction.

Generally, PN properties are divided into two classes; those dependent on the initial marking, called *behavioral properties*, and those independent of the initial marking, known as *structural properties*. Since our main concern in supervisory design is deadlock prevention, we start with the definition of *liveness* property.

Definition 6.1.4 (liveness): Petri net PN with initial marking $\mathbf{m}_0$ is *live* if there exists a firing sequence such that any transition in the PN can be fired from any marking reached from $\mathbf{m}_0$.

The notion of liveness is closely related with deadlock and circular blocking, *i.e.* live PN is deadlock free. As liveness guarantees that there always exists a sequence that fires all transitions in the PN, a system whose model is live PN cannot get into deadlock. The PN shown in Figure 6.4 is live, while those depicted in Figures 6.2 and 6.3 are not.

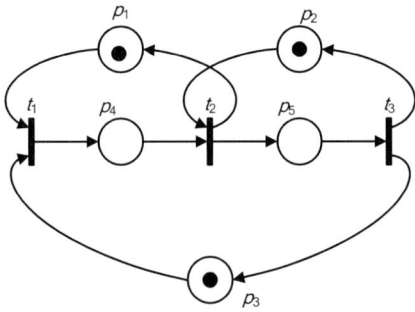

Figure 6.4. An example of a live PN

A liveness is a strong requirement, but in most cases very difficult to test. For this reason liveness is categorized with respect to transitions, so instead of checking if PN is live we consider each transition independently and say that the *transition is a live or a dead one*. There exist 4 classes of live transitions: *L1*-live – transition can fire at least once, *L2*-live – transition can fire at least *k* times, *L3*-live – transition can fire an infinite number of times, and *L4*-live – transition is *L1*-live for every **m** reached from **m**$_0$. A situation in which all transitions in PN are *L4*-live corresponds with liveness as defined in Definition 6.1.4. It has been shown in Chapter 5 that in MRF systems one dead transition is source of the system deadlock. Therefore, the PN of the controlled MRF system should be *L4*-live, *i.e.* we require PN liveness according to Definition 6.1.4.

The other property that is essential in PN analysis has already been mentioned *reachability*.

Definition 6.1.5 (reachability): A marking **m**$_j$ is reachable from marking **m**$_i$ if there exists a firing sequence $\sigma_{ij} = t_m t_r \ldots t_p$ such that it leads a marking vector from **m**$_i$ to **m**$_j$. We write **m**$_i$ [σ_{ij} > **m**$_j$.

A set of all markings reachable from **m**$_i$ is denoted by $\mathcal{R}(\mathbf{m}_i)$. Reachability is determined by a listing of all markings (states) that can be reached from a particular, usually initial, marking. Firing of enabled transition(s) produces new markings and each new marking generates even more markings. Evidently, this kind of analysis could lead to enormous number of states and it is limited to a PN with a relatively small number of places.

Reachability analysis of a PN results in a graphical structure called a *coverability tree*. For bounded PN, which we consider herein, a coverability tree becomes a *reachability tree* and it contains all reachable states of the corresponding PN. A reachability tree for the PN given in Figure 6.3 is shown in Figure 6.5.

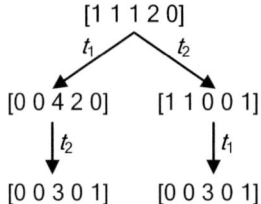

Figure 6.5. A reachability tree of the PN from Figure 6.3

The tree is constructed starting from the initial PN marking by drawing an arc for each transition that is enabled. As both transitions are enabled, two arcs (branches) should be created; firing t_1 produces marking [0 0 4 2 0], while firing t_2 leads to marking [1 1 0 0 1]. We proceed further by drawing arcs for transitions that are enabled under newly obtained markings. The process continues until all reachable markings are counted. If we treat markings as nodes then the obtained reachability tree is actually an automaton representation of the considered PN and a *set of all firing sequences*, $L(\mathbf{m}_0)$, corresponds with the language generated by this automaton. For the PN shown in Figure 6.4, an automaton equivalent to its reachability tree is depicted in Figure 6.6. Represented in this way, the analysis techniques used for automata can be used for bounded PNs as well. For example, a reachability tree node with no output branches may indicate deadlock, hence, firing of transition(s) that force undesired PN marking, corresponding to that node, should be forbidden.

It is evident from the above brief introduction that reachability analysis offers a solution to many questions posed for PNs. However, algebraic analysis, based on state equation (6.2), is proven to be more convenient for PNs. Furthermore, matrix-based modeling of manufacturing systems, presented in Chapter 3, has much in common with the state representation of Petri nets. For these reasons here we close our discussion on reachability analysis from the automata point of view and proceed with a description of a reachability test based on the algebraic equation (6.2).

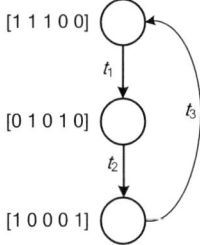

Figure 6.6. An automaton equivalent of PN from Figure 6.4

From Equation (6.2) we see that marking $\mathbf{m}_d$ could be reachable from $\mathbf{m}_k$ if there exists a transition vector **t** such that

$$\mathbf{W}^T\mathbf{t} = \mathbf{m}_d - \mathbf{m}_k \tag{6.3}$$

This equation is a *necessary* condition for reachability, that is, the existence of solution $\mathbf{t}$ does not guarantee that $\mathbf{m}_d$ is reachable from $\mathbf{m}_k$. What we know for sure is that when Equation (6.3) has no solution in $\mathbf{t}$ then there is no firing sequence that enforces $\mathbf{m}_k$ into $\mathbf{m}_d$.

An interesting result is obtained as a solution of the homogenous equation

$$\mathbf{W}^T\mathbf{t} = 0 \tag{6.4}$$

Since $\mathbf{m}_d - \mathbf{m}_k = 0$ transition vector $\mathbf{t}$ that satisfies Equation (6.4) comprises a firing sequence that returns marking $\mathbf{m}_k$ back to itself. Such a transition vector is called *t-invariant*. In close relation with t-invariant is the notion of *reversibility*.

Definition 6.1.6 (reversibility): A Petri net is said to be *reversible* if for any marking $\mathbf{m}_i$ there exists a firing sequence σ_i such that $\mathbf{m}_i [\sigma_i > \mathbf{m}_i$.

In practice, it is required for most manufacturing systems to exhibit cyclic behavior. Petri-net models of such systems should be reversible, hence, checking reversibility is an important issue for the systems we encounter in practice.

Another interesting PN structure, which plays a key role in the investigation of deadlock, is the so-called *p-invariant*, a non-negative integer place vector $\mathbf{p}$ that is a solution of

$$\mathbf{Wp} = 0 \tag{6.5}$$

As an example of PN invariants we use the net shown in Figure 6.4. Its incidence matrix is

$$\mathbf{W} = \begin{bmatrix} -1 & 0 & -1 & 1 & 0 \\ 1 & -1 & 0 & -1 & 1 \\ 0 & 1 & 1 & 0 & -1 \end{bmatrix}$$

The t-invariant is $\mathbf{t} = [1\ 1\ 1]^T$, and the p-invariants are $\mathbf{p}_1 = [1\ 0\ 0\ 1\ 0]^T$, $\mathbf{p}_2 = [0\ 1\ 0\ 0\ 1]^T$ and $\mathbf{p}_3 = [0\ 0\ 1\ 1\ 1]^T$. Of course, $\mathbf{t}_q = [q\ q\ q]^T$ is also an invariant of this PN, however, when structural properties are investigated then *minimal invariants* are of primary interest. An invariant $\mathbf{p}$ ($\mathbf{t}$) is *minimal* if there is no such invariant $\mathbf{p}_q$ ($\mathbf{t}_q$) that $p_{qi} \leq p_i$ ($t_{qi} \leq t_i$) for any vector component.

It is easy to show that the number of tokens in places that belong to p-invariant is constant. Multiplying Equation (6.2) with $\mathbf{p}^T$ from the left gives

$$\mathbf{p}^T\mathbf{m}_k = \mathbf{p}^T\mathbf{m}_{k-1} + \mathbf{p}^T\mathbf{W}^T\mathbf{t} \tag{6.6}$$

By including Equation (6.5) in Equation (6.6), and having in mind that Equation (6.2) holds for any k, we obtain

$$\mathbf{p}^T\mathbf{m}_k = \mathbf{p}^T\mathbf{m}_0 = const. \tag{6.7}$$

This equation is very important and actually confirms what we shall show later; in a PN model of an MRF system p-invariants correspond with the resource loops described in Section 5.1.2.

Let us now examine the PN shown in Figure 6.7. According to Definition 6.1.1 transitions t_1 and t_3 are enabled. However, place $\mathbf{p}_3$, which has two output arcs, is marked with only one token. Hence, firing one of these two transitions will disable the other one. This situation is known as a *conflict* and it was thoroughly discussed in previous chapters. Since we assumed that the transition fires as soon as it is enabled, marking $m(p_3)$ becomes negative upon firing of t_1 and t_3, which is not allowed. Therefore, our prime concern in PN analysis is to prevent conflict.

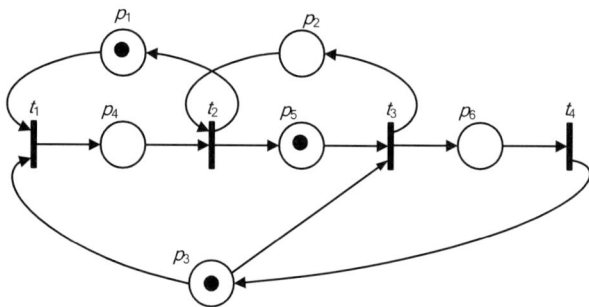

Figure 6.7. An example of a PN with conflict

An occurrence of conflict is related to Petri-net *persistency*.

Definition 6.1.6 (persistency): A Petri net is *persistent* if for any two enabled transitions firing of one does not disable the other.

This definition concludes the description of the basic behavioral properties of PN. We continue with the presentation of properties that are determined by the PN structure and do not depend on PN marking. First, let us extend the notions of *preset* and *postset* to Petri nets:

- $\bullet p = \{t \mid w(t, p) > 0\}$ – a set of input transitions of place p,
- $p\bullet = \{t \mid w(p,t) > 0\}$ – a set of output transitions of place p,
- $\bullet t = \{p \mid w(p,t) > 0\}$ – a set of input places of transition t,
- $t\bullet = \{p \mid w(t, p) > 0\}$ – a set of output places of transition t.

This notation can be extended to sets so that, for example, for $S \subset P$ one has $\bullet S = \underset{p \in S}{\cup} \bullet p$. A vector representation of a set of PN nodes remains the same as in previous chapters. To recall; a set of places (transitions) that correspond with nonzero entries in vector **s** is called the *support of* **s**, $S = sup(\mathbf{s})$. In the Petri-net literature support is usually denoted as $\|\mathbf{s}\|$. Next, we define special classes of Petri nets, called *marked graphs* and *state machines*.

Definition 6.1.7 (a marked graph): An ordinary Petri net is called a *marked graph* if $\forall p \in P, |\bullet p| = |p \bullet| = 1$, i.e. each place has one input transition and one output transition.

The Petri net shown in Figure 6.7 does not belong to that class since $|p_3 \bullet| = 2 > 1$, while the one depicted in Figure 6.4 is a marked graph. It has been proved that a marked graph is live if and only if each directed circuit in PN has at least one token under initial marking $\mathbf{m}_0$. This important result can be checked on the PN from Figure 6.4. Three directed circuits exist in this PN, $\{p_1, p_4\}$, $\{p_2, p_5\}$ and $\{p_3, p_4, p_5\}$, with initial marking $\mathbf{m}_0 = [1\ 1\ 1\ 0\ 0]^T$. Therefore, $m(p_1)=1$, $m(p_2)=1$ and $m(p_3)=1$, i.e. exactly one token is provided for each directed circuit. From Figure 6.6 we see that firing sequence $\sigma = t_1 t_2 t_3$, which returns PN in its initial marking, can be repeated an infinite number of times, thus, according to Definition 6.1.4, the PN is live.

Definition 6.1.8 (a state machine): An ordinary Petri net is called a *state machine* if $\forall t \in T, |\bullet t| = |t \bullet| = 1$, i.e. each transition has one input place and one output place.

It is easy to check the liveness property of a strongly connected state machine. Specifically, if initial marking $\mathbf{m}_0$ of a strongly connected state machine has at least one token then the state machine is live. This is a necessary and sufficient condition for state machine liveness.

In the previous chapter we have studied in detail the importance that siphons and traps have in MS analysis. The relation between an empty siphon and deadlock was explained and analytical methods for siphon determination and deadlock avoidance in MRF systems have been proposed. The definitions of siphon and trap given in Section 5.1.3 can be directly applied in Petri nets. Explicitly, in a Petri net a siphon is a set of places S such that every transition having an output place in S has an input place in S. For a set of places in trap Q every transition having an input place in Q has an output place in Q. Furthermore, the properties of these two structures hold for Petri nets as well; once a siphon becomes empty, $m(S)=0$, it remains empty for all successive markings. On the other hand, if a trap is marked under some marking it remains marked under all successive markings.

There are numerous papers published in journals and presented at conferences related to algorithms for siphon determination in PNs. Some algorithms are based on linear inequalities, while others use logical rules or algebraic equations.

However, none of these methods can be directly applied to all classes of Petri nets. Here we demonstrate a simple approach that checks each place in a PN and forms a set of inequalities [4]. As an example, let us use the PN shown in Figure 6.8. We start with the assumption that place p_1 is an element of siphon S. Then, according to the siphon definition, every transition having p_1 as an output place should have an input place in S. Hence, if $p_1 \in S$ then $p_5 \in S$. For place p_2 one has that when $p_2 \in S$ then $p_1 \in S$ or $p_4 \in S$ since $\bullet p_2 = p_1 \bullet = p_4 \bullet = \{t_2\}$. Checking of p_3, p_4 and p_5 gives the following rules; if $p_3 \in S$ then $p_2 \in S$ or $p_5 \in S$, if $p_4 \in S$ then $p_2 \in S$ or $p_5 \in S$, if $p_5 \in S$ then $p_3 \in S$ and ($p_1 \in S$ or $p_4 \in S$). This set of logical rules can be transformed into a set of inequalities written as

$$\begin{aligned}
-p_1 + p_5 &\geq 0 \\
-p_2 + p_1 + p_4 &\geq 0 \\
-p_3 + p_2 + p_5 &\geq 0 \\
-p_4 + p_2 + p_5 &\geq 0 \\
-p_5 + p_3 &\geq 0 \\
-p_5 + p_1 + p_4 &\geq 0
\end{aligned} \tag{6.8}$$

A solution of this system is a binary vector $\mathbf{s} = sup(S)$. For example, $\mathbf{s}_1 = [0\ 0\ 1\ 1\ 1]^T$ satisfies a set of inequalities, thus, $S_1 = \{p_3, p_4, p_5\}$ is a siphon. Another solution, $\mathbf{s}_2 = [1\ 0\ 1\ 0\ 1]^T$ is also a siphon, however, this siphon contains p-invariant and, as we mentioned in the previous chapter, it is not interesting for a deadlock-avoidance supervisory design in MRF systems. It is interesting to note that $\mathbf{s}_3 = [1\ 1\ 1\ 1\ 1]^T$ satisfies the above inequalities, *i.e.* the PN itself is a siphon since it is comprised of two p-invariants.

The second method we present here follows the same reasoning as the approach described above; all possible sets (combinations) of places in a PN should be checked to see if they satisfy the siphon condition. This can be done in various ways and herein we demonstrate a procedure based on a PN incidence matrix in an ordinary PN [6].

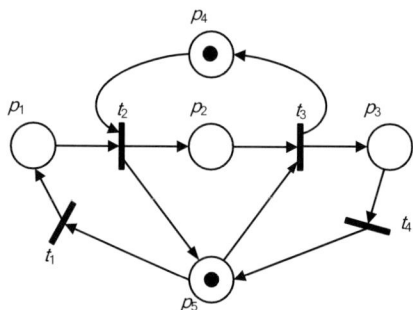

Figure 6.8. An example of a siphon in a PN

For an ordinary PN incidence matrix $\mathbf{W} = \mathbf{O} - \mathbf{I}$. Hence, an element of $\mathbf{W}$ obtains value from a set $\{-1, 0, 1\}$, with $w_{ij} = 1$ for $p_j \in t_i\bullet$, $w_{ij} = -1$ for $p_j \in \bullet t_i$ and $w_{ij} = 0$ for $p_j \notin \{\bullet t_i \cup t_i \bullet\}$. If we assume that p_j belong to siphon S, then for each $w_{ij} = 1$ there must exist $w_{ik} = -1$ with $p_k \in S$. However, when $|t\bullet| > 1$ for some transition in the PN, incidence matrix should be modified. From the PN shown in Figure 6.8 we see that, although it has 4 transitions, we needed 6 logical rules to obtain inequalities (6.7). This is due to the fact that each of transitions t_2 and t_3 has two output places. To cope with this situation we modify the incidence matrix as follows: each $w_{ij} = -1$ should be replaced with $w^*_{ij} = -r_i$, where $r_i = |t_i\bullet|$. Then, for an ordinary PN with m transitions and n places, set $S = \{p_j \mid p_j \in P\}$ is a siphon if and only if

$$\sum_j w^*_{ij} \leq 0 \text{ for all } i = 1, m \tag{6.9}$$

For the PN depicted in Figure 6.8 the incidence matrix $\mathbf{W}$ and the modified incidence matrix $\mathbf{W}^*$ are defined as

$$\mathbf{W} = \begin{bmatrix} 1 & 0 & 0 & 0 & -1 \\ -1 & 1 & 0 & -1 & 1 \\ 0 & -1 & 1 & 1 & -1 \\ 0 & 0 & -1 & 0 & 1 \end{bmatrix}, \quad \mathbf{W}^* = \begin{bmatrix} 1 & 0 & 0 & 0 & -1 \\ -2 & 1 & 0 & -2 & 1 \\ 0 & -2 & 1 & 1 & -2 \\ 0 & 0 & -1 & 0 & 1 \end{bmatrix}$$

As we have to check all combinations of places in PN we start with $S = \{p_1, p_2\}$. According to Equation (6.9)

$$w^*_{11} + w^*_{12} = 1 + 0 > 0$$
$$w^*_{21} + w^*_{22} = -2 + 1 < 0$$
$$w^*_{31} + w^*_{32} = 0 + (-2) < 0$$
$$w^*_{41} + w^*_{42} = 0 + 0 = 0$$

Since the 1st row is greater than 0, set $S = \{p_1, p_2\}$ is not a siphon. We proceed with $S = \{p_1, p_3\}$, $S = \{p_1, p_4\}$, and so on. For $S = \{p_2, p_4\}$ one has

$$w^*_{12} + w^*_{14} = 0 + 0 = 0$$
$$w^*_{22} + w^*_{24} = 1 + (-2) < 0$$
$$w^*_{32} + w^*_{34} = -2 + 1 < 0$$
$$w^*_{42} + w^*_{44} = 0 + 0 = 0$$

thus S is a siphon. When all sets containing two places are checked, the procedure continues with $S=\{p_1, p_2, p_3\}$ and other three-element sets. Applying Equation (6.9) on $S=\{p_3, p_4, p_5\}$ gives

$$w_{13}^* + w_{14}^* + w_{15}^* = 0+0+(-1) < 0$$
$$w_{23}^* + w_{24}^* + w_{25}^* = 0+(-2)+1 < 0$$
$$w_{33}^* + w_{34}^* + w_{35}^* = 1+1+(-2) = 0$$
$$w_{43}^* + w_{44}^* + w_{45}^* = -1+0+1 = 0$$

i.e. S is a siphon, which confirms the result obtained from the set of logical rules (6.8). Finally, the last set to be checked is $S=\{p_1, p_2, p_3, p_4, p_5\}$. Since the sum of elements of each row in $\mathbf{W}^*$ is ≤ 0 this set is a siphon.

The presented method can be easily converted into an algorithm. However, direct realization is time consuming since the procedure is based on the so-called "brute force" approach. On the other hand, for some classes of PNs, the method can be modified in order to reduce computation complexity. We do not elaborate on this issue in more detail since in MRF systems siphons can be determined by implementation of the results discussed in Section 5.1.3, which we illustrate later in this chapter.

There exist many interesting concepts in PN theory, such as complex-valued tokens [11] or continuous Petri nets [12], which widen the usage of PNs in fields that are beyond the scope of this book. However, two types of PNs, namely timed [7] and colored [8] PNs, are commonly used in MS analysis and design, hence, we conclude the basic definitions and properties of Petri nets with brief remarks on these two groups of PNs.

Although the PN state equation (6.2) describes movement of marking vector $\mathbf{m}$ in the state space, it does not cover the system dynamics. As we showed in Chapter 3, the concept of time is essential in performance evaluation of an MS. Therefore, the time durations of the system tasks should be included in a PN model. This can be done in two ways; we can associate time delays with PN places (*p-timed PN*) or PN transitions (*t-timed PN*). Herein we present p-timed PNs.

In principal, we follow the ideas presented in Section 3.3. An additional parameter, *p-delay*, denoted $d(p_i)$, is assigned to each place in PN. In general, $d(p_i)$ is a real number (for a fuzzy timed PN see [9]), deterministic or stochastic, which depends on the character of the modeled system. A p-delay is introduced in a PN in the form of a diagonal matrix $\mathbf{D}[d_{ii}]_{n \times n}$, which requires splitting of marking vector $\mathbf{m}$ into two vectors, one representing all tokens that are available for further propagation through the PN, the *available marking vector* $\mathbf{m}_a$, and the other showing all tokens that are delayed, the *pending marking vector* $\mathbf{m}_p$ [10]. Consequently, a place in a PN graph is split into two parts as shown in Figure 6.9a, while a PN state equation obtains the following form

$$\mathbf{m}_{p,k} = \mathbf{m}_{p,k-1} + \mathbf{W}^T \mathbf{t}$$
$$\mathbf{m}_{a,k} = \mathbf{m}_{a,k-1} + \mathbf{D} \cdot \mathbf{m}_{p,k} \qquad (6.10)$$
$$\mathbf{t} = f(\mathbf{m}_{a,k})$$

It should be noted that only the available marking vector $\mathbf{m}_a$ is used for calculation of enabled transitions. When the transition fires, the tokens are moved into the pending vector $\mathbf{m}_{p,k}$, where they stay until the delay time of a particular place expires. Then, the tokens propagate into $\mathbf{m}_{a,k}$ where they may be used to fire subsequent transitions. For marked graphs, model (6.10) can be transformed into the max-plus form.

The dynamic PN state equation (6.10) gives the correct results only when each place in PN has exactly one input transition. The reason is the same as the one described in Section 3.3. That is, each input transition requires an additional delay parameters, as shown in Figure 6.9b. In this case $\mathbf{D}$ is not a diagonal matrix and the dimensions of vectors $\mathbf{m}_p$ and $\mathbf{m}_a$ are different. Model (6.10) becomes even more complex when place p is not bounded and receives tokens with an input rate that is faster than its delay time. As we already explained in the section related to the modeling of the system dynamics, such a situation involves multiple clocks, *i.e.* each token that enters a place is associated with its own clock. When the clock expires a token is moved from $\mathbf{m}_p$ into $\mathbf{m}_a$. In fact, when an MS is modeled by a PN, pending marking vector $\mathbf{m}_p$ corresponds with vector $\mathbf{m}^s$ in Equation (3.21). Therefore, for deadlock avoidance in a p-timed PN both vectors, $\mathbf{m}_a$ and $\mathbf{m}_p$, should be considered.

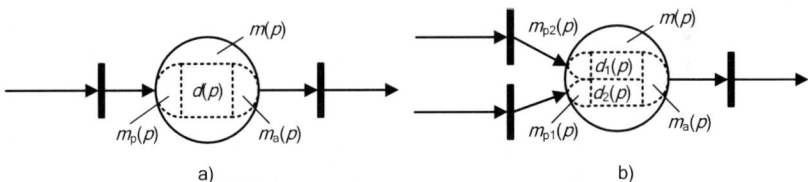

Figure 6.9. Splitting a place in a p-timed PN

Although this is not a topic of the book, it is interesting to mention the application of timed PNs in data processing. If a token is considered as data received from a sensor or some other device, then the time associated with a place could be considered as temporal-information degradation. In other words, after some time the information "value" is decreased and confidence in firing of a particular transition is reduced; if time expires, a token is removed from the place and the transition is no longer enabled.

Colored PNs are generally used for modeling of DESs in which tokens represent a particular property or type of processed part (customer) or an offered service. Various properties (or types) are characterized by different colors or differ-

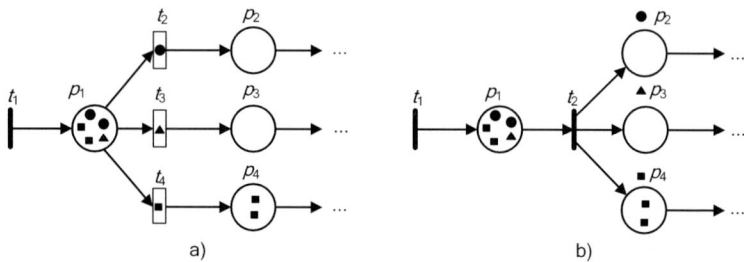

Figure 6.10. Example of colored PNs

ent shapes of tokens. Figure 6.10a shows the PN model of a customer service entry. Customers with different requests arrive into the system and each of them is routed depending on the nature of the request. There are three types of requests, marked with a circle, a triangle and a rectangle. Transition t_1 is a source transition and represents arrival of the customer. Firing of transitions t_2, t_3 and t_4 depends on the marking of place p_1 (input buffer); transition t_2 is enabled with a circle, t_3 is enabled with a triangle, and t_4 is enabled with a rectangle.

A variation of the same system is depicted in Figure 6.10b. In this PN, transition t_2 is enabled with all three types of tokens, while each place accepts only tokens of a specific type; place p_2 can receive only circular, p_3 only triangular, and p_4 only rectangular tokens. A propagation of different tokens through a PN could be associated with arcs as well. One way or the other colored PNs offer a powerful tool for modeling and analysis of complex and demanding systems. However, the final PN graph and the underlying PN state equation can be very difficult to understand.

6.2 Manufacturing Systems Modeling

In a PN model of an MS, described herein, places are associated with operations and resources, while transitions represent starting and ending of operations and tasks. Therefore, recalling the definitions given in Section 3.1, a set of places $P = P^* \cup PI \cup PO$, $P^* = R \cup J \cup P_0$, where $R = \bigcup_{k \in \Pi} R^k$ is a set of resources, $J = \bigcup_{k \in \Pi} J^k$ is a set of operations, $P_0 = \bigcup_{k \in \Pi} P_0^k$ is a set of *pallets*, and Π is the set of distinct types of parts produced (or customers served) by an MS. As we stated in Section 3.1, each part type has a predetermined sequence of operations (except for FMRF) that starts with a raw part-in operation, $J_{in}^k \in PI$, represented by a source place, and a finished product-out operation, $J_{out}^k \in PO$, represented by a sink place. We consider a source place as a token generator. That is, tokens appear in a source place according to a specified function or stochastically. On the other hand, a sink place is considered as a drain, *i.e.* a token is removed from the sink place immediately upon arrival. Pallets are used for carrying parts through the system.

PN models of non-shared and shared resources are depicted in Figure 6.11. A nonshared resource a) is represented with two places, R and J^a_R. A token in place R marks the availability of a resource, while a token in place J^a_R denotes that the resource executes the corresponding operation. The initial number of tokens is equal to the number of parts that can be simultaneously processed by the resource. A sequential shared resource b) performs more than one operation *on the same part type*; each operation is represented by one place, while resource availability is characterized with a token in place R_S. A parallel shared resource c) performs more than one operation *on different part types*. Obviously, a shared resource that executes some operations on the same part type and others on different part types can be represented as a combination of models b) and c).

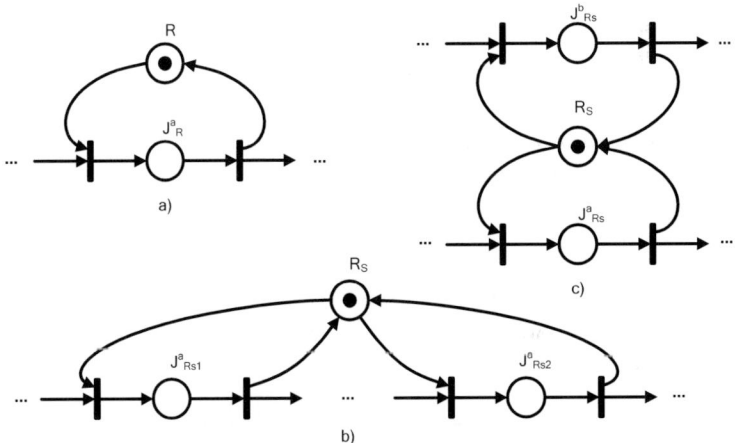

Figure 6.11. PN models of nonshared (**a**), sequentially shared (**b**) and parallel shared (**c**) resources

It is assumed that the resources shown in Figure 6.11 are released immediately upon completion of an operation. Generally, this is not the case. A resource could perform two or more operations, one after the other, as shown in Figure 6.12. In this case the last operation to be performed is the one that releases the resource (in our case resource R executes J^a_1 and then J^a_{2R}). Furthermore, it may happen that one operation requires more than one resource. In the PN depicted in Figure 6.12 operation J^a_1 requires resources R and R_A in order to be performed.

In Chapter 5 we analyzed some properties of free-choice multiple re-entrant flowlines. In the example that belongs to this class of systems and is shown in Figure 5.6 some resources cannot be described with the PN models presented so far. It can be seen that buffer B2, for example, receives parts from three machines and distributes these parts to two machines. If one considers each input individually, then the place representing the occupied buffer requires three input transitions. Additionally, each output from the buffer is represented by one transition. Thus, buffer B2 is modeled as shown in Figure 6.13. Each token in place B2A stands for an unoccupied slot, while tokens in B2P represent parts held by the buffer. This PN model is obtained by combination of the free-choice and merge prototypes shown in Figure 6.14. An assembly operation, which is commonly used in MS, could be modeled as a combination of several resource prototypes. Figure 6.15 depicts one of the possible configurations that describes the assembly of two parts, *a* and *b*, in resource R.

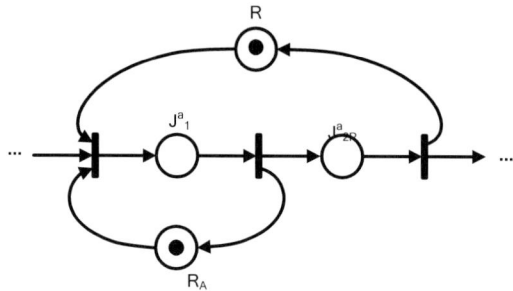

Figure 6.12. A PN model of a nonshared resource with two operations in sequence and an operation that requires two resources

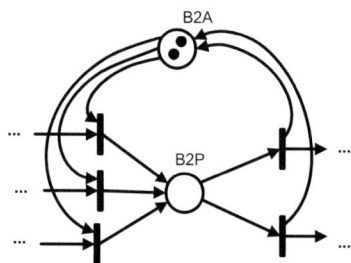

Figure 6.13. A PN model of buffer B2 in the system shown in Figure 5.6

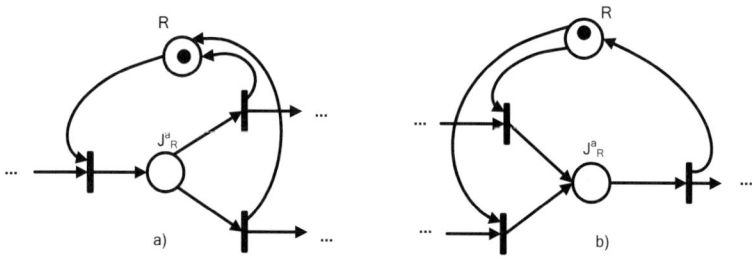

Figure 6.14. A PN model of a free choice (**a**) and merge (**b**)

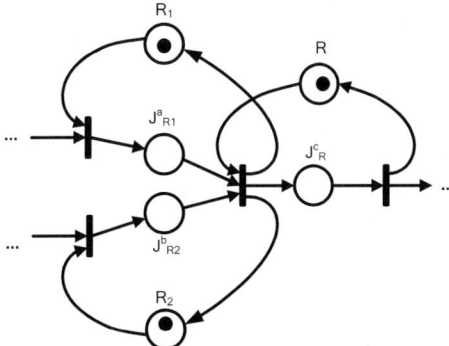

Figure 6.15. A PN model of an assembly

When the last operation on part a, J^a_{R1}, is finished and part b is ready (operation J^b_{R2} is completed), resource R takes both parts and creates a new part type c by execution of assembly operation J^c_R.

Based on the presented PN models one can conclude that, in general, physical entities of an MS, machines, robots, conveyer belts, *etc.*, could be identified directly from the PN model. Still, there are examples where some parts of the system, which in fact do not belong to the class of resources, are represented with models shown in Figure 6.11. Such an example is the multi-AGV system depicted in Figure 6.16. In this example, paths and crossing areas (sometimes called blocking areas) used by vehicles are considered as resources. Figure 6.16 shows a crossing area and its PN model that corresponds with a model of a shared resource.

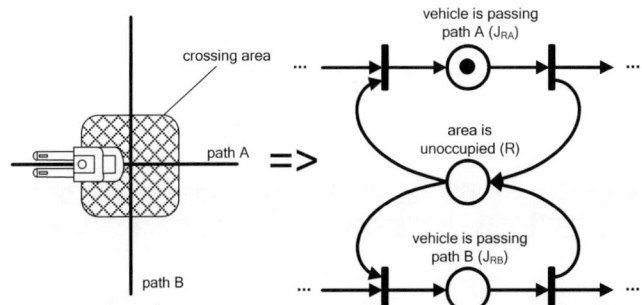

Figure 6.16. A PN model of a crossing area in a multi-AGV system

Having described resource prototypes, we can define the properties of a PN for MRF systems:

- $\forall p \in P^*, \bullet p \cap p \bullet = \emptyset$; there are no self-loops,
- $\forall k \in \Pi, t^k_1 \bullet \cap P^* \setminus J = \emptyset$ and $\bullet t^k_L \cap P^* \setminus J = \emptyset$; each part path has a well-defined beginning and an end,

230 Manufacturing Systems Control Design

- $\forall J_i^k \in J, \left|R(J_i^k)\right| = 1$ and $R(J_i^k) \neq R(J_{i+1}^k)$; each operation requires one and only one resource and the same resource cannot execute two successive operations,
- $\forall p \in J, |p \bullet| = 1$; there are no free-choice operations,
- $\forall t, |\bullet t \cap J| \leq 1$; there are no assembly operations,
- there exists at least one shared resource.

For MRF systems, for any $r \in R$, $J(r) = r \bullet \bullet \cap J = \bullet \bullet r \cap J$ and $R(J_i^k) = J_i^k \bullet \bullet \cap R = \bullet \bullet J_i^k \cap R$.

Let us now consider the assembly tree depicted in Figure 3.1. We start construction of a PN model by assigning one place with each operation in the job sequence, as shown in Figure 6.17a. Then, an extra place, representing an idle nonshared resource, is joined with a place that represents an operation performed by that particular nonshared resource (Figure 6.17b). Three places are added in the PN model; MA, B and MB, representing drilling machine, buffer and grinding machine, respectively. Next, shared resource(s), together with source (an input, PI) place for parts entering the system, and sink (an output, PO) place for parts leaving the system are added, as shown in Figure 6.17c, (the considered system has only one shared resource, thus, one place, denoted R, is added). Finally, initial marking is assigned to the PN model and transitions are denoted. As can be seen, it is assumed that resources are idle. Each machine can process one part at a time, with a buffer having two empty slots, and an input place with three parts waiting to enter the system.

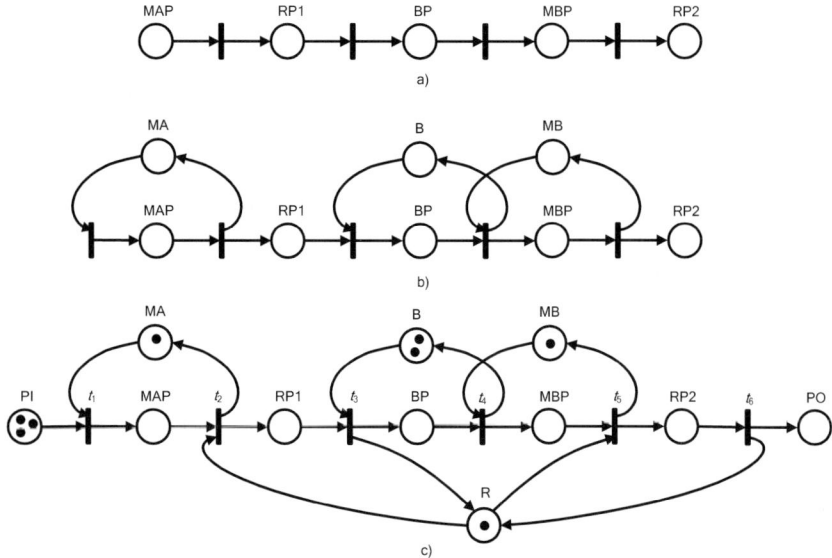

Figure 6.17. A PN model of a job sequence from Figure 3.1; (a) operations, inclusion of (b) nonshared resources, and (c) shared resource together with input and output places

The attained PN is pure and ordinary, with P = {PI, MAP, RP1, BP, MBP, RP2, MA, MB, B, R, PO}, T = {t_1, t_2, t_3, t_4, t_5, t_6}, initial marking $\mathbf{m}_0$=[3 0 0 0 0 0 1 1 2 1 0]^T, $\mathbf{M}$=[$\mathbf{O}$ | $\mathbf{I}$], $\mathbf{W}$=$\mathbf{O}$–$\mathbf{I}$, and

$$\mathbf{I} = \begin{bmatrix} 1 & 0 & 0 & 0 & 0 & 0 & 1 & 0 & 0 & 0 & 0 \\ 0 & 1 & 0 & 0 & 0 & 0 & 0 & 0 & 0 & 1 & 0 \\ 0 & 0 & 1 & 0 & 0 & 0 & 0 & 0 & 1 & 0 & 0 \\ 0 & 0 & 0 & 1 & 0 & 0 & 0 & 1 & 0 & 0 & 0 \\ 0 & 0 & 0 & 0 & 1 & 0 & 0 & 0 & 0 & 1 & 0 \\ 0 & 0 & 0 & 0 & 0 & 1 & 0 & 0 & 0 & 0 & 0 \end{bmatrix}$$

$$\mathbf{O} = \begin{bmatrix} 0 & 1 & 0 & 0 & 0 & 0 & 0 & 0 & 0 & 0 & 0 \\ 0 & 0 & 1 & 0 & 0 & 0 & 1 & 0 & 0 & 0 & 0 \\ 0 & 0 & 0 & 1 & 0 & 0 & 0 & 0 & 0 & 1 & 0 \\ 0 & 0 & 0 & 0 & 1 & 0 & 0 & 0 & 1 & 0 & 0 \\ 0 & 0 & 0 & 0 & 0 & 1 & 0 & 1 & 0 & 0 & 0 \\ 0 & 0 & 0 & 0 & 0 & 0 & 0 & 0 & 0 & 1 & 1 \end{bmatrix}$$

6.2.1 Petri-Net Controller

From the discussions in previous chapters it is clear that the notion of *state* is one of the central points in the system theory. By using various modeling techniques one is able to characterize the system behavior as movement of the state vector in the state space. Then, specifications regarding system performance may be given in the form of regions in the state space; some of these regions are preferred, while the others are forbidden. Due to its ability to capture the structural properties of the modeled system, the PN formalism is particularly convenient for implementation of this approach in the DES analysis and design. By controlling firing of transitions one can keep the system in the desired region of the state space, thus avoiding illegal states. This can be done with insertion of *control places* in an uncontrolled PN model of the system.

In this section we demonstrate how to add control places in a given PN, and how to determine their initial marking, which depends on the structure of the system and its initial state. Although many techniques for PN controller design have been proposed in the literature, we limit our discussion to a relatively simple approach based on p-invariants. Our main concern in PN controller design is the same as in the previous chapters, that is, prevention of conflict and deadlock (the

PN controller should guarantee liveness). We assume that all transitions (or at least those connected with control places) are controllable and observable.

First, let us study a conflict. We know that conflict in MS is related to the shared resources. The occurrence of simultaneous requests from two (or more) tasks that use the same resource must be handled by the supervisor, *i.e.* the decision should be made regarding a priority. One way to prevent conflict in the PN model of a shared resource R_s, is to add a control place as an input to each transition that belongs to set $R_s\bullet$, as depicted in Figure 6.18. Such a control place does not have input transitions. In other words each place is a *source* that generates tokens according to some control function, $m(u_{di}) = h_i(\mathbf{m})$. Evidently, each function $h_i(\mathbf{m})$ should be defined so that markings of control places are mutually exclusive, as for the matrix controller described in Section 3.4. Hence, the relation

$$\sum_i m(u_{di}) = 1 \qquad (6.11)$$

must be fulfilled each time conflict occurs.

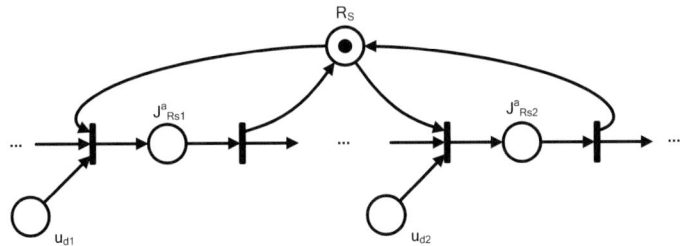

Figure 6.18. A conflict resolution in a PN, $m(u_{di}) = h_i(\mathbf{m})$

When control places are responsible only for conflict resolution, requirement (6.11) can be satisfied directly by *synchronization* of two (or more) transitions involved in a conflict, as shown in Figure 6.19. This solution is very restrictive from the resource-utilization point of view since only one token is allowed to enter the part of the PN within conflicting transitions (usage of only one control place, u_{d1}, will have the same effect). It should also be noted that initial marking of control places may be a reason for a dead PN.

Once conflict is resolved we can concentrate on the deadlock avoidance. Control of the number of tokens in a particular part of the PN is the main mechanism in the deadlock prevention [5]. This is expected since analysis of the relation between a deadlock and an empty siphon showed that the control strategy should assure that at least one place belonging to the siphon is marked at any time.

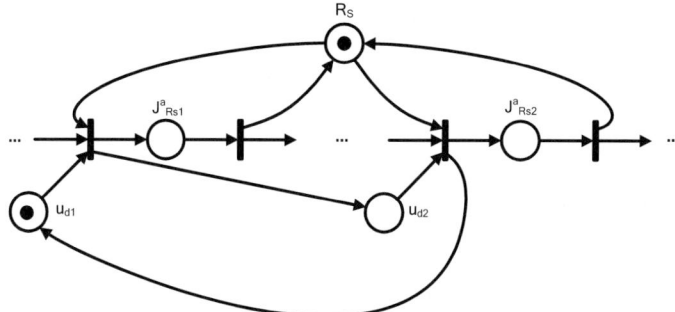

Figure 6.19. A conflict resolution in a PN by synchronization of conflicting transitions

Here we present a method that is proposed in [14]. The basic idea is to restrain the number of tokens in subsets of PN places by using linear inequalities

$$\mathbf{L} \cdot \mathbf{m} \leq \mathbf{b} \tag{6.12}$$

where $\mathbf{L}$ is an $l \times n$ integer matrix, l is the number of inequalities, n is the number of places in PN, $\mathbf{m}$ is a marking vector of an uncontrolled PN, and $\mathbf{b}$ is an integer column vector. Constraints (6.12) can be transformed into the set of linear equations in matrix form

$$\mathbf{L} \cdot \mathbf{m} + \mathbf{u}_d = \mathbf{b} \tag{6.13}$$

where $\mathbf{u}_d$ is the marking of control places added to an uncontrolled PN.

Implementation of Equation (6.13) requires determination of a) the incidence matrix of closed-loop (controlled) PN, and b) initial marking of control places, $\mathbf{u}_d(0) = \mathbf{u}_{d0}$. First, we extend marking vector $\mathbf{m}$ in order to incorporate control places, $\mathbf{m}_d = [\mathbf{m} \ \mathbf{u}_d]^T$. This extension requires a change in the closed-loop PN incidence matrix $\mathbf{W}$, which becomes $\mathbf{W}_c = [\mathbf{W} \ \mathbf{W}_d]$, where $\mathbf{W}_d$ is an unknown incidence matrix that comprises information regarding connections of control places with transitions of uncontrolled PN. Then, from Equation (6.5) it follows

$$\mathbf{W}_c \cdot \mathbf{P} = [\mathbf{W} \ \mathbf{W}_d] \cdot \mathbf{P} = 0 \tag{6.14}$$

where $\mathbf{P}$ is a p-invariant matrix formed of p-invariant vectors.

Matrix equation (6.13) should be satisfied at any time, hence

$$\mathbf{L} \cdot \mathbf{m}_k + \mathbf{u}_{dk} = [\mathbf{L} \ \mathbf{I}] \cdot \begin{bmatrix} \mathbf{m}_k \\ \mathbf{u}_{dk} \end{bmatrix} = \mathbf{b} = const. \tag{6.15}$$

By comparing this equation with Equation (6.7) we see that each row of matrix $[\mathbf{L} \ \mathbf{I}]$ in fact represents the p-invariant of a closed loop PN, *i.e.*

234   Manufacturing Systems Control Design

$$\begin{bmatrix} \mathbf{L}^T \\ \mathbf{I} \end{bmatrix} = \mathbf{P} \tag{6.16}$$

Including Equation (6.16) in Equation (6.14) yields

$$[\mathbf{W} \ \mathbf{W}_d] \cdot \begin{bmatrix} \mathbf{L}^T \\ \mathbf{I} \end{bmatrix} = 0 \tag{6.17}$$

which provides the relation for calculation of $\mathbf{W}_d$,

$$\mathbf{W}_d = -\mathbf{W} \cdot \mathbf{L}^T \tag{6.18}$$

Initial marking of control places can be directly obtained from Equation (6.13),

$$\mathbf{L} \cdot \mathbf{m}_0 + \mathbf{u}_{d0} = \mathbf{b} \quad \Rightarrow \quad \mathbf{u}_{d0} = \mathbf{b} - \mathbf{L} \cdot \mathbf{m}_0 \tag{6.19}$$

This result shows that a supervisor will impose constraints (6.12) only for those initial markings that give $\mathbf{u}_{d0} > 0$, since fulfillment of Equation (6.19) implies $\mathbf{L} \cdot \mathbf{m}_0 \leq \mathbf{b}$.

Example 6.2.1 (p-invariant-based PN controller)

We demonstrate p-invariant controller design on the workcell shown in Figure 2.12. A PN model should be developed based on a description of the system given in Example 2.2.1. First we identify the set of operations required for production of parts *a* and *b*. The PN model of both sequences is depicted in Figure 6.20.

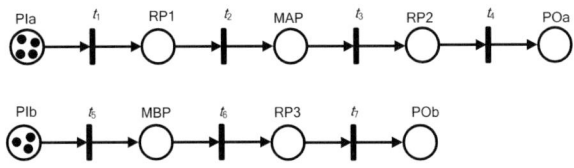

Figure 6.20. Operations sequences for the workcell shown in Figure 2.12

The next step in PN modeling is allocation of resources. The PN graph shown in Figure 6.21 is obtained by using resources prototypes described in the previous section. It is worth noting that the obtained PN model replicates a structure of the system, which is not the case with automaton representation of the same workcell (Figure 2.17). The shared resource in the system is robot R, which executes three tasks; two on part *a* path and one on part *b* path. It is assumed that both machines have the same capacity of one part at a time.

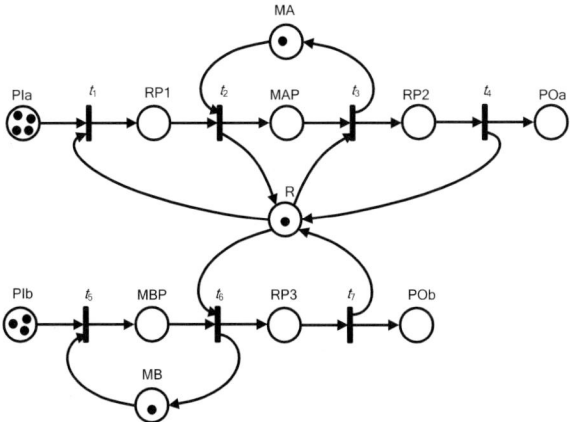

Figure 6.21. A PN model of the workcell shown in Figure 2.12

The attained PN belongs to the MRF class. It is pure and ordinary, with $P = \{$PIa, PIb, RP1, MAP, RP2, MBP, RP3, MA, MB, R, POa, POb$\}$, $T = \{t_1, t_2, t_3, t_4, t_5, t_6, t_7\}$, initial marking $\mathbf{m}_0 = [4\ 3\ 0\ 0\ 0\ 0\ 0\ 1\ 1\ 1\ 0\ 0]^T$, $\mathbf{M} = [\mathbf{O} \mid \mathbf{I}]$, $\mathbf{W} = \mathbf{O} - \mathbf{I}$, and

$$\mathbf{I} = \begin{bmatrix} 1 & 0 & 0 & 0 & 0 & 0 & 0 & 0 & 0 & 1 & 0 & 0 \\ 0 & 0 & 1 & 0 & 0 & 0 & 0 & 1 & 0 & 0 & 0 & 0 \\ 0 & 0 & 0 & 1 & 0 & 0 & 0 & 0 & 0 & 1 & 0 & 0 \\ 0 & 0 & 0 & 0 & 1 & 0 & 0 & 0 & 0 & 0 & 0 & 0 \\ 0 & 1 & 0 & 0 & 0 & 0 & 0 & 0 & 1 & 0 & 0 & 0 \\ 0 & 0 & 0 & 0 & 0 & 1 & 0 & 0 & 0 & 1 & 0 & 0 \\ 0 & 0 & 0 & 0 & 0 & 0 & 1 & 0 & 0 & 0 & 0 & 0 \end{bmatrix}$$

$$\mathbf{O} = \begin{bmatrix} 0 & 0 & 1 & 0 & 0 & 0 & 0 & 0 & 0 & 0 & 0 & 0 \\ 0 & 0 & 0 & 1 & 0 & 0 & 0 & 0 & 0 & 1 & 0 & 0 \\ 0 & 0 & 0 & 0 & 1 & 0 & 0 & 1 & 0 & 0 & 0 & 0 \\ 0 & 0 & 0 & 0 & 0 & 0 & 0 & 0 & 0 & 1 & 1 & 0 \\ 0 & 0 & 0 & 0 & 0 & 1 & 0 & 0 & 0 & 0 & 0 & 0 \\ 0 & 0 & 0 & 0 & 0 & 0 & 1 & 0 & 1 & 0 & 0 & 0 \\ 0 & 0 & 0 & 0 & 0 & 0 & 0 & 0 & 0 & 1 & 0 & 1 \end{bmatrix}$$

The system analysis, given in example 2.2.1, confirmed the existence of operation sequences that can lead the system to deadlock, which corresponds to the situation when both machines are processing parts while the robot carries part *a*. Inspection of the PN shown in Figure 6.21 reveals the existence of critical siphon $S_C=\{RP2, RP3, MA, R\}$ (one of the previously described methods for the siphon detection could be used for this purpose). A constraint that should be enforced by the supervisor must provide that $m(S_C) \geq 1$ at any time. Relation (6.12) attains the form

$$\mathbf{L} \cdot \mathbf{m} \geq \mathbf{b}$$

where $\mathbf{L} = [0\ 0\ 0\ 0\ 1\ 0\ 1\ 1\ 0\ 1\ 0\ 0]$ and $\mathbf{b} = [1]$ (note that $\mathbf{L} = \mathbf{s}_C$ with $sup(\mathbf{s}_C) = S_C$).

One control place is required since there is only one constraint. Its initial marking is obtained from

$$\mathbf{L} \cdot \mathbf{m}_0 - u_{d0} = b \;\Rightarrow\; 2 - u_{d0} = 1 \;\Rightarrow\; u_{d0} = 1$$

Matrix $\mathbf{W}_d$ is calculated from

$$\mathbf{W}_d = \mathbf{W} \cdot \mathbf{L}^T =$$

$$= \begin{bmatrix} -1 & 0 & 1 & 0 & 0 & 0 & 0 & 0 & 0 & -1 & 0 & 0 \\ 0 & 0 & -1 & 1 & 0 & 0 & 0 & -1 & 0 & 1 & 0 & 0 \\ 0 & 0 & 0 & -1 & 1 & 0 & 0 & 1 & 0 & -1 & 0 & 0 \\ 0 & 0 & 0 & 0 & -1 & 0 & 0 & 0 & 0 & 1 & 1 & 0 \\ 0 & -1 & 0 & 0 & 0 & 1 & 0 & 0 & -1 & 0 & 0 & 0 \\ 0 & 0 & 0 & 0 & 0 & -1 & 1 & 0 & 1 & -1 & 0 & 0 \\ 0 & 0 & 0 & 0 & 0 & 0 & -1 & 0 & 0 & 1 & 0 & 1 \end{bmatrix} \cdot \begin{bmatrix} 0 \\ 0 \\ 0 \\ 0 \\ 1 \\ 0 \\ 1 \\ 1 \\ 0 \\ 1 \\ 0 \\ 0 \end{bmatrix} = \begin{bmatrix} -1 \\ 0 \\ 1 \\ 0 \\ 0 \\ 0 \\ 0 \end{bmatrix}$$

As a result, the control place has transition t_1 as an output, and transition t_3 as an input. The controlled PN is depicted in Figure 6.22. It can be seen that the control place is blocking transition t_1 when one token is remaining in siphon S_C. Since t_1 draws tokens from the siphon, this mechanism prevents the siphon from becoming empty. Actually, control place u_d limits the number of tokens in places RP1 and MAP since these two places, together with u_d, form p − invariant {RP1,

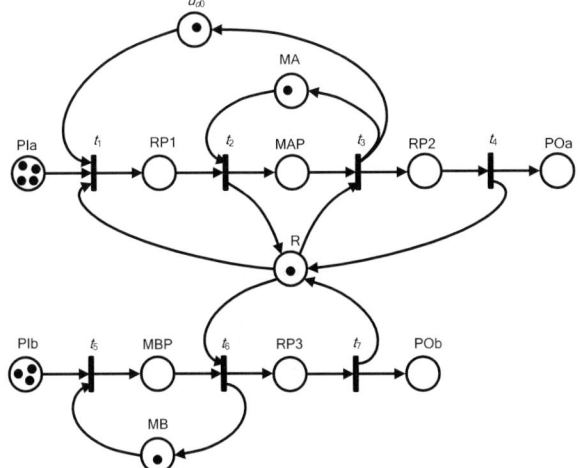

Figure 6.22. Controlled PN model of the workcell shown in Figure 2.12

MAP, u_d}. This detail is important for deadlock prevention and resources utilizations in MRF systems.

♦

Two issues, concerning siphon control and inclusion of control places in an uncontrolled PN, have to be addressed. The first one is related to observability and controllability of transitions. Each constraint stated in Equation (6.12) requires one control place that receives tokens from and dispatches tokens to transitions of the uncontrolled PN. This mechanism is feasible only when transitions that belong to $\bullet u_d$ are observable, and those belonging to $u_d \bullet$ are controllable. Constraints that generate such control places are called *admissible*. The admissibility of constraints can be tested by the following relations

$$\begin{aligned} \mathbf{W}_{uc} \cdot \mathbf{L}^T &\geq 0 \\ \mathbf{W}_{uo} \cdot \mathbf{L}^T &= 0 \end{aligned} \qquad (6.20)$$

where $\mathbf{W}_{uc}$ is an incidence matrix containing rows corresponding to uncontrollable transitions, and $\mathbf{W}_{uo}$ is an incidence matrix containing rows corresponding to unobservable transitions.

The second problem related to inclusion of control places in an uncontrolled PN lies in the fact that new places could generate new siphons. Therefore, the above method for siphon control, as well as many others, is based on an iterative procedure, *i.e.* realization of one constraint from Equation (6.12) could generate new constraint(s). More details regarding an iterative algorithm and requirements for its completion can be found in [13].

6.3 Relation Between Petri Nets and Matrix Form

In Chapter 1 it was mentioned that system matrices are closely related with Petri nets. Actually, as we shall demonstrate in this section, there is a direct relation between these two mathematical formalisms. This is expected since both tools are used for DES analysis and controller design.

The logical state vector **x** in the matrix-based approach associates logical conditions, in the form of availability of resources and parts, with consequences in the form of actions taken upon fulfillment of conditions. According to Definitions 3.1.2 and 3.1.3 matrices $\mathbf{F}_v$ and $\mathbf{F}_r$ capture conditions, while matrices $\mathbf{S}_v$ and $\mathbf{S}_r$ are responsible for actions. If we correlate components of the logical state vector with transitions in an ordinary and pure PN, then the system matrices can be directly associated with the arcs connecting transitions and places, as shown in Figure 6.23.

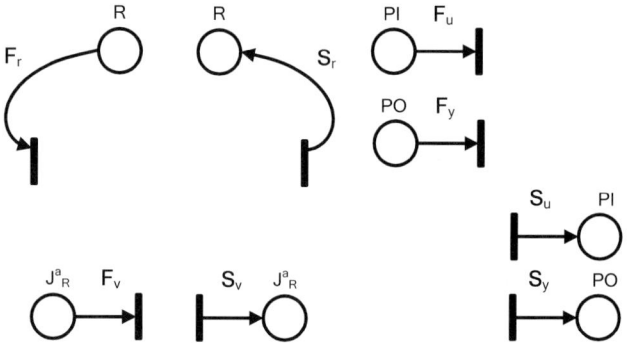

Figure 6.23. Relations between PN arcs and the system matrices

Each entry "1" in the resource-requirements matrix $\mathbf{F}_r$ is associated with an arc connecting a place, representing resource availability, with the corresponding transition; 1s in the resource-release matrix $\mathbf{S}_r$ express the connections between PN transitions and places that hold tokens when resources are idle. Correspondingly, 1s in matrices $\mathbf{F}_v$ and $\mathbf{S}_v$ represent arcs connecting transitions and places associated with operations executed by MS resources. The input matrix $\mathbf{F}_u$ portrays output arcs from input places, while output matrix $\mathbf{S}_y$ depicts input arcs to output places. Since we assume that input places are sources and output places are sinks, matrices $\mathbf{F}_y$ and $\mathbf{S}_u$ are null matrices, $\mathbf{F}_y = \mathbf{S}_u = [\mathbf{0}]$.

As a result, PN input and output incidence matrices can be obtained directly from the system matrices,

$$\mathbf{I} = [\mathbf{F}_u \quad \mathbf{F}_v \quad \mathbf{F}_r \quad \mathbf{F}_y] = \mathbf{F}$$
$$\mathbf{O} = [\mathbf{S}_u^T \quad \mathbf{S}_v^T \quad \mathbf{S}_r^T \quad \mathbf{S}_y^T] = \mathbf{S}^T \tag{6.21}$$

Even though **I** and **O** matrices define the form of a PN, they do not provide consistent and straightforward information regarding the structure of the modeled MS. By partitioning these matrices in accordance with Figure 6.23 and Equation

(6.21), one is capable of distinguishing between places that represent MS tasks and places indicating resources that perform these tasks. Moreover, the system inputs and outputs can be clearly distinguished. Now, if we include Equation (6.21) in the marking transition equation (6.2), then

$$\mathbf{m}_k = \mathbf{m}_{k-1} + (\mathbf{S} - \mathbf{F}^T)\mathbf{t} \qquad (6.22)$$

which coincides with Equation (3.12).

It is evident that the PN model, consisting of resources prototypes described in the previous section, can be constructed directly from the system matrices, which we demonstrate in the example that follows.

Example 6.3.1 (determination of PN from the system matrices)

We use matrices that describe the system analyzed in the case study in Section 5.4.

[matrices F_v, F_r, F_u, S_v, S_r, S_y shown]

The structural properties of the PN can be read from the system matrices. A number of rows of F-matrices, as well as a number of columns of S-matrices, defines a number of transitions, which in our case is 13. Matrix $\mathbf{F}_u$ has two columns, each of them corresponding to one input place, while the rows of matrix $\mathbf{S}_y$ match two output places. This information, together with the fact that $\mathbf{F}_v$ ($\mathbf{S}_v$) has no "1s" in rows (columns) in which $\mathbf{F}_u$ ($\mathbf{S}_y$) has an element equal to 1, points out that PN will have two part paths.

Let us denote part paths inputs as p_{i1} and p_{i2}, and part paths outputs as p_{o1} and p_{o2}. Furthermore, we denote places that stand for operations as $p_{v1}, p_{v2}, ..., p_{v11}$ (there are 11 columns in $\mathbf{F}_v$), and places that represent resources availability as $p_{r1}, p_{r2}, ..., p_{r8}$ (8 columns in $\mathbf{F}_r$). Then, matrix element $f_u(1,1)=1$ corresponds to $w(p_{i1},t_1)=1$, $f_v(2,1)=1$ corresponds to $w(p_{v1},t_2)=1$, $f_r(1,3)=1$ corresponds to $w(p_{r3},t_1)=1$, and so on. On the other hand, matrix element $s_v(1,1)=1$ corresponds to $w(t_1,p_{v1})=1$, $s_r(1,3)=1$ corresponds to $w(t_3,p_{r1},)=1$, $s_y(1,6)=1$ corresponds to $w(t_6,p_{o1})=1$. Following the same reasoning one is able to determine all PN arcs.

240   Manufacturing Systems Control Design

The PN graph of the system described with given matrices is shown in Figure 6.24. The model has two part paths with one parallel and one combined shared resource. Notations used in the case study are placed in parentheses.

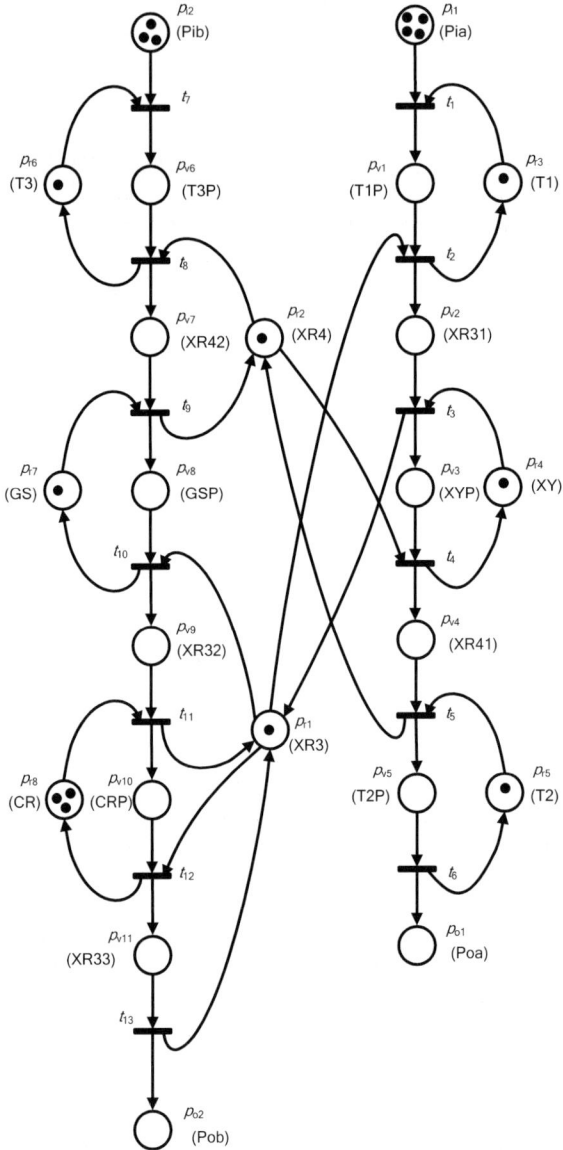

Figure 6.24. PN model of the system described in the case study in Section 5.4

It is apparent that the system analysis provided in Chapter 5 can be directly applied to a PN, given that the attained PN belongs to the MRF class. In addition, string composition presented in Chapter 4, can be used for calculation of circular paths connecting PN resource places, resulting in circular waits (as we know, critical siphons in MRF systems comprise circular waits). Alternatively, CWs can be determined directly from a PN graph. Since each circular wait includes at least one shared resource, one can move along PN arcs that connect resource places, starting with a shared resource. When a tour completes in the starting place, the executed path represents a circular wait. This can be illustrated in the PN in Figure 6.17c. The shared resource R has two output arcs, one connecting t_2 and the other one connecting t_5. Arriving in t_2 from R we can proceed along arc $t_2 \rightarrow$ MA, and then further along arc MA$\rightarrow t_1$. As there are no arcs that connect resource places with t_1, the path is completed. Evidently, the executed path is not circular. On the other hand, moving along arc R$\rightarrow t_5$ we can move further to MB and then to B. From transition t_3 we are returning to place R, which closes up a circular wait. Both paths are shown in Figure 6.25 (note the similarity with the wait relation graph in Figure 5.1).

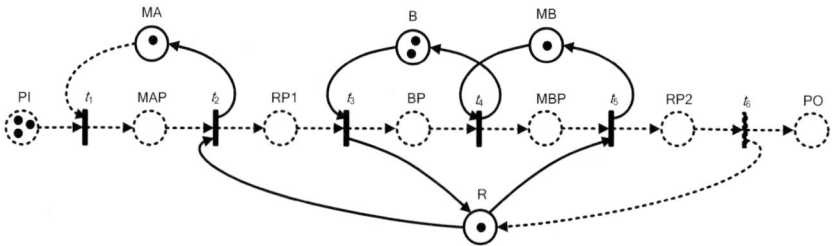

Figure 6.25. Wait relations in PN model of the workcell shown in Figure 3.2

Determined by the string composition or directly from PN graph, circular waits are starting points in the implementation of a PN controller, which could be based on the analysis given in the previous chapter. All definitions and conclusions developed therein can be applied to a PN by simple substitution of the logical state vector **x** with the transition vector **t**. Such, for example, precedent rules become precedent transitions, posterior rules becomes posterior transitions, and so on. Additionally, PN marking vector **m** is equivalent to the state vector, purposely denoted **m**, in the matrix model. Hence, most of the MS structures presented in vector form and involved in matrix equations can be recognized in the PN. Let us mention just two of them. The first one is a resource loop; Equations (5.2) and (5.43) directly associate resource loops in an MRF system with p-invariants in its PN model. The second structure is a critical subsystem; when rows of matrix **L** in Equation (6.12) are built from critical subsystem vectors $[\mathbf{v}_{OC} \; \mathbf{0}_n]$ determined by Equation (5.25) and $\mathbf{b}=\mathbf{m}_0(C)$, then the p-invariant controller, Equation (6.18), with initial conditions, Equation (6.19), provides deadlock-free behavior of the system (compare Equation (6.13) with Equation (5.48)).

6.4 Petri Nets Simulation and Implementation

There are two main groups of solution methods used in the system analysis. Methods in the first group rely on the analytical approach, while methods in the second group use simulation. Which method is used depends mainly on the character of the system and the designer's affinity. Although analytical methods offer not only accurate results but also a deep insight in the system itself, usually they suffer from complexity and may even become inapplicable in the case of large systems. Very often instead of an original method its approximation is used. This is particularly widespread in the case of analytical methods that find their applications in industry. Engineers that work onsite with real-world problems are enforced to apply approximations due to time restrictions posed on the system commissioning.

With the rapid growth of the computational power and in an industry that is cost competitive, simulation methods have become more and more popular. Their progress can be tracked in two directions; one that is related to development of faster methods that can adopt parallelism in the execution of mathematical algorithms [28], and the other that deals with the presentation of the attained results. In the previous chapters we introduced the matrix-based approach to the DES analysis that is convenient for simulation, while the last chapter of the book is devoted to the presentation of simulation results. In this section we give an insight into PN simulation together with a description of the DES simulation tool Petri.NET, which was developed in the Laboratory of Robotics and Intelligent Systems at the Department of Control and Computer Engineering, Faculty of Electrotechnics and Computation, University of Zagreb.

Petri nets, as a mathematical and graphical tool, are especially suitable for simulation. Driven by a very simple mechanism, reduced to two basic rules, from the algorithmic point of view they suggest a large diversity of solutions. This is why an extensive number of PN simulation packages is currently offered on the market [31]. Some of them are very sophisticated (and expensive) with features that allow simulation and analysis of a whole corporation on the highest, corporate, level, while others are intended to be used for small-scale systems (usually offered free of charge).

Even though all of these tools have the same purpose their differences are mainly in the operating system (OS), programming language, graphical user interface (GUI), simulation capabilities and analytical capabilities.

Today, most of the tools work on a Windows platform, but only ten years ago Unix systems were predominant together with DOS. Most of the early applications were programmed in C and C++, but with development of Java, an interpreted, object-oriented, portable, and multithreaded programming language, applications became independent of OS. Some of them even evolve in a way that provides writing of new features that can be incorporated in existing code [29]. At the same time extensive use of XML speeds up data transfer. A further step ahead is the appearance of open-source applications [30].

One of the benefits of PN is their graphical capability, which is extensively used in GUI design and presentation of results. Almost all PN simulation tools are more or less attractive and user friendly, GUI with drag-and-drop ability. Some kind of graphical editor is used for model definition with a token game as a result of simulation. Generally, features such as backward simulation, step-by-step simulation and pause, are integral parts of applications.

The main differences between PN simulation tools are their analytical capabilities. Many of them do not provide any analysis of the PN model. Some of them do provide analysis of the reachability tree (its construction and representation) together with determination of liveness and boundedness. Additionally, some of the simulation tools have statistics analysis, such as the number of times a transition fires, the average number of tokens in a place, *etc*. Investigation of structural properties, such as p and t invariants is rarely included in applications.

In the rest of the section we describe the PN simulation tool Petri.NET. This tool, written in .NET for a Windows platform, incorporates features that are typical of most of PN simulators. Additionally, it comprises some specific properties required for analysis MRF systems and implementation of MS supervisory controller.

The main window of the Petri.NET GUI is shown in Figure 6.26. It comprises three tabs (central part of the screen): PetriNet Editor, Description and Response, and four dockable frames: Toolbox, Document Explorer, Properties and Rules Editor.

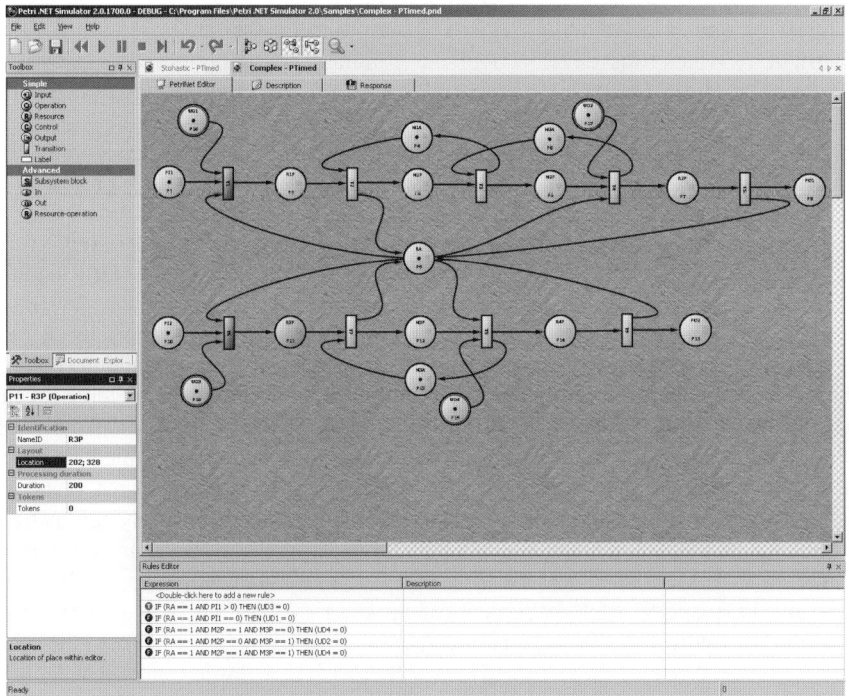

Figure 6.26. The main window of Petri.NET

Toolbox is a special TreeView control containing all objects that can be dragged to the editor: simple objects like places and transitions and more complex resource prototypes. Document Explorer shows the objects tree of the currently active PN model. It helps in navigation of the objects hierarchy. The properties window is used to display and edit all properties of objects (places and transitions) that are part of the Petri-net model. The properties of other objects (labels, subsystem blocks,...) used in the application, can be displayed and edited as well. The Rules Editor is used to add/edit/remove rules that are applied to the currently active PN model. It contains a collection of rules that define the activities of the control places included in the PN model.

A PN model is built with PetriNet Editor by a simple drag-and-drop principle. Since Petri.NET is primarily designed for simulation and analysis of MS, the Toolbox window contains five types of places: Input, Operation, Resource, Control and Output. Some properties are common to all types (NameID), while others are specific and depend on the type of the place. An input, for example, as a source place can receive tokens with predefined, fixed or stochastic, frequency. A resource on the other hand, has a unique property related to release times (Figure 6.27).

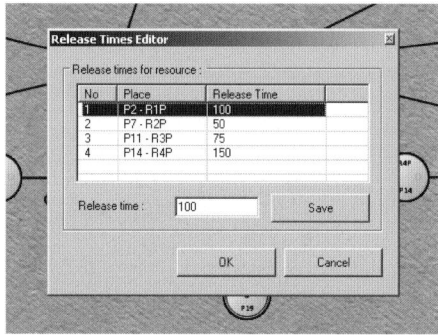

Figure 6.27. Release Times Editor in Petri.NET

As we described in Section 6.2.1 there are two basic ways in which to control how places are related with other places in the PN graph; they can receive tokens according to some control function, or they can have input transitions. In Petri.NET the control function has the form of rules and it is defined in Rules Editor, depicted in Figure 6.28.

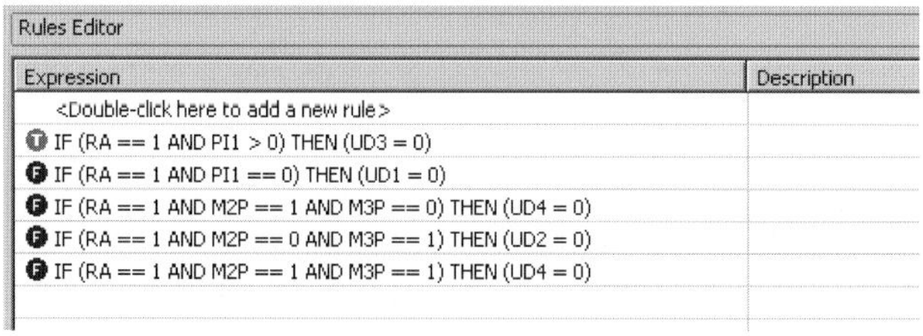

Figure 6.28. Rules Editor in Petri.NET

A rule has the following syntax:
IF (Expr1 **AND** Expr2 **AND** ... **AND** ExprN) **THEN** (Assign1 **AND** Assign2 **AND** ... **AND** AssignN)

where:
Expr: NameID1/const1 **op** NameID2/const2 **op** ... **op** NameIDN/constN
             RELOP NameID1/const1 **op** NameID2/const2 **op** ... **op** NameIDN/constN

Assign: NameID = const

op – arithmetic operators: '+' or '-'
RELOP – relational operators: ==, !=, <, <=, >, >=

Upon definition of the model, Petri.NET can simulate time-invariant and p-timed PN. Simulation can be tracked by the selection of a token game, while in the case of p-timed PN a pie object that indicates the remaining time, appears inside a place (Figure 6.29).

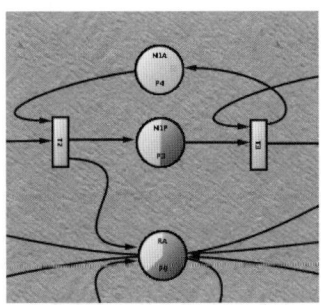

Figure 6.29. Pie objects in Petri.NET (indication of remaining time in a p-timed PN)

Once simulation is finished, by using the Response tab the user selects a type of presentation of simulation results. Two types of presentations are available, Spreadsheet and Oscilogram. When the PN model belongs to the MRF class, Petri.NET provides basic system analysis; determination of circular waits, transitions in conflict, the system matrices and the wait relation matrix, and calculation of resources utilizations. All these options are available in Description tab.

We conclude this section with a description of another Petri.NET feature, an automatic PLC code generator (Figure 6.30), which makes this application different from most PN simulation tools.

The PLC code generator executes two functions; first the PN model is transformed in generic PLC code, and then a parser is used to create a file that is readable by the target PLC. Currently, the code generator supports the Siemens S7-200 PLC family, but due to its modular design, Petri.NET provides a very simple method for insertion of additional parsers. Nevertheless, due to the large variety of PLCs some other options should be investigated. The OPC standard is one of the solutions, since almost all PLC manufacturers provide programming tools that allow PLC to connect to an OPC server as a client. Then, Petri.NET as another OPC client, would be able to exchange information with a PLC through the OPC server.

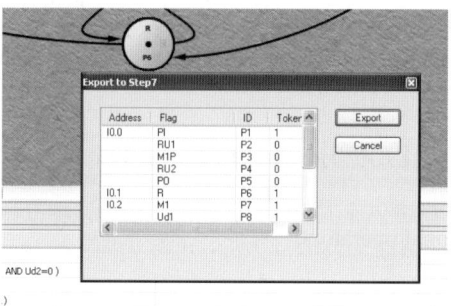

Figure 6.30. Main frame of the automatic PLC code generator in Petri.NET

Let us now return to the automatic PLC code generation. Transformation of an ordinary PN graph in a ladder logic diagram is based on several rules as stated below:

- each place is associated with one PLC variable; a Boolean is assigned to the place with $K(p)=1$, a counter is assigned to the place with $K(p)>1$,
- each transition is associated with a Boolean variable and PLC outputs connected with tasks that should be started when a transition fires,
- a Boolean variable, associated with a place that represents a task (operation, resource release), is "set" on a positive edge of PLC input, connected with a corresponding task-completion sensor,
- a Boolean variable, associated with a control place is "set" on a positive edge of the variable that represents its input transition, or upon fulfillment of its control function,
- a counter, associated with a place that represents a task (operation, resource release), is increased on a positive edge of PLC input, connected with a corresponding task-completion sensor,
- a counter, associated with a control place is increased on a positive edge of the variable that represents its input transition, or upon fulfillment of its control function,
- a Boolean variable representing a transition is "true" when all conditions for firing the corresponding transition are met,
- a Boolean variable, associated with a place that represents a task (operation, resource release) or control place, is "reset" on a positive edge of the variable associated with its output transition,
- a counter, associated with a place that represents a task (operation, resource release) or control place is decreased on a positive edge of the variable that represents its output transition,

A slight change of these rules should be made in order to provide code generation for a general type of a PN graph. We assumed that tasks-completion sensors and tasks-start drivers are connected with PLC digital I/Os. Usually this is not the case (see Section 5.4). However, there should not be a problem to follow given rules even if places and transitions are associated with variables that are changed by some communication protocol. An example of automatic PLC code generation is given in the section that follows.

6.5 Validation of Implemented Petri Nets

For the last three decades PLCs have had a leading role in industrial automation. From process industry to assembly lines they serve as a main part of various control loops. Having a modular hardware concept and user-friendly programming software, PLCs were, and still are, used for implementation of simple logic tasks as well as for very complicated control schemes that includes thousands of signals and requires a whole network of controllers.

As the requirements for control quality and safety increase, implementation of complex control algorithms becomes a problem. Methods used by engineers who transfer complex algorithms into a PLC program are not able to cope with the complexity problem. Furthermore, most of the information related to the control problem has an informal character, thus making PLC programming even harder. This is why in recent years a lot of work has been done in the field of applying formal methods in PLC programming. As stated in [15], three steps in the control design process may be identified: a) formalization and reinterpretation, b) synthesis and c) implementation. In the case study, given in Chapter 5, all three steps have been demonstrated and, as a result, a matrix-based controller was successfully implemented in PLC by using an automatic code generator.

Even though large efforts have been made in this direction there is still no unique solution for transformation of a general PN in PLC code. One of the reasons is, as we already mentioned, the large variety of PLCs. Although almost all PLCs are programmed with standard programming languages, each of them has some exclusive feature or particular programming syntax, which makes a general solution very difficult to achieve. In the previous section we presented a PN simulation tool with the ability to generate program code for PLC S7-216. In [16] – [18] and [27] methods for implementation of PN in PLC by using structured text (ST), an instruction list (IL) and a ladder diagram (LD) have been proposed. In [19] SIMULINK®, high-level timed Petri nets and functional block diagram (FBD) are used for design and analysis of control systems. All these methods offer more or less straightforward and convenient procedures for PN transformation into generic PLC code, but when it comes to target PLC code generation they lack suitable solutions. It should be mentioned that in 1975 GRAFCET appeared as a "missing link" between PN and PLC code [25], [26]. In 1988. IEC announced "Sequential Function Chart" as an international standard for PLC programming based on GRAFCET.

The other two problems encountered by PLC programmers are verification and validation (V&V) of implemented algorithms. As today's engineers apply many various strategies in PLC programming, V&V procedures differ one from another depending on approach, formalism and the method used in software development. A V&V based on coupling of so-called interpreted Petri nets of the controller (SIPN) and the process (PIPN) is described in [20]. In [21] it is shown how PLC code, written in IL, can be translated into a Petri net. Then, by using standard PN analysis (reachability tree, boundness check, *etc.*), the PLC program is checked for possible errors. The other approach, which also deals with IL, is described in [22]. In [23] the condition/event (C/E) model of a process is connected with sequential function chart (SFC) control software, thus making a closed-loop system. The set of reachable states is then compared with the set of forbidden states providing insight into system behavior under various conditions. Another model checker, which is developed for LD control logic, is presented in [24].

In this section we present a method for verification and validation of PLC control algorithms developed from PN models. Due to the existence of a direct relation between PN and the system matrices, a matrix-based MS controller can be tested as well. Based on super blocks, designed in SIMULINK®, and by using MATLAB® Real Time Workshop (RTW), the method provides an efficient tool for real-time investigation of various dispatching policies as well as the influence of manufacturing system parameters on the behavior of the control system. This approach is convenient for small-size PLCs, since their programming software usually does not include online simulators.

The main components of the testbed are shown in Figure 6.31. Since the model of the uncontrolled system is built in SIMULINK®, the PC should have installed MATLAB® with RTW. Furthermore, a board with digital I/Os has to be included in the PC hardware configuration. Inputs and outputs of a SIMULINK® model of an uncontrolled process are connected to modules, which communicate with the I/O board.

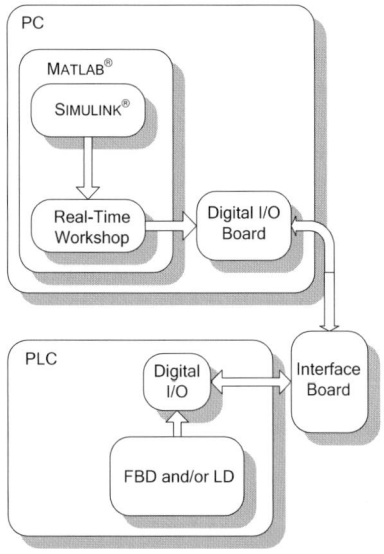

Figure 6.31. Main components of the V&V testbed

Since the levels of signals on the I/O board and PLC are different (TTL versus 24 V) the board sends/receives signals to/from PLC through an interface. A PLC configuration mainly depends on the system to be controlled and the control algorithms to be implemented. Complex systems with numerous states require PLCs with high computational power and a large number of I/O units.

The SIMULINK® model used for verification and validation of the control algorithm is made of basic PN components (prototypes), which have the form of predefined super blocks. There are four different prototypes: Input Place, Nonshared Resource, Shared Resource and Output Place. Every super block is determined by its inputs, outputs and parameters. Inputs of a super block are associated with the PC I/O card and connected with the PLC controller outputs. On rising edge of the PLC output signal an operation that corresponds with that signal is started. Outputs of super blocks can be separated into two groups. The first group comprises outputs that illustrate the state of the prototype, such as number of parts that are currently processed and/or the number of idle resources. These signals are used for online MS analysis. The second group includes logical outputs (0 or 1), which are used by the PLC controller. These signals can be associated with sensors planned to be installed in real MS. There are three types of signals:

- error – signal is set to "1" if an error occurs (machine failure, number of WIP is negative, *etc*.),
- resource available – signal is set to "1" if corresponding resource is idle,
- operation completed – signal is set to "1" if corresponding operation is finished.

Super blocks that represent Input Place and Output Place are shown in Figure 6.32.

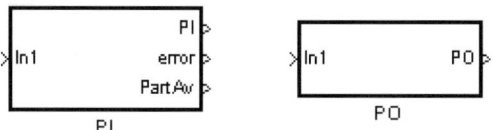

Figure 6.32. Input Place and Output Place super blocks

As its name implies, Input Place super block describes the input of the system. The superblock has one input and three outputs. Input "In1" is a trigger signal; transition from 0 to 1 decreases the number of parts in Input Place. Output "PI" is an integer that represents the current number of parts in Input Place, while output "error" is set to 1 if the number of parts becomes less than zero or larger than maximum number of parts allowed. Output "PartAv" is set to 1 if the number of parts in Input Place is positive, otherwise is set to 0. Input Place and Output Place configuration masks are shown in Figure 6.33. The Input Place mask comprises four parameters: "Initial condition" – initial number of parts in input place, "Limit" – maximum number of parts allowed, "Period" – time delay (in seconds) between parts arrival, "Range" - if "Random" is checked, then the time delay is a random variable in the range between 0 and Range. In this case the value entered in "Period" is ignored.

Output Place super block describes the system output. It has one input and one output. Input "In1" is a trigger signal connected with the PLC controller. Output "PO" is an integer that represents the number of parts in Output Place. The initial number of parts in Output Place can be defined in the configuration mask.

The super blocks that represent typical MS resources are shown in Figure 6.34. The Shared Resource prototype is used to model the resource that performs more than one operation, while Nonshared Resource represents a resource with one task only.

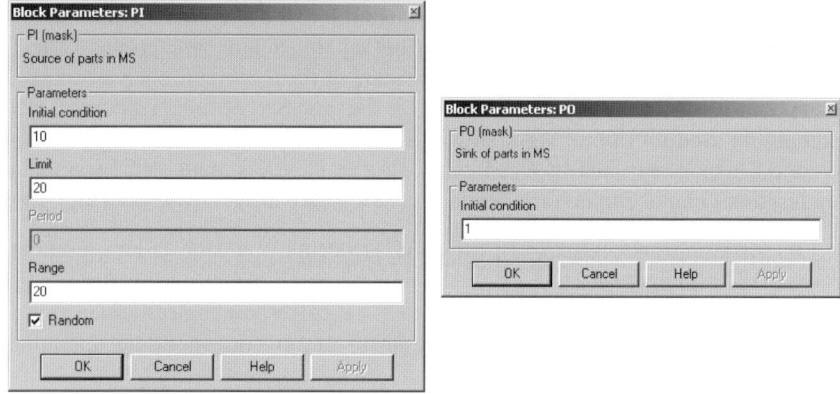

Figure 6.33. Input Place and Output Place configuration masks

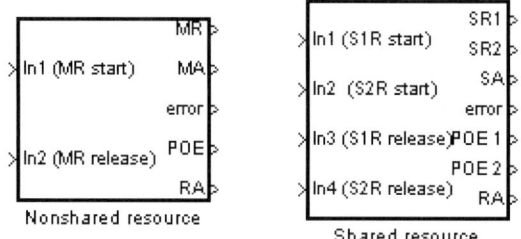

Figure 6.34. Nonshared Resource and Shared Resource super blocks

The super block that represents Shared Resource has four inputs, all of them connected with a PLC controller:

- In1 – input that starts operation 1,
- In2 – input that starts operation 2,
- In3 – input that starts resource release after operation 1,
- In4 – input that starts resource release after operation 2.

Shared Resource super block has three outputs, generally used for MS analysis:

- SR1 – number of parts processed by operation 1,
- SR2 – number of parts processed by operation 2,
- SA – number of available slots.

Besides these outputs, the super block has logical outputs connected to and used by a PLC:

- error – error signal,
- POE1 – is set to "1" if operation 1 is completed,
- POE2 – is set to "1" if operation 2 is completed,
- RA – is set to "1" if resource is idle.

The Shared Resource configuration mask is shown in Figure 6.35. The configuration mask has fields for definition of all parameters required for simulation of the shared resource dynamics (p-timed PN, Figure 6.9). The duration of operations and the duration of resource-release tasks can be set by the designer. The initial number of parts processed in operations and the initial number of idle slots are defined in a form of SIMULINK® vector. When a shared resource with more than two operations is required, a new prototype may be designed by following a simple procedure implemented a for two-operations shared resource.

Nonshared Resource super block has two inputs:

- In1 – input that starts operation,
- In2 – input that starts resource release,

and three logical outputs:

- error – error signal,
- POE – is set to "1" if operation on part is completed,
- RA – is set to "1" if resource is idle,

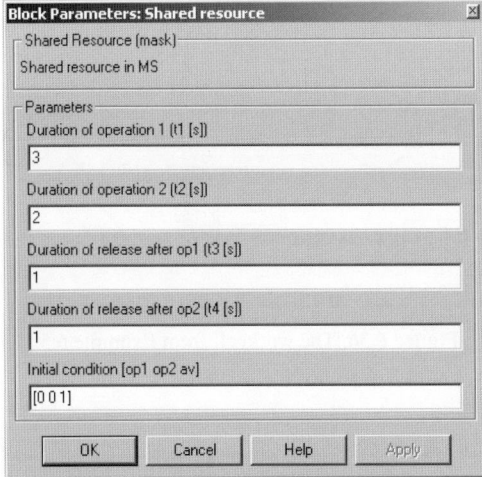

Figure 6.35. Shared Resource configuration mask

all connected with a PLC. Two outputs used for MS analysis are:

- MR – number of currently processed parts,
- MA – number of available slots.

Nonshared Resources are configured through a configuration mask similar to the one shown in Figure 6.35. Since Nonshared Resource has only one operation to perform, the difference between two masks is only in the number of parameters required for resource definition.

Example 6.5.1 (validation of PN implemeted in PLC)

We consider the workcell depicted in Figure 6.36. The workcell, comprised of three machines and two robots, processes two part-types. Its PN graph with control places is shown in Figure 6.37. The implemented control policy restricts the number of parts in path a (control place u_{d1}), and path b (control place u_{d2}) (we leave thorough analysis of the system to the reader). Conflicts are resolved by sequential execution of ladder diagram networks; when robot R1 part a has priority over part b, whilst robot R2 gives priority to part b.

Since the PN is ordinary and pure, and all places, except control place u_{d1}, initially have only one token, places are associated with markers in PLC memory. Place u_{d1} is associated with a counter. The symbol table is depicted in Figure 6.38. As may be seen, markers M1–M3 are used for places, M4 is used for a control signal u_{d2}, M5 and M6 are assigned to transitions, while control signal u_{d1} is assigned to counter C0. We assume that each place corresponds with one PLC input and one PLC output, *i.e.* the input and output interface functions are of the form one-to-one (see the case study in Section 5.4).

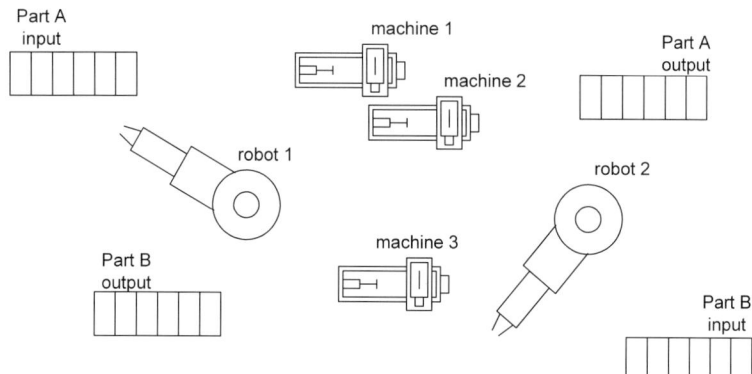

Figure 6.36. The workcell from Example 6.5.1

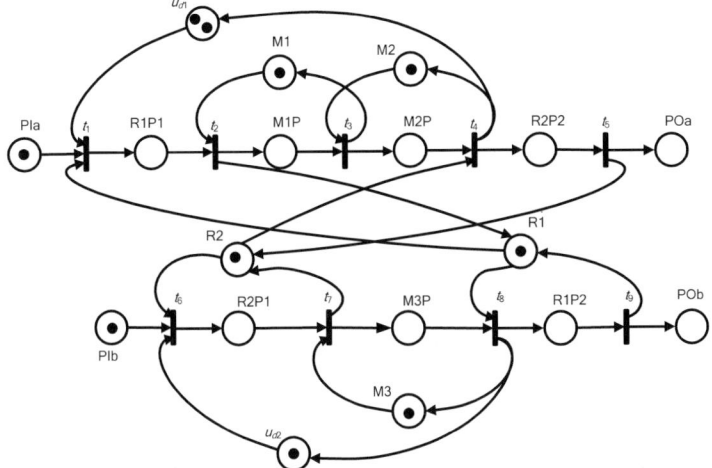

Figure 6.37. Controlled PN of the workcell shown in Figure 6.36

	Name	Address	Comment
1	R1P1	M1.0	robot 1 operation 1
2	M1P	M1.1	machine1 operation
3	M2P	M1.2	machine2 operation
4	R2P2	M1.3	robot 2 operation 2
5	R2P1	M1.4	robot 2 operation 1
6	M3P	M1.5	machine3 operation
7	R1P2	M1.6	robot 1 operation 2
8	M1	M2.0	machine1 idle
9	M2	M2.1	machine2 idle
10	M3	M2.2	machine3 idle
11	R1	M2.3	robot1 idle
12	R2	M2.4	robot2 idle
13	Pla	M3.0	part a available
14	Plb	M3.1	part b available
15	POa	M3.2	part a out
16	POb	M3.3	part b out
17	ud2	M4.0	control place ud2
18	ud1	C0	control place ud1
19			
20	tr1	M5.0	transition 1
21	tr2	M5.1	transition 2
22	tr3	M5.2	transition 3
23	tr4	M5.3	transition 4
24	tr5	M5.4	transition 5
25	tr6	M5.5	transition 6
26	tr7	M5.6	transition 7
27	tr8	M5.7	transition 8
28	tr9	M6.0	transition 9
29	reset	M7.0	system reset

Figure 6.38. The symbol table of S7-216 PLC for controlled PN in Figure 6.37

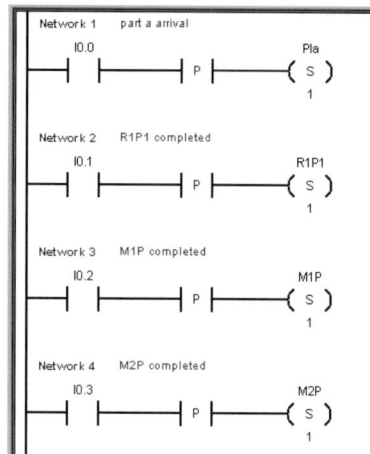

Figure 6.39. First four networks of the ladder diagram subroutine for reading inputs

The PLC code for S7-216 is obtained by a Petri.NET code generator. Part of the ladder diagram subroutine for digital inputs acquisition is shown in Figure 6.39. As may be seen, a particular PN place is set to "true" on the positive edge of the corresponding digital input. This action matches up with a token entering the place.

Upon completion of the input subroutine, the PLC starts to execute a subroutine that calculates the (•t) part of PN. A fraction of that subroutine is depicted in Figure 6.40. By comparing this subroutine with the PN graph shown in Figure 6.37, conditions for firing the first four transitions can be clearly recognized from the ladder networks.

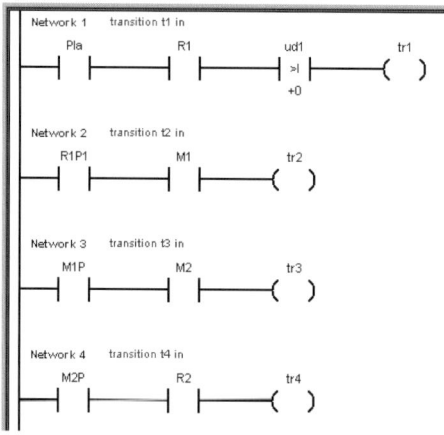

Figure 6.40. First four networks of the ladder diagram subroutine for PN execution (•t)

At the end of the PLC cycle, a subroutine that sets PLC outputs and resets markers associated with PN places is executed (Figure 6.41). This action corresponds with token withdrawal in the PN.

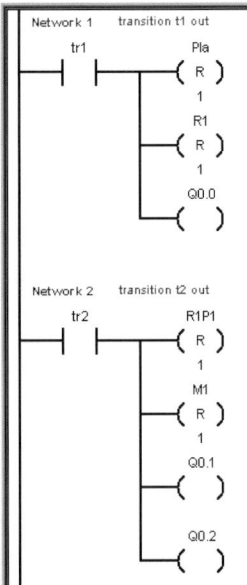

Figure 6.41. First two networks of the ladder diagram subroutine for PN execution (t•)

As we mentioned, implementation of the proposed dispatching strategy requires one counter for tracking the number of tokens in control place u_{d1}. This counter is realized in ladder network 5 (Figure 6.42). The positive edge of variable tr4 increases, while the positive edge of variable tr1 decreases the counter value, which corresponds with activities in the PN graph.

The SIMULINK® model of the workcell is given in Figure 6.43. Super blocks that represent resources are connected with the PLC's inputs and outputs by using MATLAB's® RTW and Advantech PC I/O card.

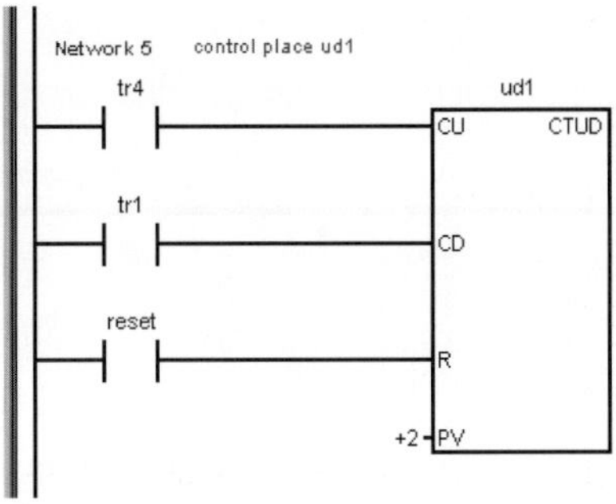

Figure 6.42. Ladder network with counter for control place u_{d1}

256 Manufacturing Systems Control Design

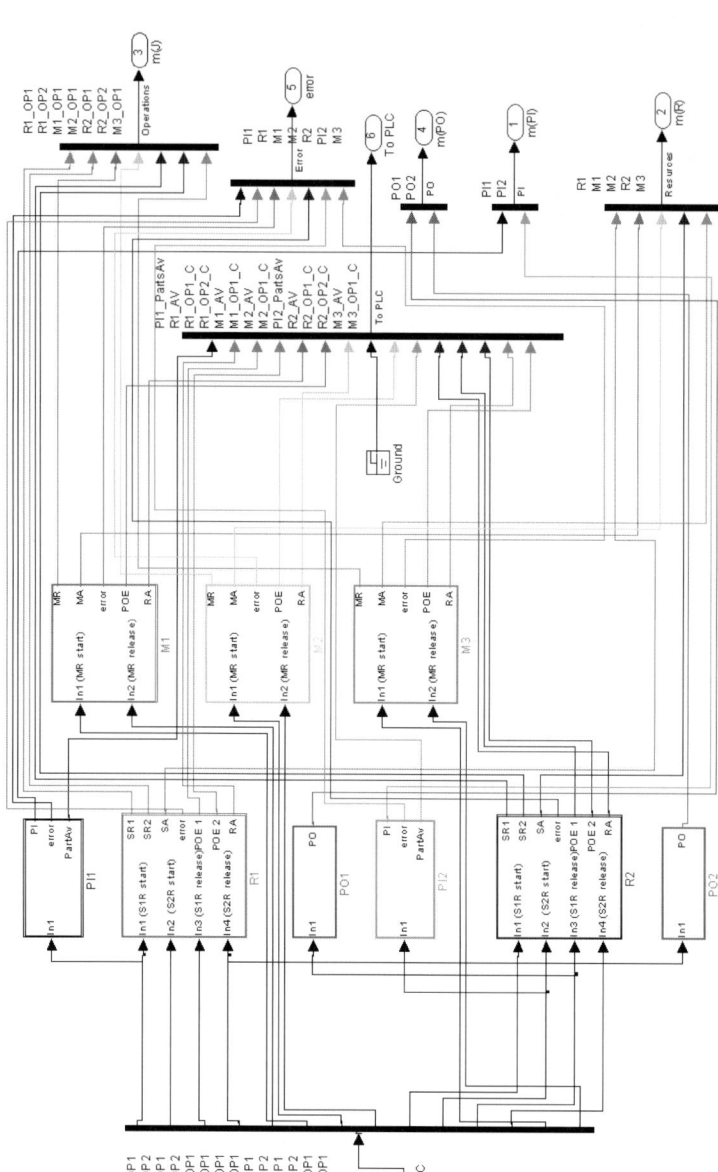

Figure 6.43. SIMULINK® model of the workcell shown in Figure 6.37

References

[1] Petri CA. Kommunikation mit Automaten, Bonn: Institut für Instrumentelle Mathematik, Schriften des IIM Nr. 2, 1962, 2nd edn:, New York: Griffiss Air Force Base, Technical Report RADC-TR-65—377 1966;1.
[2] Proth JM, Xie X. Petri Nets: A Tool for Design and Management of Manufacturing Systems. Chichester: Wiley, 1996.
[3] Zhou MC, Venkatesh K. Modeling, Simulation and Control of Flexible Manufacturing Systems: A Petri Net Approach. Singapore: World Scientific, 1998.
[4] Murata T. Petri nets: properties, analysis and applications, Proc. IEEE 1989;77;4:541–580.
[5] Kumar PR, Meyn SP. Stability of queuing networks and scheduling polices, IEEE Trans. Aut. Contr. 1995;40:251–260.
[6] Amer-Yahia C, Zerhouni N, El Moundi A, Ferney M. On finding deadlocks and traps in Petri nets, System Analysis, Modeling and Simulation 1999;34:495–507.
[7] Wang J. Timed Petri Nets. Boston: Kluwer, 1998.
[8] Jensen K. Colored Petri Nets: Basic Concepts, Analysis Methods and Practical Use. Berlin: Springer., 1992.
[9] Pedrycz W, Camargo H. Fuzzy timed Petri Nets, Fuzzy Sets and Systems 2003;140:301–330.
[10] Tacconi DA, Lewis FL. A New Matrix Model for DES: Application to Simulation, IEEE Contr. Sys. Mag. 1997;October: 62–71.
[11] Desrochers AA, Deal TJ, Fanti MP, Complex-Valued Token Petri Nets, IEEE Trans. Aut. Sci. Eng. 2005;2;4:309–318.
[12] Alla H, David R. A modeling and analysis tool for discrete event systems: Continuous Petri net, An Int. J. on Performance Evaluation 1998;33:175–199.
[13] Moody JO, Antsaklis PJ. Supervisory Control of Discrete Event Systems Using Petri Nets. Boston: Kluwer Academic Publishers, 1998.
[14] Iordache MV, Moody JO, Antsaklis PJ. Automated Synthesis of Deadlock Prevention Supervisors Using Petri Nets, Technical report of the ISIS Group at the University of Notre Dame 2000, ISIS-2000-003.
[15] Frey G, Litz L. Formal methods in PLC programming, Proc. of the IEEE SMC00 2000;2431–2436.
[16] Cutts G, Rattugan S. Using Petri nets to develop programs for PLC systems, Proc. of Application and Theory of Petri Nets - Springer 1992;368–372.
[17] Stanton MJ, Arnold WF, Busk AA. Modelling and control of manufacturing systems using Petri nets, Proc. of 13th IFAC World Congress 1996;329–334.
[18] Uzam M, Jones AH, Khan AH, Karimzadgan D, Kenway SB, A general methodology for converting Petri nets into ladder logic: the TPLL methodology, Proceedings of the 5th International Conference on Computer Integrated Manufacturing and Automation Technology - CIMAT96 1996;357–362.
[19] Baresi L, Mauri M, Monti A, Pezzè M. PLCTOOLS: Design, Formal Validation, and Code Generation for Programmable Controllers, Proc. of the IEEE SMC'00; 2000.
[20] Frey G, Litz L. Verification and validation of control algorithms by coupling of interpreted Petri nets, Proc. of the IEEE SMC98;1998:7–12.
[21] Mertke T, Menzel T. Methods and tools to the verification of safety-related control software, Proc. of the IEEE SMC'00;2000.
[22] Canet G, Couffin S, Lesage JJ, Petit A, Schnoebelen Ph. Towards the automatic verification of PLC programs written in Instruction List, Proc. of the IEEE SMC'00; 2000.
[23] Kowalewski S, Preusig J. Verification of sequential controllers with timing functions for chemical processes, Proc. of 13th IFAC World Congress 1996; 419–424.

[24] Moon I. Modelling programmable logic controllers for logic verification, IEEE Control Systems;1994:14:6:53–59.
[25] Baracos P. GRAFCET Step by Step, Famic Inc.,1992.
[26] David R, Alla H. Petri nets and GRAFCET: Tools for Modeling Discrete Event Systems. New York London: Prentice-Hall, 1992.
[27] Lee GB, Zandong H, Lee JS. Automatic generation of ladder diagram with control Petri net, Journal of Intelligent Manufacturing 2004;15:245–252.
[28] Chiola G, Ferscha A. Distributed simulation of Petri nets, IEEE Parallel & Distributed Technology 1993;1:3:33–50.
[29] http://www.informatik.hu-berlin.de/top/pnk
[30] http://parsys.informatik.uni-oldenburg.de/~pep
[31] http://www.informatik.uni-hamburg.de/TGI/PetriNets

7

Virtual Factory Modeling and Simulation

Manufacturing systems (MSs) are assembled from elements such as robots, machine tools, fixtures, buffers, rotary tables, belt conveyers, pallets, *etc.* that are connected and supervised through a local area network. Using today's classification of systems, MSs can be treated as hybrid systems that contain a mixture of various dynamic behaviors—continuous and discrete control loops, Boolean variables related to process states, and discrete events, all embraced by a usually hierarchical decision-making overhead. This means that an MS structure contains both hard and soft technology, first focused on the product fabrication, assembly and distribution, while later the focus is on the support and coordination of manufacturing operations.

The MS's hard technology is split into several levels – from the factory level via the operating center, workcell and robotic station levels to a particular manufacturing process level. The accompanying soft technology is also split into several levels – from the highest strategy level, via lower planning, supervisory, and manipulating levels to the basic manufacturing task level.

Today, virtual models provide a very inexpensive and convenient way for complete factory design. Instead of building real systems, a designer first builds new factory layouts and defines resource configurations in the virtual environment and refines them without actual production of physical prototypes. Allowing clear visualization of all potential problems caused by the layout, virtual modeling and dynamic simulation of manufacturing processes has traced a completely new route to analysis and design of MSs [1–3].

A factory layout design, physical modeling, control synthesis, performance analysis, dynamic simulation and visualization of robotized manufacturing systems have become much easier and more effective with specialized programs for virtual-factory modeling and simulation. Some virtual-factory simulators originated from the academia [4–8], but most of them are sophisticated products of leading robot manufacturers and independent companies [9–11]. In this chapter we briefly portray several tools such as Grasp2000 from BYG Systems Ltd., eM-Plant from Tecnomatix, RobotStudio from ABB, CimStation Robotics from Silma, and Cosimir from FESTO. Then we describe FlexMan – a virtual-factory simulator with an integrated matrix-based MS controller [12].

A typical structure of a virtual factory simulator is shown in Figure 7.1. The aim of virtual modeling is to create an experimental MS by combining a tentative factory layout with existing or newly created virtual models of constituent MS objects. Usually, MS objects and layouts can be loaded from the corresponding libraries of objects and layouts, but they can also be imported from other CAD software or created as new entities within the simulator itself.

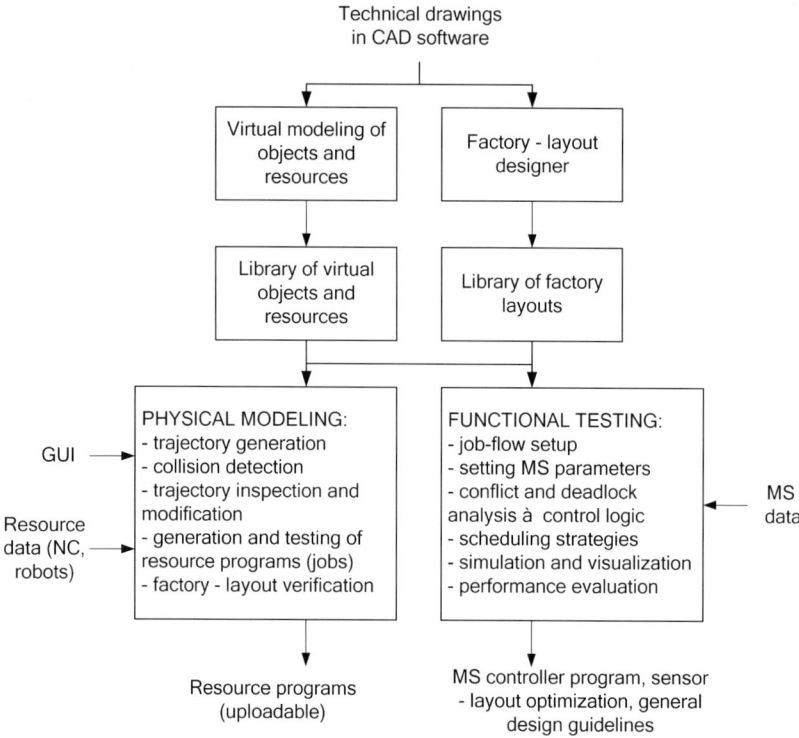

Figure 7.1. A typical structure of a virtual-factory simulator

MS simulation consists of multiple tasks that are highly interdependent. As shown in Figure 7.1, there are two main groups of tasks related to physical modeling and functional testing of the simulated MS. Physical modeling is mainly concerned with resources that play an active role in the manufacturing process, especially with robots and numerically controlled (NC) machine tools. The trajectory generation for these resources is closely related to circumventing the inverse kinematics problems (*e.g.* joint limits, singularity points), working-space constraints, and particularly to prevention of collisions with surrounding MS objects. In order to achieve reliable and precise collision detection, exact physical measures of all virtual models and their postures are needed. Positive collision tests lead to consecutive trajectory or factory-layout modifications. Physical modeling allows the designer to generate and test single manufacturing jobs performed by a corresponding MS resource, but the main goal of physical modeling is verification of the simulated factory layout. After successful validation of the simulated virtual MS setup, most commercial MS simulators generate programs executable in controllers of particular active resources.

Functional testing has a goal to connect a physical setup with the plan of the simulated MS. As shown in Figure 7.1, functional testing is concerned with a job-sequence definition, setting of MS parameters, conflict and deadlock analysis at the local and global level (at the robot workcell or robot station, and at the whole MS level), synthesis of control logic, study of different job-scheduling strategies, simulation and visualization of dynamic phenomena during MS operation. Having a plan of a manufacturing process and all necessary MS data, functional testing should help the MS designer to reach a reliable and objective MS performance evaluation.

In most cases, MS control depends on the states of sensors installed in the system. Therefore, a successful functional testing generates two outputs: the executable MS controller program, and the optimized sensor layout. Based on the acquired designer's experience, virtual simulators may serve as efficient design accelerators and trustworthy sources of implementation guidelines.

7.1 3D Modeling of Manufacturing Systems

A factory-layout design is primarily a hard-technology-related task, whose goal is to establish an optimal arrangement of individual MS elements, viewed from the spatial and operational point of view. In the very recent past, factory-layout design was a job that had to be done before the onsite MS construction could start. Today, three-dimensional (3D) modeling serves to define the physical shape information of a particular MS object prior to its physical creation. One more complex 3D model, like the model of a palletization robot work cell shown in Figure 7.2, actually represents a combination of primitive 3D shapes – cuboids, cylinders, prisms, polygons, and lines, combined together in a hierarchical (so called parent–child) order and characterized by different material properties, textures, colors, shininess etc. In most cases, a parent–child relationship means that a group of subordinate objects ("children") is translated, rotated and scaled with respect to the superimposed ("parent") coordinate frame. Initially, all 3D objects on the scene are positioned at the origin of the virtual environment, and then by using suitable commands, are put in spatial relations. Such a 3D model is further used for display on the computer screen and for calculations carried out as defined by the simulation context.

The model with more details is computationally more demanding. The complexity of the 3D model is usually dictated by the required precision of the model. For example, when collision avoidance is explored then a more detailed 3D model is preferred. On the other hand, logical testing of operations in the simulated factory layout can be achieved by using models with fewer details.

A proper 3D model requires precise physical dimensions – *e.g.* height, width and depth of the primitive shapes. This information is usually obtained by measuring the object, or it is taken from the original technical drawings. 3D models can be created using ISO standards 3D file formats such as X3D

(Extensible 3D), its predecessor VRML (Virtual Reality Modeling Language) or by using CAD programs (*e.g.* AutoCAD™, Catia™ or 3D-Studio™). CAD programs generate different file formats (*e.g.* DXF, 3DS, IGES, STEP, VRML), and support conversion from one format to another.

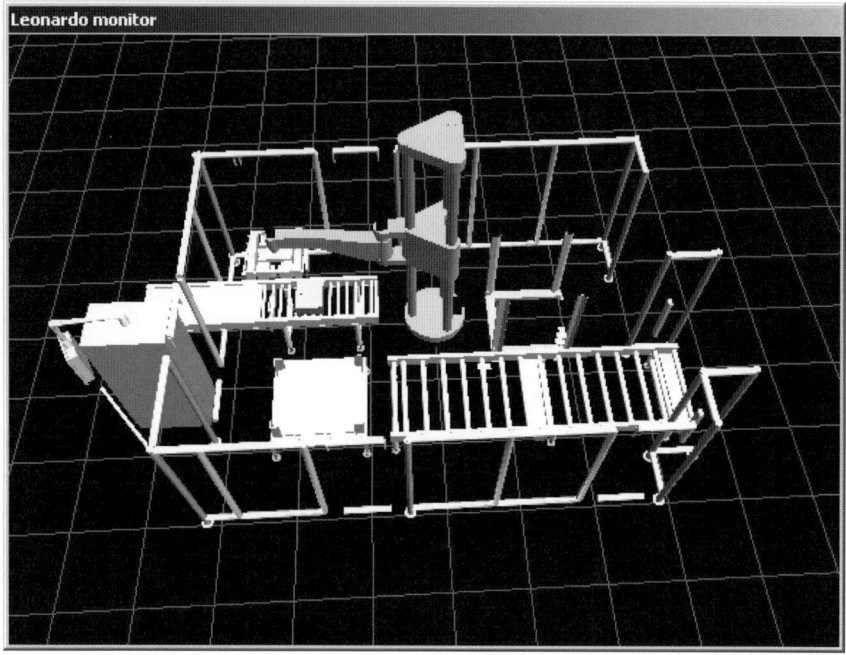

Figure 7.2. The 3D model of a palletization work cell (Courtesy of Euroimpianti s.p.a)

7.2 Modeling FESTO FMS in VRML (X3D) Format

As mentioned above, 3D models of solid objects can be created in many ways and many tools are at the designer's disposal for this purpose. One way is modeling in VRML format (or in X3D, which is the successor to the VRML), which has become an international standard established in 1994 for description of 3D shapes and environments suitable for World Wide Web program applications. Besides the creation of virtual environments, VRML enables introduction of 3D motion, sound and other dynamic features [13]. Virtual objects modeled in VRML can be visualized in independent VRML viewers or in popular web browsers providing that some VRML viewer plug-in has been previously installed. X3D improves upon VRML with new features, advanced application programmer interfaces, additional data-encoding formats, stricter conformance and a componentized architecture that allows for a modular approach to supporting the standard [14].

7.2.1 Basic VRML Features

Basic programming elements in VRML are nodes and fields that together with the header and comments form a VRML file (extension *.wrl). Nodes may be interpreted as "commands" that designate different geometric shapes, materials, light, spatial transformations, *etc*. Fields describe node features that can change. Dimensions in VRML are normalized. For example, the shape *box* with dimensions {10, 10, and 10} may have a 10 mm, 10 m or 10 km long edge, depending on the metric measure defined by the user.

Basic geometric shapes (primitives) are Box, Cone, Cylinder, and Sphere. The group of geometric primitives is extended with the two-dimensional VRML object Text representing a particular text. Geometric shapes are created with the node *Shape*, which has two fields – appearance and geometry:

```
Shape {
        appearance ...   – defines color and object texture
        geometry ......  – defines form or structure
        }
```

All VRML objects are initially positioned in the origin of the VRML environment. In order to place the objects at different positions, a node *Transform* is used. This node is a grouping node, which enables simultaneous translation, rotation and scaling of a group of subordinate objects, so-called children. In fact, all children objects tied to this new coordinate frame are translated, rotated and scaled with respect to the superimposed or so-called parent coordinate frame.

The syntax of the *Transform* node is defined in the following way:

```
Transform {
        translation  dx dy dz        # position
        rotation  rx ry rz delta     # orientation (in radians)
        scale  sx sy sz              # scaling
        children [ ....... ]         # subordinate objects
        }
```

Variables *dx*, *dy* and *dz* denote displacements of all children objects with respect to the global coordinate frame. In terms of homogeneous coordinates regularly used in robotics, translation is represented with the following homogeneous transformation matrix:

$$\mathbf{T} = \begin{vmatrix} 0 & 0 & 0 & dx \\ 0 & 0 & 0 & dy \\ 0 & 0 & 0 & dz \\ 0 & 0 & 0 & 1 \end{vmatrix}$$

Variables *rx*, *ry* and *rz* assume values 0 or 1, depending on about which axis rotation is going to occur (the other two variables get the value 0). The counterclockwise rotation is assumed positive. Assuming that rotation is defined around one of the axes, the homogeneous coordinate transformation attains the form:

$$\mathbf{R} = \begin{vmatrix} r_{11} & r_{12} & r_{13} & 0 \\ r_{21} & r_{22} & r_{23} & 0 \\ r_{31} & r_{32} & r_{33} & 0 \\ 0 & 0 & 0 & 1 \end{vmatrix}$$

In order to achieve different scaling factors for each axis, in calculations one must take into account a multiplication with a scaling matrix:

$$\mathbf{S}_v = \begin{vmatrix} sx & 0 & 0 \\ 0 & sy & 0 \\ 0 & 0 & sz \end{vmatrix}$$

The orientation in the VRML environment is defined as shown in Figure 7.3. One can see that the orientation of the *x–y–z* coordinate frame in the VRML environment does not coincide with the usual orientation representation in the Cartesian space, also shown in Figure 7.3. The difference between two orientation representations must be taken into account in all coordinate transformations and related calculations.

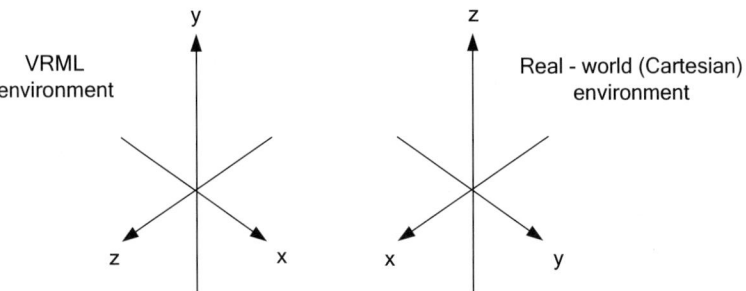

Figure 7.3. Orientation representation in the VRML and Cartesian worlds

It should be noted that transformations of children objects at every parent-children level are always relative to the involved parent coordinate frame.

Having in mind, for example, that robots move and work thanks to coordinated motion of their prismatic and revolute joints, a so-defined hierarchical structure in

VRML simplifies virtual modeling of robotic manipulators and other similar complex solid objects very much. By branching and nesting of *Transform* nodes, attachment of new coordinate frames to each robot joint or any other robot part (*e.g.* working tool) becomes easy and straightforward. Since transformations defined by *Transform* nodes are relative, the change in the outermost *Transform* node (*e.g.* in rotation) will affect all subordinate coordinate frames and objects. The coincidence with the way how robots move, viewed from the robot base to the working tool, is more than obvious.

7.2.2 FESTO FMS VRML Model

Let us describe VRML-based modeling of the FESTO FMS laboratory setup at the Faculty of Electrical Engineering and Computing, University of Zagreb. The aim of the FMS is to produce several types of cylinders assembled from the bodies, pistons, springs, and caps varying in shape and color. FESTO FMS is composed of four PLC-controlled work stations connected via the Profibus network: the distribution station, testing station, processing station, and assembly station. The flexibility in the assembly line is increased by using the five degrees of freedom rotational robot Mitsubishi Movemaster EX RV-M1 (see Figure 7.4).

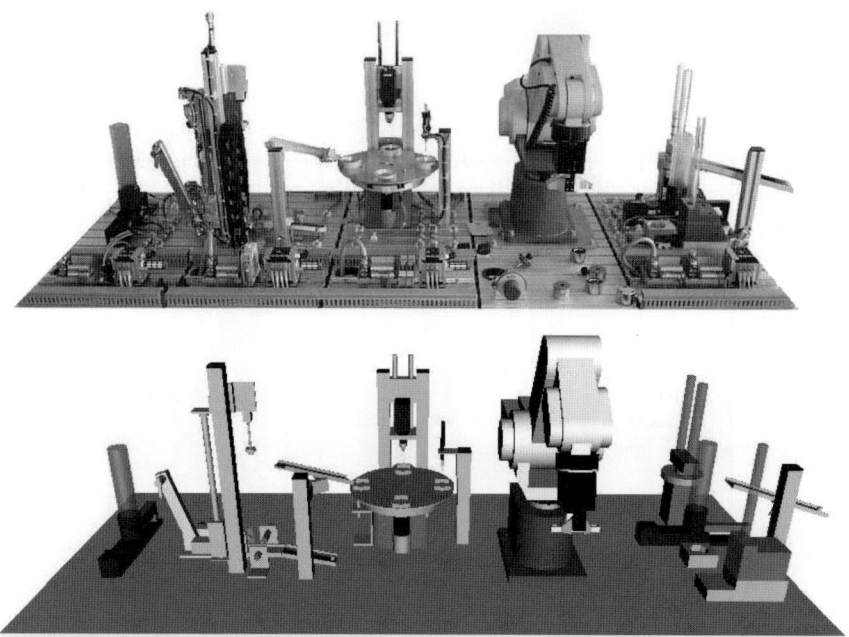

Figure 7.4. FESTO FMS: Laboratory set-up (above), virtual model (below)

The aim of virtual modeling is to prepare the modeled FMS for the functional testing. The richness of the model details is determined by the function of the particular FMS components. Virtual models of work stations contain system modules that have an active role in the manufacturing process. Besides the stack magazine module, the separating module, the testing module, the changer module, the spring magazine module, and the drilling module shown in Figure 7.5, important FESTO FMS parts are also the rotary indexing table module, the lifting module, the measuring module, the air-cushioned slide module (5 objects capacity), two slide modules (4 and 6 objects capacity), the cap magazine module (10 objects capacity), the place of assembly, and the robot arm.

The FESTO FMS VRML model is shown in Figure 7.4 together with the real system [15]. One can see that the outlook and layout of the virtual FMS fully resemble the outlook and layout of the real FMS. All insignificant details from the functional point of view are omitted (*e.g.* models of pipes, wires, connectors, gauges, some construction details, *etc.*).

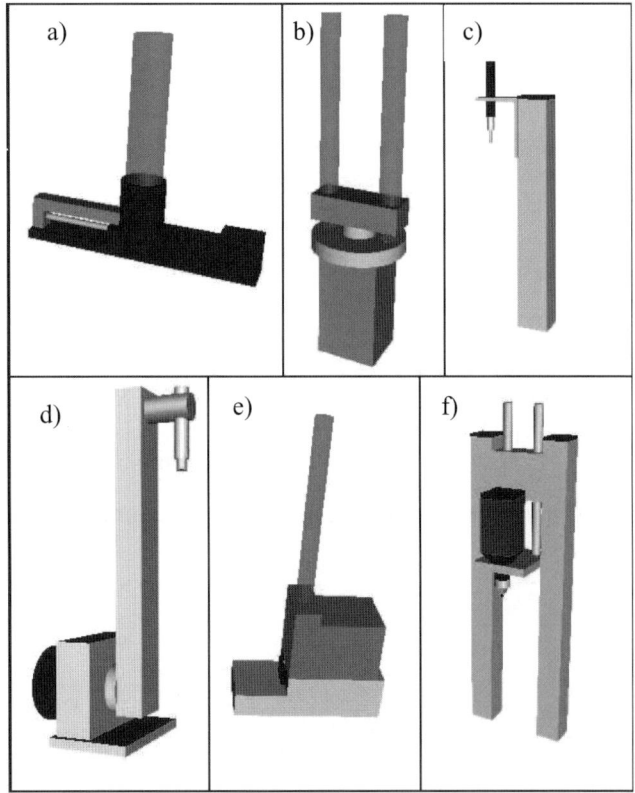

Figure 7.5. Virtual models of FESTO FMS components: **(a)** stack magazine module, **(b)** separating module (pistons), **(c)** testing module, **(d)** changer module, **(e)** spring magazine module, and **(f)** drilling module

The next step in FMS modeling and simulation is the generation of a functional model. This model, which describes operations and operating rules, is used for the creation of system matrices that are later used for the matrix-based FMS controller design. The reader can find more about functional modeling and simulation of

FESTO FMS in Section 7.9, which describes the usage of the Internet-based modeling and simulation tool FlexMan [12].

7.3 Modeling in LISA

Let us illustrate the use of virtual models in another simulator called LISA – a C++ and OpenGL^{TM}-based software for simulation and 3D modeling of complex kinematic configurations [16]. Models used in LISA are first created in the CAD software, then described in the XML (eXtensive Markup Language) and thereafter imported as an *xml* file into LISA. Applying this procedure, the palletization workcell from Figure 7.2 is displayed in LISA in the way shown in Figure 7.6. One can see from this example that 3D models can be used in different program applications without loss of model quality.

3D models in LISA are polygonal structured, *i.e.* polygons form a closed manifold, hierarchical nonconvex models. Polygons are made entirely of triangles because hardware accelerated rendering of the triangles is commonly available in the graphic hardware. Triangle meshes can be used for extraction of all geometric parameters including, for example, robot joint positions, link lengths, *etc*. Every virtual object is composed of an arbitrary number of links that form a parent–child hierarchy. There is no limit on the number of child links for a parent, so complex kinematics configurations can be formed, like the articulated robot arm shown in Figure 7.7.

Figure 7.6. The 3D model of a palletization work cell displayed in LISA

Frames (coordinate systems) are assigned to the links sequentially and may be either static or dynamic (see Figures 7.8 and 7.9). Namely, a 3D object can have an active or passive role on the scene. Active objects consist of static and dynamic frames, while passive objects are built only from the static ones. Dynamic link frames undergo rigid-body transformations during a simulation in a virtual environment [17].

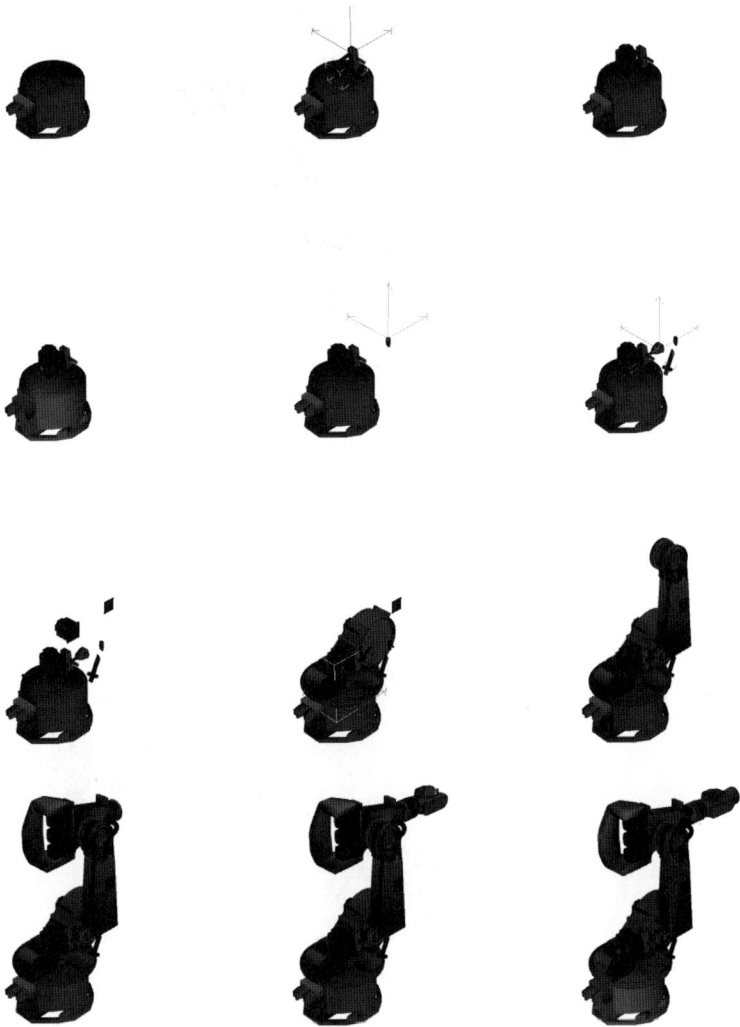

Figure 7.7. The creation of the 3D robot model

Virtual Factory Modeling and Simulation 269

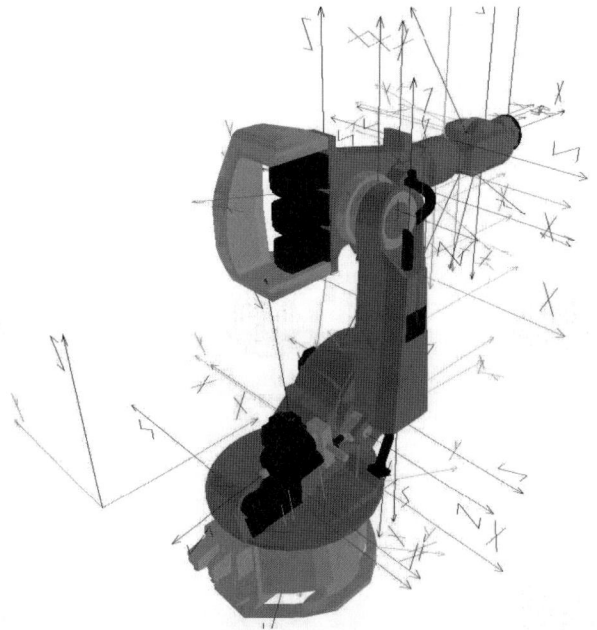

Figure 7.8. The 3D model of a KUKA robot (Courtesy of Kuka Roboter) – static and dynamic frames

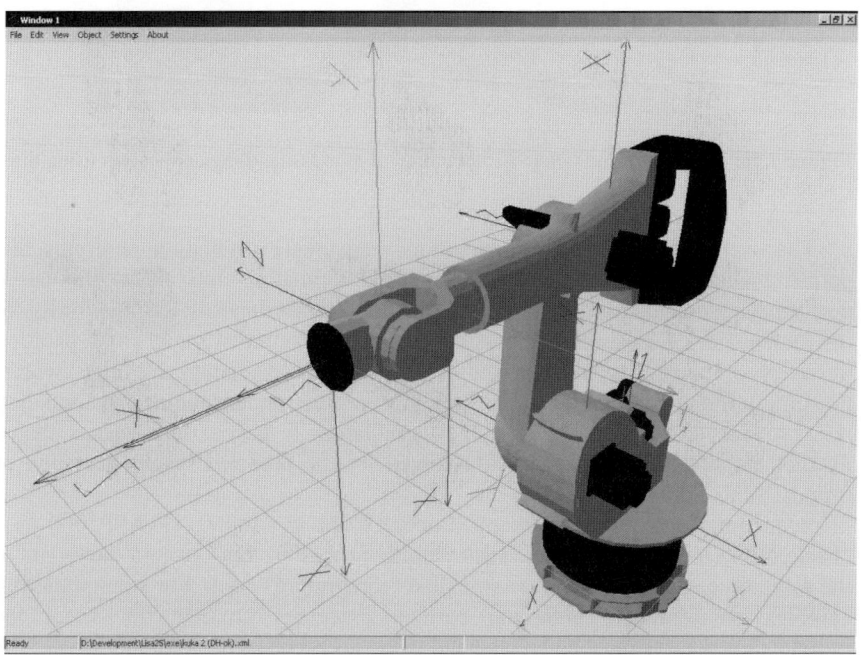

Figure 7.9. The 3D model of a KUKA robot – dynamic frames

7.4 GRASP2000 (BYG Systems Ltd, UK)

GRASP2000 is a program tool that integrates a time-based simulation system with advanced 3D graphics capabilities. The user is able to create virtual models of arbitrary complexity including all types of manufacturing systems, robotic and kinematic structures, production systems and AGV routing systems. The software allows up to 24 revolute or prismatic joints for each individual mechanism.

Figure 7.10 taken from [18] shows the example of the GRASP2000 3D model of a brick-handling application using three Fanuc M410iHW robots.

Figure 7.10. Example of the GRASP2000 model of a brick-handling application using three Fanuc M410iHW robots

The user can use a set of instructions and create simulation programs that permit "what-if" type of analysis using 3D animation and exact time-based performance calculations. The result of simulation depends on the order of instructions and the way they are used within one simulation track, so preparation of every simulation scenario requires considerable planning. In other words, meaningful results can be obtained only with a clear understanding of simulation requirements and the model on which simulation is based. This means that a detailed knowledge of the process involved for the modeled system (existing or proposed) is required [18].

In manufacturing systems many processes run in parallel. GRASP2000 uses "background tracks" to simulate such parallel processes. Background tracks run at the same time as the "current" or so-called "foreground" track. For simulation of a complex environment containing parallel processes separate tracks for the individual processes must be created and then "invoked" as background tracks. The aim of a foreground track is to control the simulation. When the foreground track is running, any tracks that have been invoked as background tracks run as well, starting at the same time. Synchronization of parallel processes (tracks) can be achieved in two ways; by inserting delay (PAUSE) instructions, and by waiting for an event. As described in [18], an event may be that the simulation clock has reached a certain time, it may be the arrival of an object, or it may be a variable being set to a particular value. Waiting for an event can be achieved using the WAIT command. The foreground and background tracks execute in the same manner, using the simulation clock to control the synchronization between all the tracks.

Regarding generation of robot tracks, GRASP2000 generates tracks for all robot models contained in the BYG robot library. The tracks are supplied with a complete set of configuration rules for the robot, with meaningful names that are understood by the target robot controller conversion program.

Among different commands, GRASP2000 also includes functionality to allow factory and process simulation using discrete event systems (DES) tools.

7.5 Robot Studio (ABB, Sweden)

ABB's RobotStudio is a simulation and "true" offline programming software due to the ABB VirtualRobot™ Technology, whose main characteristic is that the actual robot system software controls the robot simulation. In this way the successfully tested robot program can be downloaded as a whole to the real system without any further translation.

As for other concurrent simulation programs, RobotStudio can import data in major CAD formats including IGES, STEP, VRML, VDAFS, ACIS and CATIA. Having a CAD model of the part to be processed, RobotStudio allows the user to automatically generate the robot positions needed to follow the path curve, significantly shortening the time usually spent for manual programming of such a task. Standard robot programming in RobotStudio is done with a program editor ProgramMaker shown in Figure 7.11. The basis for programming in RobotStudio is ABB's robot programming language RAPID.

The software is characterized by several optimization features, such as path optimization and AutoReach™ computation. RobotStudio can automatically detect and warn about programs that include motions in close vicinity to singularities, so that measures can be taken to avoid such conditions. Simulation Monitor is a visual tool for optimizing robot movement. Red lines indicate what targets can be improved to make the robot move in the most effective way.

As shown in Figure 7.12, tool-center position (TCP) speed, acceleration, singularity or axes can be optimized to gain cycle time [19]. AutoReach automatically analyzes reachability while moving the robot or the work piece around until all positions become reachable. This allows quick verification and optimization of the workcell layout. Also, integrated collision detection helps to identify possible collisions among concerned objects and modify critical paths.

Event Tables is a tool used in RobotStudio for debugging and verifying the program structure and logic. As the program executes, the user can observe the I/O

272   Manufacturing Systems Control Design

states of the analyzed workcell. The I/O lines can be wired to simulation events allowing simulation of the robot and all equipment in the robot station.

RobotStudio provides the possibility of using Visual Basic to adapt and expand RobotStudio's functionality for various applications. This enables the user to create different add-on modules, macros or customized user interfaces.

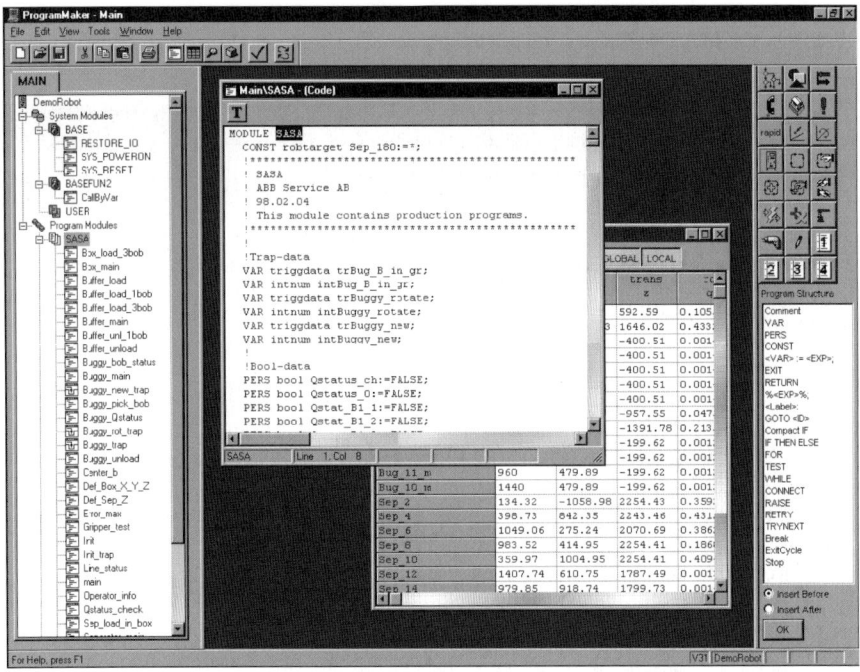

Figure 7.11. RobotStudio programming editor ProgramMaker

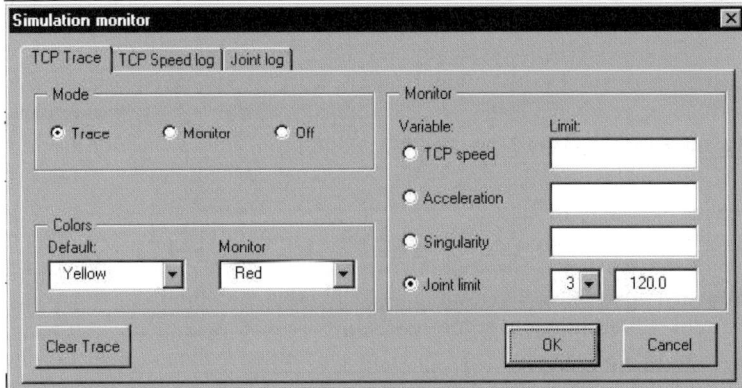

Figure 7.12. Path optimization by tracing the TCP position, speed and acceleration

Based on the use of Visual Basic for Applications (VBA) RobotStudio offers optimized solutions for applications such as arc-welding, press break tending, spot-welding, CalibWare (absolute accuracy), blade grinding, and BendWizard (press brake tending). Figure 7.13 shows the 3D model of one such application - the spot-welding robot work cell of the Volvo Cars "body-in-white" manufacturing line [20].

Figure 7.13. The example of the ABB's RobotStudio model of a Volvo body-in-white manufacturing line using ABB industrial robots (Courtesy of ABB)

7.6 Tecnomatix eM-Plant (UGS, USA)

Tecnomatix is a suite of software applications intended to support so-called digital manufacturing (also known as manufacturing process management). As discussed in [21], digital manufacturing is a combination of software and manufacturing methods that transforms manufacturing processes and manufacturing-related business initiatives. Besides process planning, digital manufacturing has a goal to optimize production operations by allowing the production planner to compare the process plan to how well that plan is actually executing.

Tecnomatix provides a broad range of applications for manufacturing management of both parts and assemblies. These solutions enable the designer to define and verify product-assembly sequences, create assembly-line layouts, simulate specific operations and material flows to optimize the process, allocate the required time for each operation, verify line performance and perform line balancing, analyze product and production costs, virtually commission and program production lines using digital planning data, execute and continually manage a production process, track and trace specific customer orders according to the materials included and the processes they undergo, and feed back real-time process changes, as executed, into manufacturing process plans [21].

eM-Plant is a Tecnomatix application that enables the simulation and optimization of production systems and processes [22]. Like other concurrent products, eM-Plant enables the designer to explore the production systems' characteristics and to optimize its performance.

Basic features of eM-Plant enable the user to simulate complex production systems and control strategies; use object-oriented, hierarchical models of plants, encompassing business, logistic and production processes; use dedicated application object libraries for fast and efficient modeling of typical scenarios; generate graphs and charts for analysis of throughput, resources and bottlenecks; use comprehensive analysis tools, including Automatic Bottleneck Analyzer, Sankey diagrams and Gantt charts. Software provides 3D online visualization and animation, which allow the user to see all system phenomena in a genuine way.

eM-Plant also has some advanced features, such as integrated neural networks and experiment handling, genetic algorithms for automated optimization of system parameters, open system architecture supporting multiple interfaces and integration capacities (ActiveX, CAD, Oracle SQL, ODBC, XML, Socket, *etc.*).

Using the eM-Plant virtual (digital) model of the manufacturing system, the user can run experiments and what-if scenarios to note critical situations and determine optimal solutions that work best. Tecnomatix software can be used for various industrial applications, and Figure 7.14 shows one such creation of the virtual expansion of the existing manufacturing facility [21].

Figure 7.14. The virtual expansion of the factory created in Tecnomatix eM-Plant

7.7 CIMStation Robotics (AC&E, UK)

CimStation Robotics is a 3D graphics program tool that enables designers to quickly and easily design, simulate and offline program robotic workcells (Figure 7.15). The software allows engineers to visualize and evaluate automation concepts to determine the cost, feasibility and performance of a proposed robotic system, long before the equipment is purchased or a part prototype is available.

Based on close collaboration with industrial users, CIMStation Robotics offers specialized application solutions tailored to the requirements of a particular robotic task. Thus the software provides advanced functionality and ease of use for painting, spot welding, arc welding, polishing, assembly and press operations.

Figure 7.15. The virtual model of the flexible manufacturing system created in CIMStation Robotics [23]

7.8 COSIMIR (FESTO, Germany)

COSIMIR is the 3D-simulation program that can be used to plan robotized workcells before they are actually built. The program allows the designer to check the reachability of all positions, develop programs for robots and controllers, and to optimize the workcell layout.

Virtual models of robots, machinery, tools, conveyer belts, part feeders, *etc.*, taken from the library of virtual models, just-created new models, or models imported from other CAD programs, can be combined to create arbitrarily complex robot-based workcells. COSIMIR allows the designer to check the developed robot programs against possible collisions and to optimize cycle times. Sensor simulation is a very useful COSIMIR's feature that extends the program's capability to simulate complete work cells. Since the program solutions for each robot in the workcell is written in the robot-compatible programming language, the direct download of tested programs and positions into the robot controller is supported. COSIMIR provides an automatic face-oriented trajectory generation suitable for applications like coating and ablation processes [24]. An industrial PLC simulation is an additional feature that makes the program adjusted for simulation of real system conditions.

Figure 7.16 shows the COSIMIR user interface for a selected robot workcell layout.

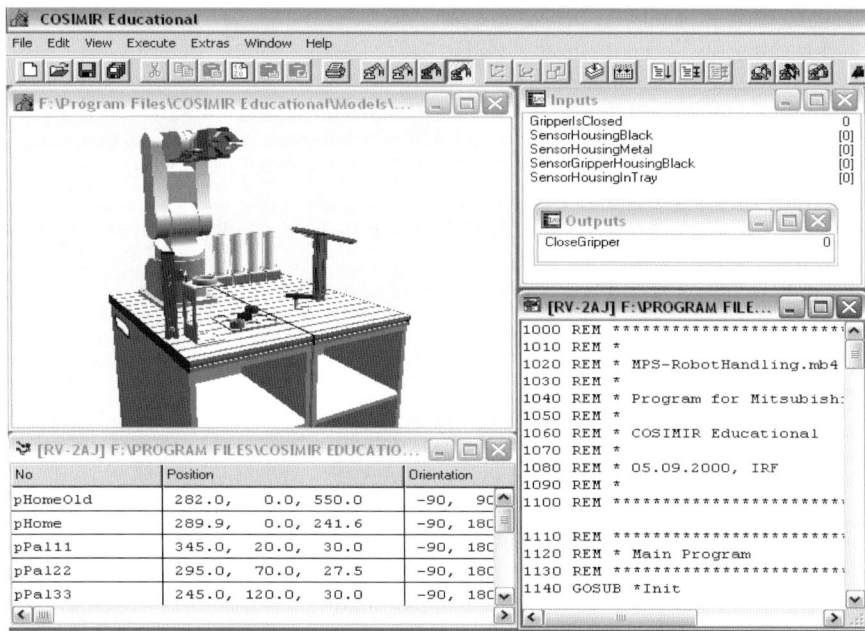

Figure 7.16. The user interface for programming and testing a virtual model of the robot work cell created in COSIMIR (Courtesy of FESTO)

7.9 FlexMan (LARICS, University of Zagreb, Croatia)

In general, all 3D simulation programs have many advanced features, including true "offline" robot programming and direct download of developed controller programs, but the problem arises when the results of analysis and (re)design performed in the virtual environment must be converted into actual real-time algorithms that should control the real system as a whole. Most of the aforementioned design and simulation programs do not offer such an elegant way that would allow transfer of system supervisory control algorithms from the virtual to the physical world.

The aim of the MS design tool FlexMan presented in this chapter is to make this step forward and show how effective the analysis of MS dynamic behavior can be, with the usage of virtual models and accompanying matrix-based dynamic models, and how straightforward it is from a matrix-based supervisory controller used in the simulator to the program for supervisory PLC in the real MS. FlexMan is a web-based virtual modeling and simulation tool using virtual models in the VRML 3D file format (see Section 7.2). The interested reader can use FlexMan and so learn more about it by visiting the FlexMan web address [25].

The usage of virtual-reality models in conjunction with the Internet-related technologies has made a significant advance in visualization of complex physical systems such as robotic systems and FMS [26].

FlexMan fulfills some basic requirements: it provides the user with a GUI for easy creation of FMS simulation prototypes including FMS layout, description of operations and operation rules, generates automatically a matrix model of the FMS as a basis for running a dynamic simulation, integrates a tool with a suitable user interface for web-based task/robot-dependent trajectory planning with embedded algorithms for solving direct and inverse kinematics problems for a user-defined type of manipulator, displays virtual FMS elements by using advanced 3D graphics and animation routines, and finally, provides status information for the job-schedule evaluation criteria.

7.9.1 FlexMan Structure

The FlexMan structure is shown in Figure 7.17. It is based on the client – server architecture. Communication between server and client(s) uses TCP/IP protocol, while all transferred and stored data is in the standard XML format [27].

Any work in FlexMan starts first with a user authorization. For different types of users, different program functions are enabled, and the work of every user is tracked and stored on the server for easier supervising and administration. This can be very convenient for training purposes, as trainees can do their work from any remote location (home, computer lab, Internet cafè), and the tutor can easily review the data about the trainee's work being stored on the server.

As shown in Figure 7.17, FlexMan's client side contains three major parts: Scene Builder, Web Trajectory Planner (WTP) and Visualization Client (VC). These three components are implemented as a single Java applet inserted in an HTML page together with a VRML plug-in that provides visualization of a 3D scene. VRML 2.0 standard defined external authoring interface (EAI) as an interface between the virtual world and the external environment. EAI defines the functionality of the VRML browser that the external environment can access, and it enables a Java applet to fully control and modify a VRML scene [13]. A new ISO standard X3D file format, which is the improved VRML format, opens new possibilities for tools like FlexMan to become more efficient and reliable.

At the server side, FlexMan has three major parts: the trajectory planner tool LEONARDO, FMS Controller, and Database.

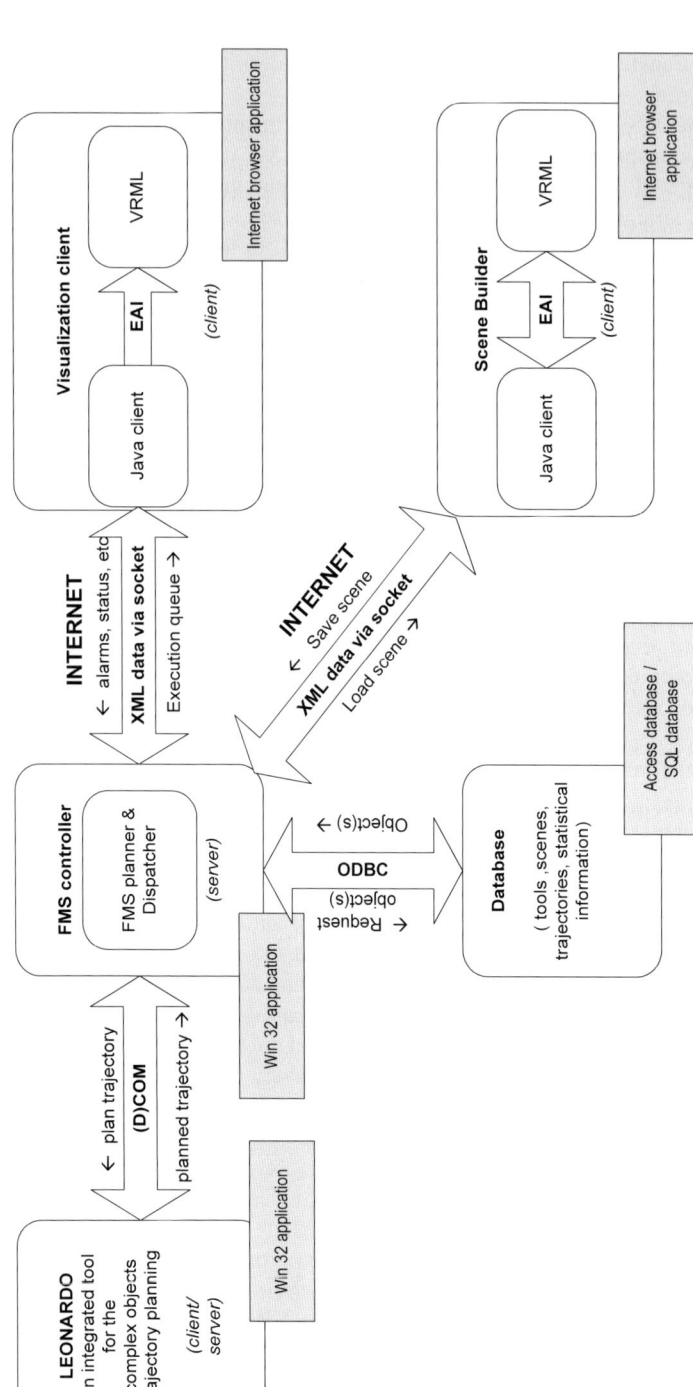

Figure 7.17. The structure of FlexMan

7.9.2 Database

Database contains information such as libraries of VRML prototypes, user information, saved work, statistical information, planned trajectories, and simulation logs. It allows new virtual FMS elements to be easily imported. Once a VRML prototype of a new resource is made, there is no need for any programming intervention in the application itself. A new element is simply added to the database and linked to the appropriate library that determines its scope of use.

7.9.3 Virtual FMS Modeling

The Scene Builder is the component that serves for modeling of the FMS layout in the virtual 3D environment and for definition of FMS functional properties.

A virtual FMS is modeled using predefined models (prototypes) of objects like robots, machines, conveyers, buffers, *etc.* Libraries of these prototypes are stored in a database on the server. Depending on the user's status and permissions, different libraries of FMS elements are available. A desired element is selected from the list of available objects (shown in the main layout of the Scene Builder (Figure 7.18)), and after setting its designation, position, orientation and scaling factor, an appropriate 3D model appears in the virtual scene. With this pick-and-place approach, even the creation of the most complex layouts is very easy, and straightforward. Figure 7.18 shows the layout of the two-robot FMS described in detail in the matrix-based controller design example in Chapter 5.

7.9.4 Functional Modeling of FMS

After the visual layout of the FMS is set, the functions and behavior of these elements are described by defining a list of operations for each element, the nature and duration of each operation, and initial system conditions. This is done with the operations editor (Figure 7.19). In order to visualize FMS operations in the virtual world as if they were real, we need an active algorithm in the background [12, 28] whose input and output must be connected to the elements of the virtual model. Providing that the resources and operations they perform are defined, the final step in FMS modeling by using FlexMan is definition of FMS operation sequencing and behavior. The part of Scene Builder named Rule Editor (Figure 7.20) serves that purpose. From the previously defined objects and their tasks the user builds a set of IF-THEN rules (see Section 3.2) that describe the sequencing of operations in the FMS.

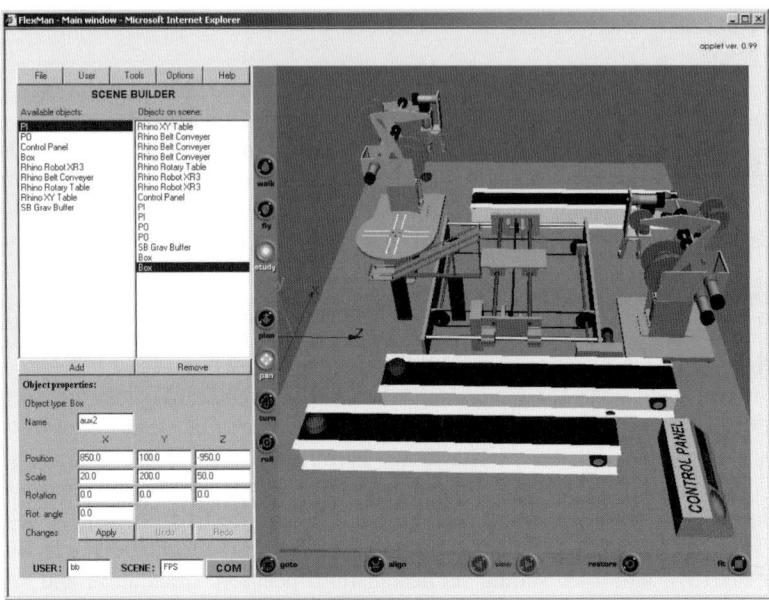

Figure 7.18. FlexMan client in a Microsoft Internet Explorer browser (the layout of the virtual model of the two-robot FMS from a case study in Section 5.4)

Based on these rules and object properties, the FMS model matrices needed for dynamic simulation are calculated automatically. The output from the Rule Editor is a set of matrices S_r, S_v, S_y, S_u, F_r, F_v, F_u, and F_y that are explained in Section 3.1. Matrices F_y, F_r, F_u and F_v are created from the antecedent (IF) part of the rule, and matrices S_y, S_r, S_u and S_v are created from the consequent (THEN) part of the rule.

7.9.5 Generating Trajectories in FlexMan

In FlexMan, trajectories for resources with one degree of freedom are generated online, but trajectories for resources with two or more degrees of freedom (*e.g.* robots) are planned with a FlexMan component – Web Trajectory Planner (WTP).

Virtual Factory Modeling and Simulation   281

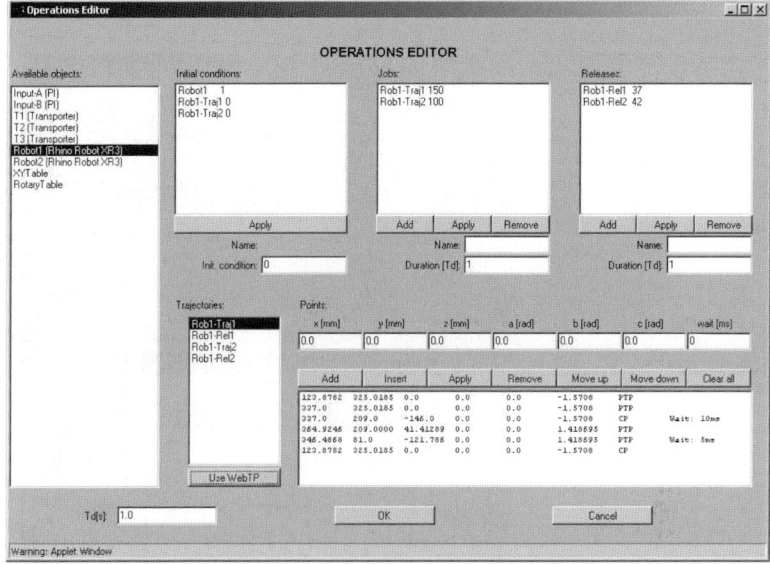

Figure 7.19. Operations Editor Window

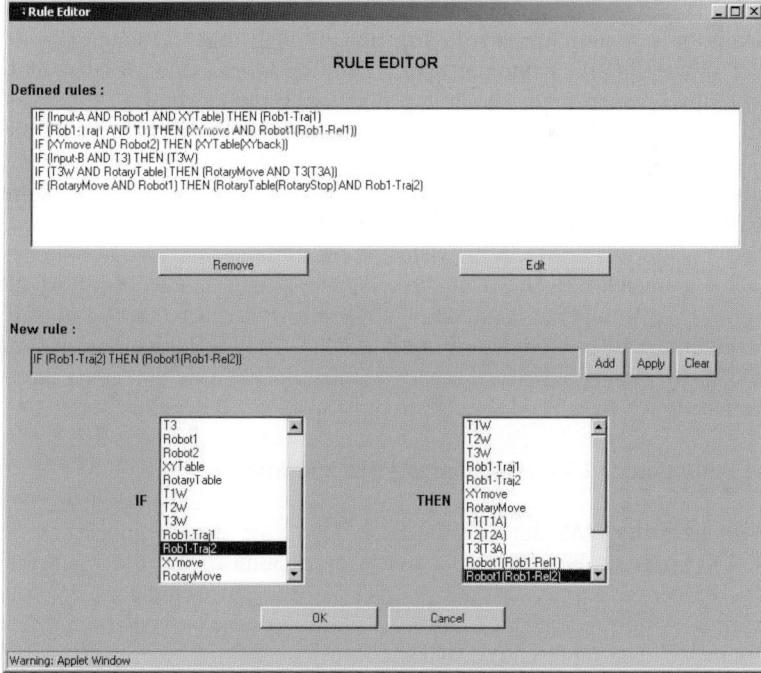

Figure 7.20. Rule Editor Window

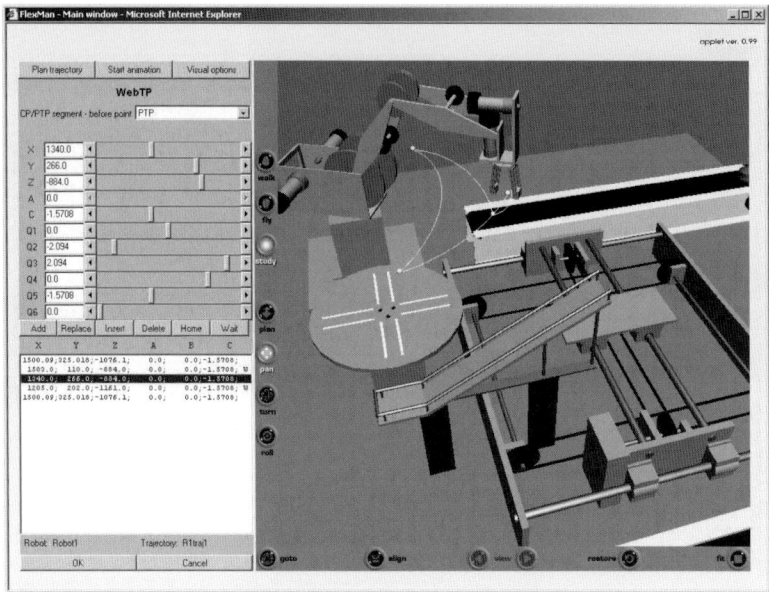

Figure 7.21. Web Trajectory Planner

In WTP (Figure 7.21), the user moves the robot from one position to another, adds key tool-tip coordinates to a list, assigns wait times and movement types (point-to-point (PTP) or continuous-path (CP)) for desired trajectory segments, and finally sends a request to the server to plan the resulting trajectory.

At the server side the trajectory planner tool LEONARDO accepts requests from FlexMan's WTP and plans combined CP/PTP motions with a given error tolerance [29]. It returns the planned trajectory points to the client and stores the trajectory for future use by FMS Controller during dynamic simulation. The planned trajectory is drawn in the virtual scene at the client side, and the user can then view animated movement of the robot along the planned trajectory (Figure 7.21).

When this phase of modeling is completed, the FMS is defined both structurally (via the VRML formatted virtual scene) and functionally (via the matrix model and the planned trajectories), and simulation of its work can proceed.

7.9.6 Simulation and Visualization of FMS operation

VC visualizes the FMS during simulation. Upon the start of simulation (Figure 7.22), VC sends to the server a complete description of the FMS generated by Scene Builder – visual layout, matrix model, and references for planned trajectories. After processing of information and necessary calculations, the server returns to VC data representing states of every element on the virtual scene in a given time frame. Through EAI, VC constantly updates the virtual scene, and thus a realistic 3D simulation of the FMS behavior is achieved, clearly depicting what is going on during the simulated manufacturing process.

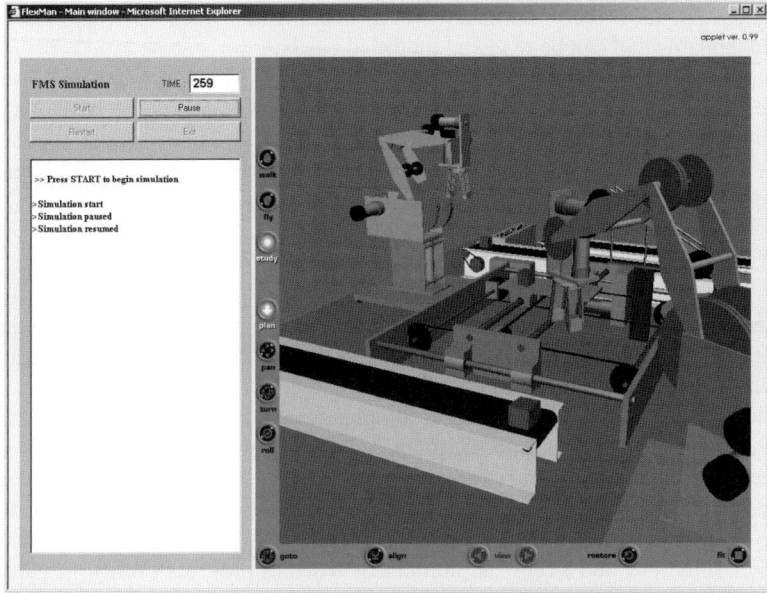

Figure 7.22. Visualization Client – enables 3D simulation of FMS operation

7.9.7 Internet-based Multiuser FMS Control with FlexMan

FMS Controller is the core of FlexMan. It is a server application that handles complete server-client communication in FlexMan, database access, client requests towards LEONARDO, user and file management, and on top of that, it executes FMS simulation. FMS Controller uses different protocols to communicate with other components: TCP/IP socket for client connections, ODBC for database, and COM/DCOM for LEONARDO access (Figure 7.17).

When a new client connects, a new communication thread is instantiated. Within this thread, a separate thread is started in which simulation is performed. After client authorization is made, FMS Controller gives the client the access to appropriate VRML prototype libraries and previously stored user files like saved models and planned trajectories. All data transferred between the server and clients is in XML format. FlexMan's XML document for scene description is used both for saving defined scenes and as input data for FMS simulation, because it contains all the necessary information for both purposes. The file size of these documents is minimal, a vital demand in any Internet-based application. This feature is a consequence of prototyped virtual scenes building, which enables full description of a VRML scene only by defining references to the required VRML prototypes and their parameters, instead of saving the data about the complete 3D model.

It must be noted that the increasing number of equal objects in the scene will only slightly increase the size of the XML file and will not affect the size of the VRML file at all. Matrix model FMS description is also very convenient for XML formatting and provides a complete functional description of the modeled FMS. It all adds up to a very compact XML document, which is very clear and understandable and can be read and modified easily.

Upon a request for simulation sent by the client, FMS Controller processes the scene description document received from the client, loads LEONARDO's robot trajectories referenced in that document, calculates trajectories for the one-degree-of-freedom resources, and starts the dynamic simulation by using a timed matrix-based model of FMS [30].

7.9.8 A Selection of an FMS Control Method

Shared resources in FMS may cause conflict situations when conditions for starting more than one concurrent job are satisfied. In that case, FMS Controller uses system matrices, finds the rules that lead to the conflict situations, and solves the problem by generating suitable control signals according to a desired dispatching policy that must be added into the model. Control signals are automatically added as prerequisites in the critical rules.

Users may choose, for example, from LBFS, FBFS, and MAXWIP dispatching policies. As described in Section 5.3, MAXWIP dispatching policy resolves conflict situations and keeps the number of work in progress (WIP), in particular FMS subsystems, at the maximum allowed level in order to avoid deadlock.

In every sampling interval, the current state of each resource is sent to the VC that updates the virtual scene. If the state of one resource has not been changed, updating for that resource is omitted to reduce the data flow and prevent communication lags.

Any problems caused by the FMS layout or by the manufacturing plan (MS data) can be easily observed, critical operations or production rules can be modified and simulations can be rerun until a suitable FMS behavior is achieved.

Example 7.9.1 (the FESTO FMS modeling and simulation with FlexMan)

Let us use the FESTO FMS laboratory setup described in Section 7.2.2 (Figure 7.4) as the target system for the matrix-based controller design. As depicted in Figure 7.23, the manufacturing task of the FMS is to assemble a cylinder by putting together four components: a body, a piston, a spring, and a cap [31]. The body colors of a cylinder can be red, silver or black. There are also two types of pistons that vary in color (black and gray) and shape (see Figure 7.23). The assembly process is organized according to the assembly specifications from Table 7.1.

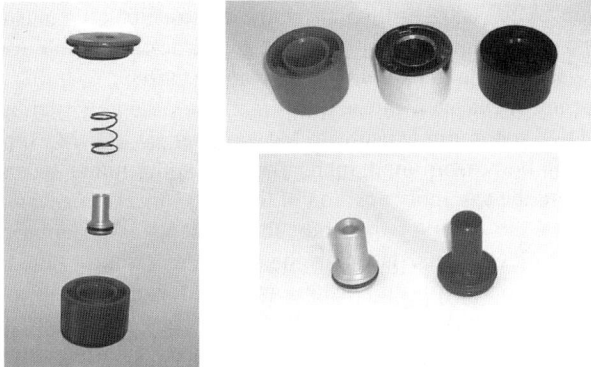

Figure 7.23. Assembly of the cylinder: a body, a piston, a spring, and a cap

Table 7.1. Assembly specifications for a cylinder

Cylinder color	red	black	silver
Cylinder material	plastic	plastic	metal
Cylinder height, [mm]	25	22.5	25
Piston color	black	gray	black
Piston radius, [mm]	20	16	20

As described in Section 7.2.2, four work stations participate in the assembly process [31]. The *distribution station* separates cylinder bodies from the *stack magazine module*, whose capacity is limited to 8 bodies. The number of bodies in the magazine is detected with a through-beam sensor. A pneumatic cylinder pushes out the bodies, one by one, and the *changer module* grips the body using a suction cup. Another sensor, a vacuum switch, checks whether the cylinder body has been picked up. The *transfer unit*, driven by a rotary drive, conveys the body to the *testing station*, which is next in the line.

The testing station determines the characteristics of inserted cylinder bodies. Different sensors serve that purpose: the sensing module identifies the color of a body and a capacitive sensor detects the body irrespective of its color. A diffuse sensor identifies silver (metallic) and red (plastic) bodies, but not the black (plastic) ones. The analog sensor of the *measuring module* determines the height of the body. The output signal is either digitalized (via a comparator with adjustable threshold value) and connected to the digital I/O of a PLC, or fed directly to the PLC analog I/O. A retro reflective sensor checks whether the working area above the body retainer is free before the body is lifted by the *lifting module*. A linear cylinder guides the correct cylinder body to the *processing station* by means of the *air cushioned slide module*. Other nonfitting bodies are sorted on the lower *slide module*.

In the processing station, cylinder bodies are positioned, processed (drilled), and then tested on a DC motor-driven *rotary indexing table*. The table has a capacity of four body places that are positioned 90° apart from each other. A solenoid actuator with an inductive sensor checks that the bodies are inserted in the correct position. After each drilling the table rotates 90° CW and the processed bodies undergo the drilling quality test. After each test, the table rotates 90° CW and the body waits for the transfer to the *assembly station*.

As we already mentioned in Section 7.2.2, the assembly station is equipped with the five DOF robot arm Mitsubishi Movemaster EX RV-M1 (see Figure 7.4), which fetches the body from the transfer position at the rotary indexing table and moves the body to the assembly position of the *assembly retainer module*.

Depending on the color of the body the robot takes an appropriate type of piston from the pallet and inserts it into the body. According to the assembly plan shown in Table 7.1, black (plastic) pistons are used for red and silver bodies, while gray (metallic) pistons are used for black bodies. Then the piston spring is taken from the *spring magazine module* and inserted. Finally, the robot picks up a cap at the *cap magazine module*, establishes the orientation of the cap and places it in the correct orientation on the body. The finished cylinder is placed on a slide, which is the end of the assembly cycle.

Having clearly defined task sequencing and ready-to-use virtual models of all the physical components of the FESTO FMS, we can use FlexMan to create a complete virtual and functional model, which together with an automatically generated matrix model will enable simulation and 3D visualization of the system.

A virtual model shown in Figure 7.4 is built of the following elements – resources (a symbolic notation for each resource is given in the parentheses): stack magazine module (SM), pneumatic pusher (PP), the transfer unit (TU), the measuring module (MM), the lifting module (LM), the rotary indexing table module (RT), the drilling module (DM), the testing module (TM), the pistons separating module (PSM), the cap magazine module (CMM), the spring magazine module (SMM), the robot arm (RA), the air-cushioned slide module (ASM), two gravitational (slide) modules (GM1 and GM2), and the place of assembly (A). In order to preserve the characteristics of the MRF line, during each new assembly job the place of assembly is treated as a new resource (that is, A converts to places A1, A2, A3 and A4). The list of jobs and releases of the above-mentioned resources, along with their symbolic notation and duration, is displayed in Table 7.2. This list can be entered using FlexMan's Operation editor. The durations of each job and release are initially determined by the actual duration of jobs and resource releases in the real FESTO FMS. These parameters can be varied during simulations in order to examine different dispatching techniques that would satisfy different manufacturing quality criteria (maximal product throughput, optimal resource utilization, minimum energy consumption, *etc.*).

All resources except the robot and the rotary table have only one job to do. The rotary table has three jobs that are done simultaneously. The only shared resource in the FMS is the robot, whose tasks are to transfer cylinder bodies from the testing station to the assembly station and fetch parts needed for assembly.

The next step is the definition of IF-THEN rules that explain the sequence of jobs and conditions, which must be fulfilled to start or finish a particular job according to a selected control strategy. Control places (CP1–CP5) have an important role in the creation of operational rules, as the state of control places dictates, in conflict situations, which job of the shared resource (that is, the robot) will be done first. This job is done with FlexMan's Rule editor.

Table 7.2. The list of jobs and releases of the FESTO FMS resources [15]

Job/Release	Resource action (movement)	Symbol	t [s]
J	A pneumatic cylinder pushes out the body	PPw	0.7
R	A pneumatic cylinder retracts	PPr	0.7
J	Transfer unit conveys the body to the testing station	TUw	2.1
R	Transfer unit retracts	TUr	1.4
J	The body is lifted by the lifting module, and a linear cylinder guides the body to the processing station via the air cushioned slide module	LMw	2.8
R	The lifting module retracts	LMr	2.8
J	The rotary indexing table module rotates to the drilling position	RTw1	1.2
J	The rotary indexing table module rotates to the drilling testing position	RTw2	1.2
J	The rotary indexing table module rotates to the transfer position	RTw3	1.2
J	The drilling module is going down	DMw	0.9
R	The drilling module is going up	DMr	0.9
J	The testing module is going down	TMw	0.9
R	The testing module is going up	TMr	0.9
J	The piston separating module pushes out the piston	PSMw	1.2
R	The pistons separating module retracts	PSMr	1.2
J	The cap magazine module pushes out the cap	CMMw	0.7
R	The cap magazine module retracts	CMMr	0.7
J	The spring magazine module pushes out the cap	SMMw	0.7
R	The spring magazine module retracts	SMMr	0.7
J	The robot fetches the body and moves it to the assembly place and the assembly place is occupied	RAw1 & A1w	4.3
R	The place of assembly is "virtually" released	A1r	0
J	The robot picks up the piston and inserts it into the body and assembly place is occupied	RAw2 & A2w	3.8
R	The place of assembly is "virtually" released	A2r	0
J	The robot picks up the spring and inserts it into the body and the assembly place is occupied	RAw3 & A3w	6.2
R	The place of assembly is "virtually" released	A3r	0
J	The robot picks up the cap, puts it onto the body, and twists it on and the assembly place is occupied	RAw4 & A4w	4.7
R	The place of assembly is "virtually" released	A4r	0
J	The robot grasps the cylinder and moves it to the slide module	RAw5	4.1
R	The robot parks in home position	RAr	2.4
J	The body is sliding down the air-cushioned slide module to the rotary indexing table module	ASMw	1.3
J	The assembled cylinder is sliding down the slide	GM2w	1.2

The jobs and releases of resources shown in Table 7.2 represent the foundation for the creation of rules. Table 7.3 shows the list of 32 rules created in the Rule editor. In order to provide a realistic 3D visualization of FMS dynamics, some auxiliary resource releases are used, such as the release of the air-cushioned slide that feeds the rotary indexing table module, and the release of the rotary table itself. These releases are instant ($t = 0$ s), as they only serve to free the resource once they have delivered the work piece to the downstream resource. Symbols PI1–PI4 and PO, which are used in the rules, denote inputs and output for the system work pieces (PI – Part In and PO – Part Out).

Table 7.3. The list of operation rules of the FESTO FMS [15]

Rule	Rule definition
1	IF (PP AND PI1) THEN (PPw)
2	IF (TU AND PPw) THEN (TUw AND PPr)
3	IF (LM AND TUw) THEN (LMw AND TUr)
4	IF (ASM AND LMw) THEN (ASMw AND LMr)
5	IF (RT AND ASMw) THEN (RTw1 AND ASMr)
6	IF (DM AND RTw1) THEN (DMw AND RTr1)
7	IF (RT AND DMw) THEN (RTw2 AND DMr)
8	IF (TP AND RTw2) THEN (TPw AND RTr2)
9	IF (RT AND TPw) THEN (RTw3 AND TPr)
10	IF (PSM AND PI2) THEN (PSMw)
11	IF (SMM AND PI3) THEN (SMMw)
12	IF (CMM AND PI4) THEN (CMMw)
13	IF (RTw3=0) THEN (CP1)
14	IF (PSMw=0) THEN (CP2)
15	IF (SMMw=0) THEN (CP3)
16	IF (CMMw=0) THEN (CP4)
17	IF (Aw+A1w+A2w+A3w+A4w=0) THEN (CP5)
18	IF (RA AND RTw3 AND CP5) THEN (RAw1 AND RTr3)
19	IF(A1 AND RAw1) THEN (A1w AND RAr1)
20	IF (RA AND PSMw AND CP1) THEN (RAw2 AND PSMr)
21	IF(A2 AND RAw2) THEN (A2w AND RAr2)
22	IF (RA AND SMMw AND CP1 AND CP2) THEN (RAw3 AND SMMr)
23	IF(A3 AND RAw3) THEN (A3w AND RAr3)
24	IF (RA AND CMMw AND CP1 AND CP2 AND CP3) THEN (RAw4 AND CMMr)
25	IF(A4 AND RAw4) THEN (A4w AND RAr4)
26	IF (A1w > 0) THEN (CP1)
27	IF (A2w > 0) THEN (CP2)
28	IF (A3w > 0) THEN (CP3)
29	IF (A AND A1w AND A2w AND A3w AND A4w) THEN (Aw AND A1r AND A2r AND A3r AND A4r)
30	IF (RA AND Aw) THEN (RAw5 AND Ar)
31	IF (GM2 AND RAw5) THEN (GM2w AND RAr5)
32	IF (GM2w) THEN (GM2r AND PO)

System resources with more than one degree of freedom, the robot and the lifting module, require planning of trajectories. This is done by using FlexMan's Web Trajectory Planner, which invokes a trajectory planning tool Leonardo on the server side. Having all trajectories planned, simulation can start and the process of cylinder assembly can be examined. Figure 7.24 shows several instants of the assembly process captured during animated 3D visualization in FlexMan's VC.

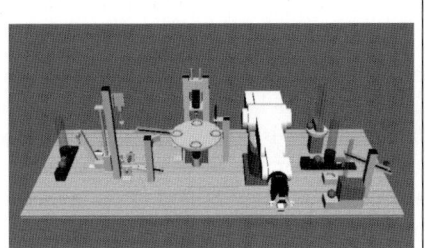	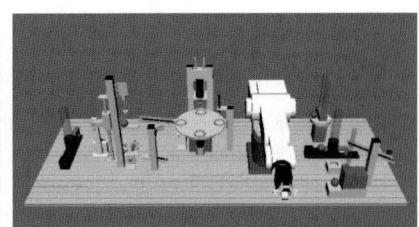
All cylinder components are present and assembly can start	The body of a cylinder is inspected by the measuring module
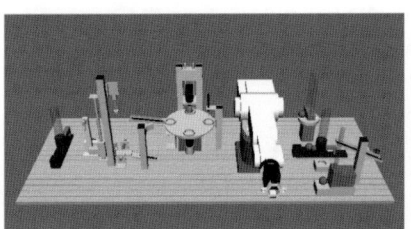	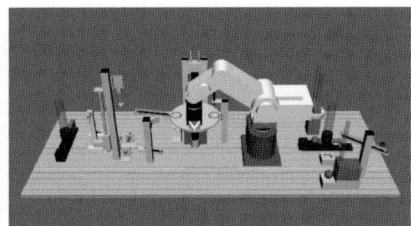
The drilling module is drilling the body.	The robot fetches the body from the transfer position of the rotary table.
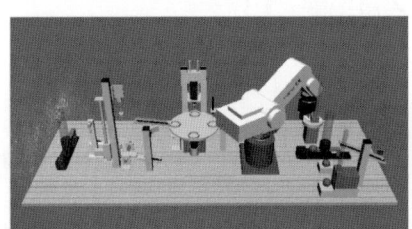	
The robot picks up the piston before inserting it into the body.	The robot puts the fully assembled cylinder on the output slide module.

Figure 7.24. The phases of the simulated assembly process visualized in the FlexMan Visualisation Client

7.10 Exercise

Figure 7.25 shows a layout of the laboratory workcell that contains an educational robot Rhino XR-3, two belt conveyers, one transporter and two pistons [32]. A processed part visits several resources on its way through the system. The system has a shared resource, *i.e.* a conflict-resolution algorithm by using a matrix-model approach should be implemented.

For the given system layout, define operational times and specify the number of sensors and their positions. The part that is processed is put into the system by piston 1. When the part gets to the end of the conveyer 1 the robot transfers it to the conveyer 2. At the end of the conveyer the robot picks the part and places it on the transporter. Once the part is close to the piston 2, it is moved out of the system.

By using FlexMan [25], create a virtual model of the FMS and describe the functions and behavior of the system elements by defining a list of operations for each element. Define the nature and duration of each operation and initial system conditions. For this purpose use FlexMan's Operations Editor (Figure 7.26). Create operation rules for a selected job-scheduling strategy with Rule editor, plan the robot trajectories with Web Trajectory Planner (Figure 7.26). Start the simulation and watch the FMS dynamic behavior while 3D animation of the FMS operation is displayed in the Visualization Client (Figure 7.27).

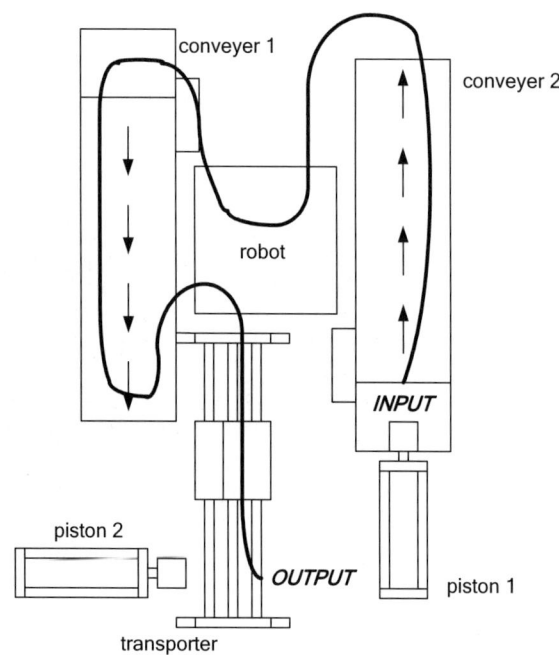

Figure 7.25. Example of the laboratory FMS layout

Virtual Factory Modeling and Simulation 291

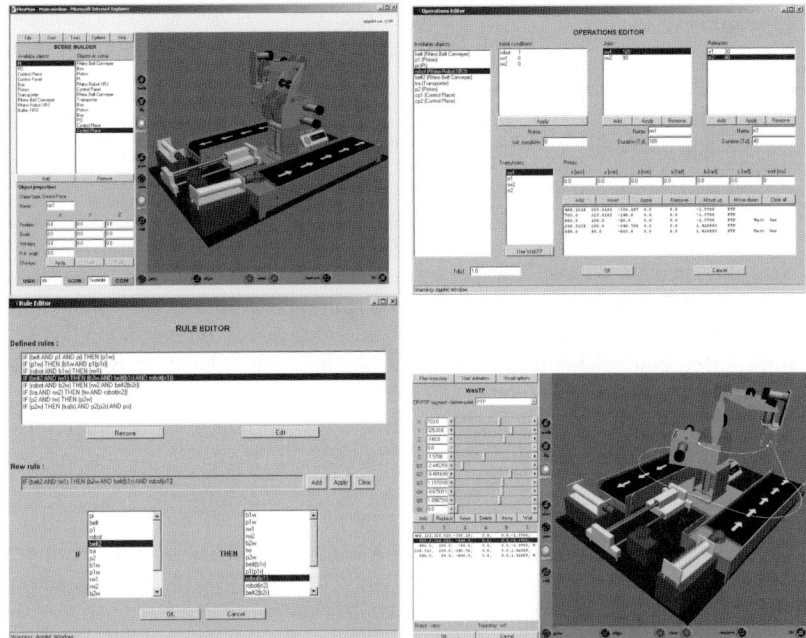

Figure 7.26. Steps of FMS control design in FlexMan for the example of the FMS layout

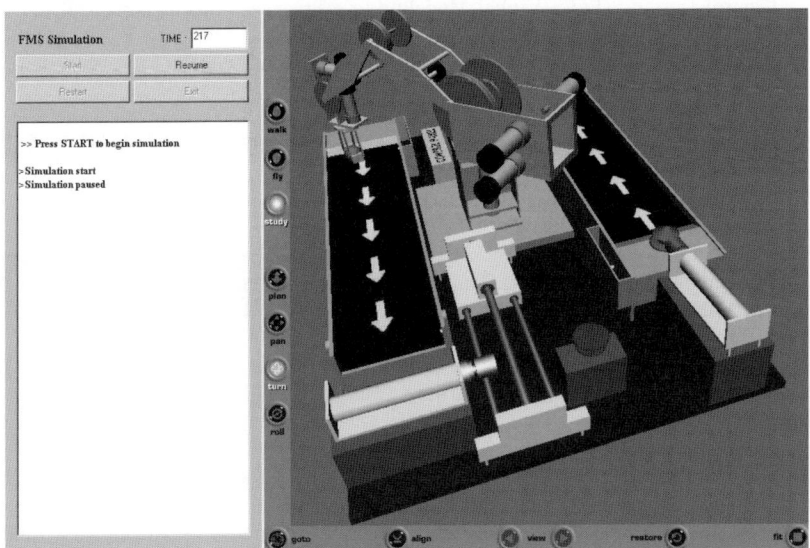

Figure 7.27. 3D visualization as a final result of the FMS control design in FlexMan

References

[1] Viswanadham N, Narahari Y. Performance Modeling of Automated Manufacturing Systems. New Jersey: Prentice Hall, 1992.
[2] Vince J. Virtual Reality Systems. Reading, MA: Addison-Wesley, 1995.
[3] Mayr H. Virtual Automation Environments – Design, Modeling, Visualisation, Simulation. New York Basel: Marcel Dekker, 2002.
[4] Gertz M W, Khosla P K. Onika: A Multilevel Human-Machine Interface for Real-Time Sensor-Based Robotics Systems, Proc. of SPACE 94: The 4th International Conference and Exposition on Engineering and Construction, 1994.
[5] Nethery J, Spong M W. Robotica: A Mathematica Package for Robot Analysis, IEEE Rob. Aut. Mag. 1994; 1: 1: 13–20.
[6] Ge S S, Lee T H, Gu D L, Woon L C. A One Stop Solution in Robotic Control System Design, IEEE Rob. Aut. Mag. 2000;7:3:42–54.
[7] Corke P. Robotic Toolbox for Matlab, CSIRO Manufacturing Science and Technology, http://www.cat.csiro.au/cmst/, visited 2005.
[8] Choi B, Park B, Ryu H Y. Virtual Factory Simulator Framework For Line Prototyping, J. of Advanced Man. Sys., World Scientific Publishing Company 2004;3:1:5–20.
[9] Sly D. Object-oriented factory layout in AutoCAD, Proceedings of the 1998 Winter Simulation Conference, 1998, 275–277.
[10] Heinicke M U, Hickman A. Eliminate bottlenecks with integrated analysis tools in eM-Plant, Proceedings of the 2000 Winter Simulation Conference, 2000, 229–231.
[11] Li Y F, Ho J, Li N. Development of a physically behaved robot work cell in virtual reality for task teaching, Rob. and Comp-Integr. Manuf. 2000;16: 91–101.
[12] Bogdan S, Kovačić Z, Smolić-Ročak N, Birgmajer B. A Matrix Approach to an FMS Control Design – From Virtual Modeling to a Practical Implementation, IEEE Rob. Aut. Mag. 2004;11:4:92–109.
[13] Jacobs K, Lemay L (eds.), Murdock K, Couch J. Laura Lemay's Web Workshop: 3D Graphics and VRML 2, Sams Publishing, 1996.
[14] Web3D Consortium web page: http://www.web3d.org/x3d/, visited 2005.
[15] Tomić M. Modeling, Simulation and Control of FESTO FMS, Diploma Thesis, University of Zagreb, 2005.
[16] Reichenbach T. Collision Avoidance in Virtual Robotized Plants, Masters Thesis, University of Zagreb, 2005.
[17] Lin M C, Gottschalk S. Collision detection between geometric models: a survey, Proceedings of the 8th IMA Conference on the Mathematics of Surfaces (IMA-98), ser. Mathematics of Surfaces (R. Cripps, ed.) 1998;VIII:37–56.
[18] GRASP 2000 User Manual, BYG Systems Ltd., 2nd edn, 2002.
[19] RobotStudio Features, ABB Information for System Partners CD-ROM, ABB, 2001.
[20] RobotStudio^{TM} – Industrial IT Software, Datasheet, ABB, 2002.
[21] CXOs: Meet your new core competency – digital manufacturing, White paper, UGS, 2005.
[22] Tecnomatix eM-Plant – eMPower for manufacturing process management, Fact sheet, UGS, 2005.
[23] CIMStation Robotics News Update 5, Applied Computing & Engineering Limited web page: http://www.acel.co.uk/, visited 2005.
[24] Karras U. COSIMIR Educational User Guide, FESTO Didactic Gmbh, 2000.
[25] LARICS FlexMan web page: http://flrcg.rasip.fer.hr/flexman, University of Zagreb, updated 2005.
[26] Hirukawa H, Hara I. Web-Top Robotics: Using the World Wide Web as a Platform for Building Robotic Systems, IEEE Rob. Aut. Mag. 2000;7:2: 40–45.

[27] Paradi W J. XML in action, Microsoft Press, 1999.
[28] Bogdan S, Lewis F L, Kovačić Z, Gurel A. New Matrix Formulation for Supervisory Controller Design in Practical Flexible Manufacturing System, Proceedings of the IEEE International Symposium on Intelligent Control ISIC 1999; 144–149.
[29] Kovačić Z, Bogdan S, Petrinec K, Reichenbach T, Punčec M. LEONARDO - The Off-line Programming Tool for Robotized Plants, CD-ROM Proceedings of the 9th Mediterranean Conference on Control and Automation 2001;WM2-C.
[30] Mireles J Jr, Lewis F L. Intelligent Material Handling: Development and Implementation of a Matrix-Based Discrete Event Controller, IEEE Trans. Ind. Electr. 2001; 48:6:1087–1097.
[31] FESTO Modular Production System - Distribution station, Testing station, Processing Station, Assembly station, User manuals, FESTO Didactic Gmbh, 2000.
[32] Kovačić Z, Bogdan S, Smolić-Ročak N, Birgmajer B, Teaching Flexible Manufacturing Systems by Using Design and Simulation Program Tools, Proceedings of the IEEE Region 8 EUROCON 2003 International Conference on COMPUTER AS A TOOL 2003;47–51.

Index

active event function, 36
activity-completion matrix, 64
activity-start matrix, 64
adjacency matrix, 104, 108, 112, 125, 151, 185
and/or algebra, 16, 58, 151
arc, 37, 98, 103
arc adjacency matrix, 185
assembly line, 4, 17, 247, 265
assembly tree, 6, 55, 230
asynchronous events, 34
automaton, 36

bill of materials (BOM), 6, 53
binary loop, 156

circle (cycle), 101, 116
circular blocking, 2, 73, 150, 168, 178
circular (cyclic) path, 129, 150, 241
circular wait relation, 15, 150
circular waits (CWs), 113, 150
clock, 70, 225
conflict, 12, 54, 73, 172, 180, 193, 220, 232
conflict resolution, 80, 94, 141
conflict-resolution matrix, 79
conflicting-rules vector, 80
content of CW, 159, 169
control function, 26, 48, 78, 90, 202, 232, 244
controllability, 26, 46, 237
coordination level, 91
critical CCW, 170
critical circuit, 101
critical resources, 169

critical siphon, 158, 165, 201, 236
critical subsystem, 164, 175, 241
critical traps, 158
CW adding rules, 159, 174
CW clearing rules, 159, 174
cycle mean, 101, 129
cyclic circular wait, 152, 169, 177
cyclic posterior rules, 170
cyclic precedent rules, 170

deadlock, 9, 13, 41, 47, 73, 83, 148,, 159, 178, 184, 195, 216, 236
deadlock avoidance, 9, 91, 150, 178, 184, 195
deadlock detection, 9
deadlock prevention, 9, 41, 87, 216, 232
delay matrices, 68, 141
deterministic automata, 37
digraph (directed graph), 10, 37, 99, 106
discrete event controller (DEC), 4, 15
dispatching, 2, 8, 16, 89, 147, 178, 248, 255, 284
dispatching-control input, 16, 78
dispatching matrix, 79, 201
dispatching problem, 5, 54, 86, 156
dispatching vector, 78, 90, 141
dispatching vector release matrix, 83, 141
dispatching, first-buffer-first-serve (FBFS), 8, 12, 77, 180
dispatching, last-buffer-first-serve (LBFS), 9, 12, 48, 180
downstream node, 100, 150

edges, 98
empty string, 47, 112

296   Index

event-driven state, 30
event-driven system, 10, 34, 45, 132
event graphs, 121, 133, 141

feedback, 16, 26, 45, 59, 70, 79, 127
final node, 101
flexible manufacturing systems (FMS), simulation tools, 17, 242, 259
free-choice multiple re-entrant flowlines (FMRF), 147, 170, 184, 226

graph, 3, 14, 37, 47, 98, 107, 123, 151, 212

hold while waiting, 53, 149
hybrid matrix model, 64
hybrid systems, 30

idle-resource vector, 59
implementation level, 92
incidence matrix, 105, 215, 223, 233
input matrix, 57, 238
irregular system, 169

job-completed vector, 59, 168
job-sequencing matrix, 6, 54
job-start equation, 60, 77
job-start matrix, 56, 71
job-start vector, 59, 204
job vector, 16, 58, 156

kanban, 162, 180
key resource, 148, 169, 183

language, 45
lifetime, 67, 79, 141
livelock, 41, 186
logical state vector, 54, 59, 77, 90, 141, 148, 181, 238

marked language, 47
marked states, 36, 47
material handling buffer (routing resources), 171
mathematical programming, 3, 16
matrix-based DE controller (supervisor), 15, 52, 77, 86, 91, 178, 202, 247, 277
maximization, 121, 132
maximum cycle mean, 101, 129, 140
max-plus, 120, 132, 225
mean weight of a path, 101

multipart re-entrant flowline (MRF), 86, 148, 169, 178, 217, 229
mutual exclusion, 10, 53, 149

neutral job set, 163, 179
neutral rules, 164, 174
nodes, 14, 98, 111, 150, 185, 212
nonshared resources, 54, 68, 156, 227, 249
NP-(nonpolynomial)-completeness, 16

observability, 27, 46, 132, 237
one-step look-ahead, 147, 178
origin, 99, 116, 185
output matrix, 57, 238
overlap, 184, 192

parallel composition, 42, 111, 122
parallel sharing, 54, 227
part path, 4, 60, 157, 184, 229
path length, 101
Petri nets, colored, 225
Petri nets, input incidence matrix, 214, 238
Petri nets, output incidence matrix, 214, 238
posterior rules, 164, 241
postset, 100, 148, 159, 174, 220
precedent rules, 164, 241
predecessor, 100, 111, 131
pre-emption, 53
prefix, 47
prefix-closed, 47
preset, 100, 148, 159, 174, 220
Programmable logic controller (PLC), 10, 48, 68, 199, 245

reachable, 102, 217, 248
re-entrant flowline, 4, 150
resource cycle, 130
resource job set, 54
resource loop, 54, 156, 178, 220, 241
resource-release equation, 15, 61, 77
resource-release matrix, 56, 71, 238
resource-release vector, 59, 204
resource requirements matrix, 2, 8, 54, 78, 238
resource utilization, 65, 130, 181, 208
resource vector, 16, 58, 156

second-level deadlock, 148, 169, 182
self-loop, 98, 105, 212, 229

sequential sharing, 54, 227
series composition, 111, 122
shared resources, 5, 54, 61, 81, 133, 149, 171, 227, 249
shift (delay) operator, 68, 141
siphon, 158, 179, 201, 221, 236
siphon job set, 161, 175
siphon job vector, 161
siphon vector, 159
siphon-trap job set, 162, 175
sorted vector, 187
state transition diagram, 37
strictly adding rules, 174
strictly neutral job set, 163
strictly siphon job set, 163
strictly trap job set, 163
string, 46, 110, 123, 130
string matrix, 112
strongly connected graph, 102, 129
substring, 47, 111
successor, 100, 199
suffix, 47
supervisory controller, 1, 15, 45, 52, 77, 120, 180, 199, 216, 236, 277
support, 58
system
system state, 23, 32, 45, 98, 108, 137
system vector, 63, 79
system, inputs, 22, 57, 129, 143
system, outputs, 22, 27, 134, 143, 250
system, time-driven, 22, 34, 125
system, time invariant, 22
system, time-variant, 22

task-sequencing matrix, see also job sequencing matrix, 2, 16
temporary system vector, 90, 159, 204
3D Modeling, 261
time driven system, see system, time-driven

time vectors, 186, 194
time windows, 178, 189
time-windows overlap, 184, 193
timed sequence, 35
token game, 11, 243
tokens, 11, 14, 212, 224, 232
transition function, 36
trap, 158, 221
trap job set, 162, 175

uncontrollable events, 31, 46, 237
unique production cycle, 129
unreachable, 31
upstream node, 100, 150

vector negation, 58
vertices, 98
virtual-factory simulators, 259
virtual-factory simulators, CimStation Robotics, 275
virtual-factory simulators, Cosimir, 275
virtual-factory simulators, eM-Plant, 273
virtual-factory simulators, FlexMan, 267, 276
virtual-factory simulators, Grasp2000, 270
virtual-factory simulators, RobotStudio, 271
virtual node, 114
VRML, 262
VRML, basic features, 263

wait relation graph, 14, 151, 241
weight, 99, 106, 129
weight of path, 101
weighted adjacency matrix, 106, 121, 129

X3D, 262

Other titles published in this Series (continued):

Analysis and Control Techniques for Distribution Shaping in Stochastic Processes
Michael G. Forbes, J. Fraser Forbes, Martin Guay and Thomas J. Harris
Publication due August 2006

Process Control Performance Assessment
Andrzej Ordys, Damien Uduehi and Michael A. Johnson (Eds.)
Publication due August 2006

Adaptive Voltage Control in Power Systems
Giuseppe Fusco and Mario Russo
Publication due September 2006

Advanced Fuzzy Logic Technologies in Industrial Applications
Ying Bai, Hanqi Zhuang and Dali Wang (Eds.)
Publication due September 2006

Distributed Embedded Control Systems
Matjaž Colnarič, Domen Verber and Wolfgang A. Halang
Publication due October 2006

Modelling and Analysis of Hybrid Supervisory Systems
Emilia Villani, Paulo E. Miyagi and Robert Valette
Publication due November 2006

Model-based Process Supervision
Belkacem Ould Bouamama and Arun K. Samantaray
Publication due February 2007

Continuous-time Model Identification from Sampled Data
Hugues Garnier and Liuping Wang (Eds.)
Publication due May 2007

Process Control
Jie Bao, and Peter L. Lee
Publication due June 2007

Optimal Control of Wind Energy Systems
Iulian Munteanu, Antoneta Iuliana Bratcu, Nicolas-Antonio Cutululis and Emil Ceanga
Publication due November 2007